Chemistry Made Clear

GCSE EDITION

R Gallagher
P Ingram

Oxford University Press

Oxford University Press, Walton Street, Oxford OX2 6DP

Oxford New York
Athens Auckland Bangkok Bombay
Calcutta Cape Town Dar es Salaam Delhi
Florence Hong Kong Istanbul Karachi
Kuala Lumpur Madras Madrid Melbourne
Mexico City Nairobi Paris Singapore
Taipei Tokyo Toronto

and associated companies in
Berlin Ibadan

Oxford is a trade mark of Oxford University Press

First published 1984

ISBN 019 914267 X

GCSE edition 1987
Reprinted 1989, 1990, 1991, 1992, 1993, 1994 (twice)

Typeset by Tradespools Ltd, Frome, Somerset
Printed in Great Britain by The Bath Press, Avon

Acknowledgements

The Publisher would like to thank the following for permission to reproduce photographs:

Aerofilm, p.26 (top); Aga-Rayburn, p.66 (top); Airco Industrial Gases, p.100; Air Products Limited, p.11; Alcan Aluminium Limited, pp.139 (top centre), 144(1,2,3,4,5); Alfa Romeo GB Limited, p.71 (top left); Aluminium Association, p.145 (top); Aluminium Company of America, p.141; American Optical, p.198 (bottom); Heather Angel, p.98 (bottom); Ardea, p.98 (top); Ashmolean Museum, Oxford, p.148 (top right); Associated Press, p.37 (top right); Astra Pyrotechnics Limited, p.65 (top); Barnaby's Picture Library, pp.88, 130 (left), 131 (left), 139 (bottom), 141 (left); BBC Hulton Picture Library, pp.26 (bottom), 30 (bottom), 54, 108, 174 (bottom); Dr. Hans Bibelreither, p.171 (top); Biophoto Associates, pp.6 (bottom left, centre, and right), 184 (centre); Black and Decker, p.76 (right); British Oxygen Company Limited, pp.33, 101 (top, centre, and bottom); B.P. Chemicals Limited p. 193; B.P. Oil, pp.86 (bottom centre), 196 (top left, and right); G.B. Bremmer, Department of Biological Sciences, Portsmouth Polytechnic, p.105 (centre); Paul Brierley, p.48 (left); British Alcan Aluminium Limited, pp.145 (bottom), 149 (bottom); British Gas, p.167 (left, centre, and right); British Quarrying and Slag Federation Limited, p.123 (bottom); British Steel Corporation, pp.72, 86 (top right), 142 (bottom left), 142 (bottom right), 146, 147 (top and bottom), 148 (top centre, and bottom); Vernon Brooke, pp. 24 (top left, and right), 30 (left and centre), 40 (left and centre), 46, 47, 48, (right), 49 (right), 50 (bottom), 56, 67, 83, 86 (top left), 87, 111 (top and bottom), 116, 117, 118, 119 (top), 122, 131 (centre), 148 (top left), 173; 194 (top); Camera Press, pp.27, 102 (bottom), 109, 139 (top right), 152 (bottom), 154 (left), 166 (right); Camping Gaz GB Limited, p.191; Jonathan Cape, p.131 (right); Cement and Concrete Association, pp.70 (centre), 86 (top centre); Central Electricity Generating Board, p.77 (bottom); Central Office of Information, p.36; Cheze-Brown, p.112, Chubb Fire Security Limited, p.71 (bottom); CLEAR, p.103 (bottom); Thomas Cook Limited, p.152 (top); T.A. Cutbill and Partners Ltd, p.195 (top); John Day, p.154 (centre); De Beers Consolidated Mines Limited, pp.24 (top centre), 51 (top); Derby Art Gallery, p.24 (bottom); The Distillers Company, p.182 (top); Zoe Dominic, p.9 (top); Eggs Information Bureau, p.65; Electrolux, p.91; Elizabeth Photo Library, p.35 (bottom); Esso Petroleum Company Limited, pp.192, 197; Fiat Auto UK Ltd, p.189 (right); Ian Fraser, pp.7, 14, 16, 18 (bottom) 19, 42, 49 (left), 77 (left, centre and right), 90, 119 (bottom), 120 (bottom), 126, 135 (top), 156, 176; Ford Motor Company, p.153; RoseMarie Gallagher, pp.6 (top left, and right), 10, 15 (top), 69 (top), 99, 103 (top), 106 (top), 110 (bottom), 142 (top right, and bottom centre), 155, 160 (bottom centre), 181 (top), 184 (top left, and bottom), 194 (bottom), 199 (top); Garrard and Company, p.138 (bottom); Colin George, p.40 (right); Geoscience Features/Dr. Basil Booth, pp.64, 170 (bottom, left and right); Griffin and George, p.18 (top); Harringtons, p.184 (top right); H.J. Heinz Company Limited, p.149 (top); Hoover plc., p.110 (bottom); ICI p.172 (bottom); ICI Agricultural Division, pp.70, 95, 127, 154 (right), 158, 159, 160 (bottom left), 161 (top, centre, bottom left and right); ICI Mond Division, pp.9 (bottom), 32 (bottom); ICI Plant Protection Division p.162 (top); Image in Industry, p.86 (bottom left); Indusfoto, p.110 (top); Industrial Diamond Information Bureau, p.180 (top); Institute of Geological Sciences, p.180 (bottom centre); Lead Development Association, p.142 (top centre); L & G Fire Appliance Company, p.182 (bottom); Leeds University, Dept. of Food Science, p.171 (bottom right); London Tourist Board p.180 (bottom left); Los Alamos National Laboratory p.153; Lyons Bakery, p.181 (bottom); Manchester Town Hall, p.58; Dr. John Martin, p.50 (left); Mark Mason, pp.44, 69 (top); Medical Research Council, p.35 (top and centre); Milk Marketing Board, p.142 (top left); Ministry of Agriculture, Fisheries and Food/David Bucke, p.112 (bottom); Dr G.R. Millward, p.51 (bottom); National Coal Board, p.184 (bottom); Natural History Photographic Agency, p.123 (top); Network, p.104; Oxfam, p.160 (top); Oxford Scientific Films, p.160 (bottom right); Pictor International, p.163 (top right); Polydor Records, p.32 (top); Popperfoto p.113 (top left); Prestige Group, p.130 (right), 199 (bottom), Raphol Perrard, p.188 (right); Reed Paper and Board UK Ltd., p.171 (left); Phelim Rooney, p.196 (bottom); Royal Mint, p.136; RSPCA, p.113 (bottom); Scottish Tourist Board, p.138 (top); Shell, p.198 (top); Space Frontiers, pp.66, 144(8); Sulphur Institute, p.170 (right); Syndication International, pp.102 (top), 113 (top right); Thames Water Authority, pp.196 (top centre), 107; Thermit Welding, p.134; Times Newspapers, p.86 (bottom right); John Topham Picture Library pp.140 (bottom), 174 (top), 183, 186 (bottom); Trust House Forte, p.194 (centre); UKAEA, pp.31, 37 (left, centre right, and bottom right); Derek West, p.105 (bottom); C. James Webb, pp.6 (top centre), 105 (top), 166 (left, and centre), 184 (top centre); The Wind Energy Group, p.188; Yorkshire Post Newspapers Limited, p.71 (top right); Zefa Picture Library (UK) Ltd, Front cover; Backcover. Additional photography by Chris Honeywell. Additional diagrams by Illustra Graphics Ltd. Cartoon by Peter Joyce.

Introduction

What science is about
Science is about asking questions. You can ask questions about anything – you ask *scientific* questions when you are reasonably sure that the answers you get can be trusted.

What chemistry is about
Chemistry is the science that asks questions about materials, the differences between them, how they react with one another and how heat or other forms of energy affect them. What is water made of? What happens when hydrogen burns? How are plastics made? Why are some acids very dangerous, and other acids (like orange juice) not dangerous? All these questions are of interest to chemists.

What Chemistry Made Clear is about
Many experiments have been done to find the answers to questions like those above. This book will tell you some of the answers. It explains, as clearly as possible, the facts about chemistry that you will need to know when studying chemistry for the GCSE examination.

How to use this book
Everything in this book has been organized to help you find out things quickly and easily. This is why it is written in two-page units. Each unit is on a topic you are likely to study. This is how to find and use the information you want:

Use the contents page
If you are looking for information on a large topic, look it up in the contents list on the next two pages. But if you cannot see the topic you want, then:

Use the index
If there is something small you want to check up on, look up the most likely word in the index. The index gives the page number where information about that word is given.

Use the questions
Asking questions and answering them is a very good way of learning. To help you learn, there are questions at the end of every topic. At the end of each chapter, you can test yourself using the exam level questions. The answers to numerical questions are at the back of the book.

Chemistry is important
Chemistry is an important and exciting subject. We hope that this book helps you with your studies, that you enjoy using it, and that at the end of your course, you agree with us!

R. Gallagher
P. Ingram

Contents

1.1 Everything is made of particles

Everything around you is made up of very tiny pieces, or **particles**. Your body is made of particles. So is your desk, your chair, and this book.

A penny is made up of about 34 000 000 000 000 000 000 000 particles.

A small raindrop contains about 1 000 000 000 000 000 000 000 particles.

Take a deep breath – and you will breathe in about 40 000 000 000 000 000 000 000 particles from the air.

These particles are so tiny that it is impossible to pick up just one of them, and look at it. However, there is a machine that is powerful enough to take pictures of *groups* of particles. It is called a **scanning electron microscope**. Below are some photographs of a needle, taken with this machine.

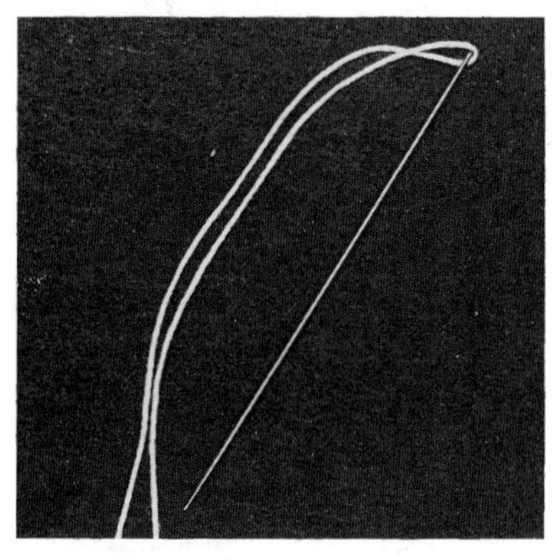

This photograph shows the needle as you would see it.

This is eye of the needle, magnified about 130 times. You can see that the surface is really quite rough.

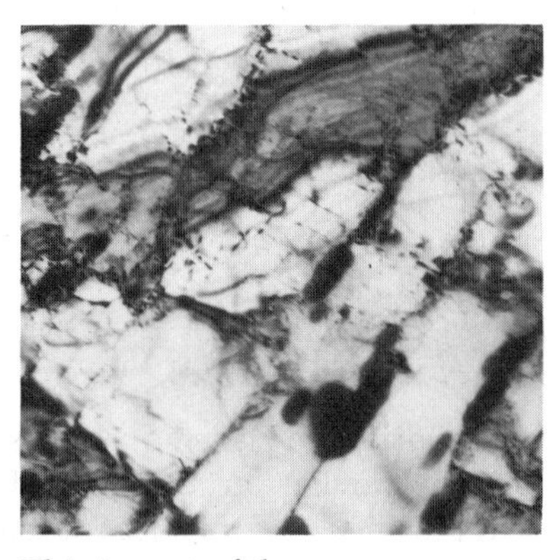

This is part of the eye, magnified 60 000 times. The photograph shows that it is made up of clusters of particles.

The scanning electron microscope is very powerful, but still not powerful enough to show just one particle. Each shape in the last picture above contains millions of smaller particles.
Your school laboratory does not have a scanning electron microscope. However, even without one, you can still find evidence that things are made of particles. Some examples are given on the next page.

A crystal dissolving

When a crystal of potassium manganate(VII) is placed in a beaker of water, the water slowly turns purple.

Explanation Both the crystal and water are made of particles. The colour spreads because purple particles leave the crystal and mix with water particles.

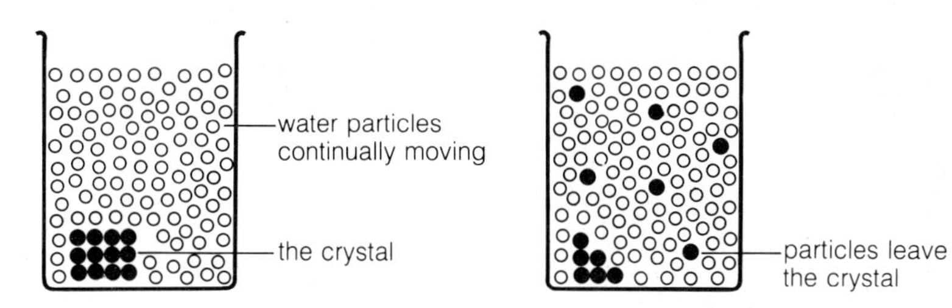

This movement of different particles among each other, so that they become evenly mixed, is called **diffusion.**

Two gases mixing

Air is colourless. Bromine vapour is a red-brown colour and is heavier than air. When a jar of air is placed upside-down on a jar of bromine vapour, the red-brown colour spreads up into it. After a few minutes, the gas in both jars looks the same.

Explanation Both air and bromine are made of tiny moving particles. These collide with each other and bounce about in all directions, so that they become evenly mixed.
This is another example of diffusion.

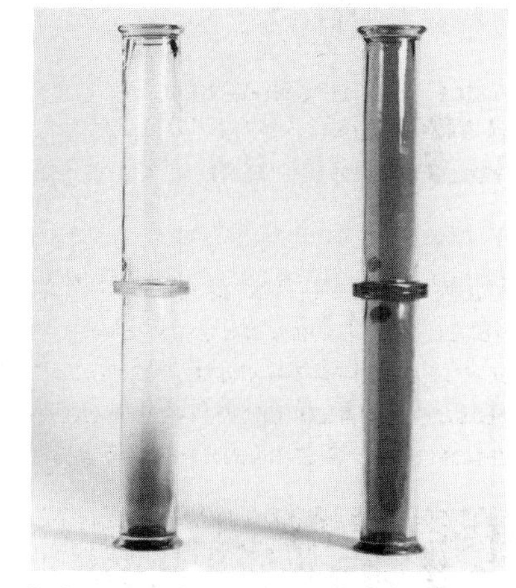
Before and after: the diffusion of bromine in air.

The movement of smoke

Smoke can be examined by trapping some in a small glass box, shining a light through it sideways, and looking at it with a microscope. The smoke specks show up as bright shiny spots that dance around jerkily. They are never still.

Explanation The smoke specks move because they are knocked about by moving particles of air. The air particles are themselves invisible under the microscope, because they are very small.
In the same way, pollen dances about on the surface of water, because it is bombarded by tiny moving water particles. The warmer the water, the faster the pollen moves. The movement of smoke and pollen was first discovered by a scientist named Robert Brown, about 150 years ago, so it is called **Brownian motion.**

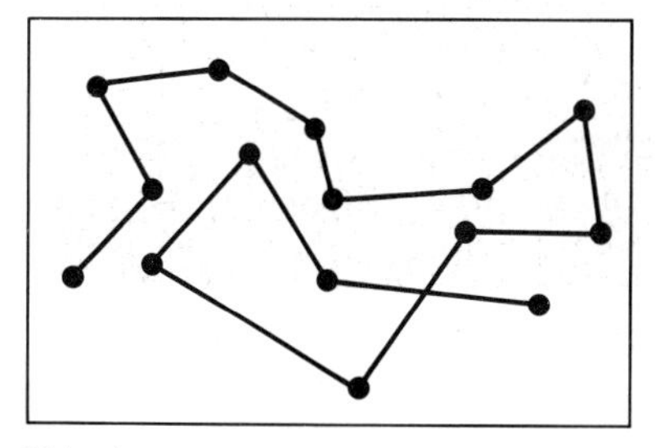
This shows the path of a single smoke speck. It changes direction because it is bombarded by air particles.

Questions

1 Why does the purple colour spread, when a crystal of potassium manganate(VII) is placed in water?
2 What is diffusion? Explain how bromine vapour diffuses into air.
3 What is Brownian motion? Sketch the movement of a particle during Brownian motion.
4 On a sunny day, you can sometimes see dust dancing in the air. Explain why the dust moves.

1.2 Solids, liquids and gases

It is easy to tell the difference between a solid, a liquid and a gas:

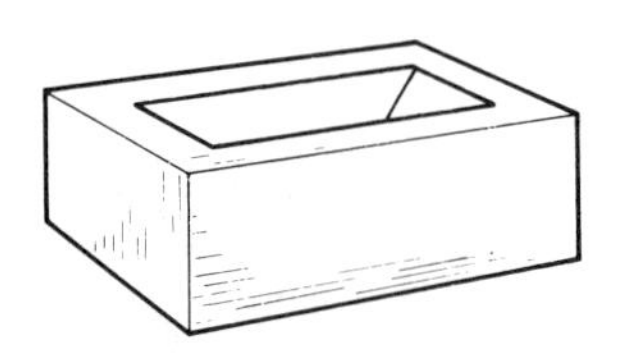

A solid has a definite shape and a definite volume.

A liquid flows easily. It has a definite volume but no definite shape. Its shape depends on the container.

A gas has neither a definite volume nor a definite shape. It completely fills its container. It is much lighter than the same volume of solid or liquid.

Water: solid, liquid and gas

Water can be a solid (ice), a liquid (water) and a gas (water vapour or steam). Its state can be changed by heating or cooling:

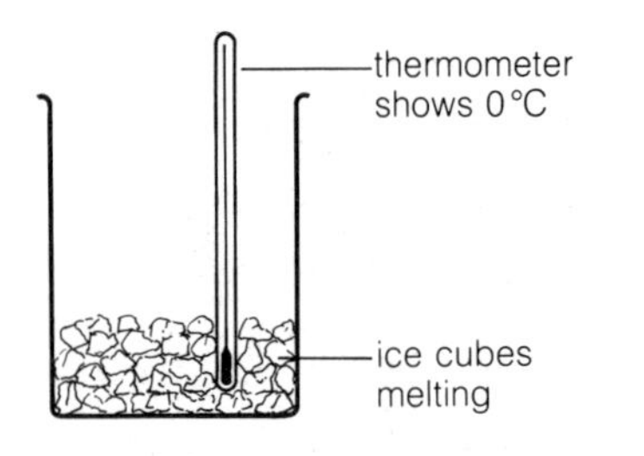

1 **Ice** slowly changes to **water**, when it is put in a warm place. This change is called **melting**. The thermometer shows 0 °C until all the ice has melted, so 0 °C is called its **melting point.**

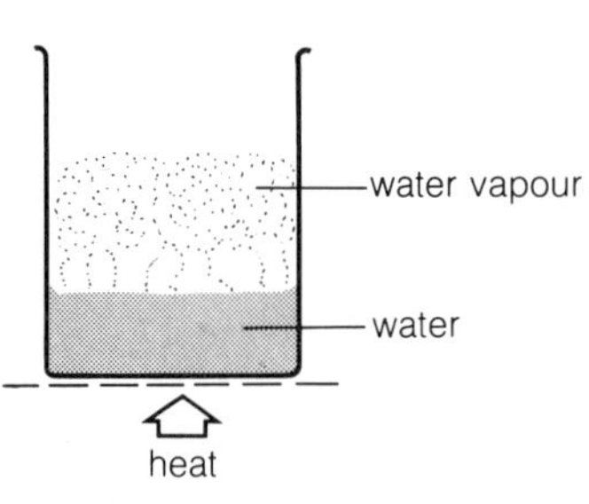

2 When the water is heated its temperature rises, and some of it changes to **water vapour**. This change is called **evaporation**. The hotter the water gets, the more quickly it evaporates.

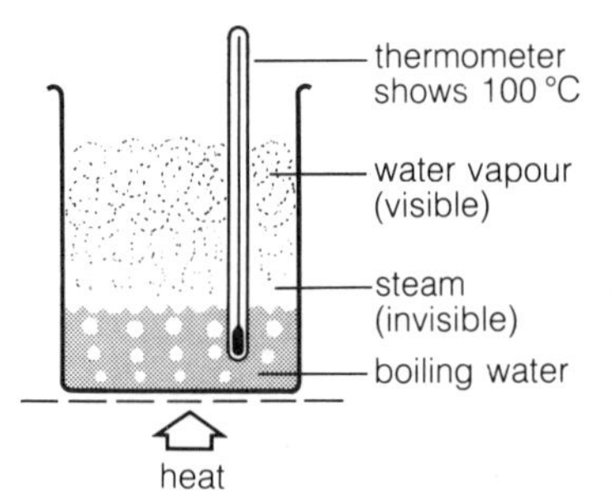

3 Soon bubbles appear. The water is **boiling**. Water vapour forms faster. It is now called **steam**. The thermometer shows 100 °C until all the water has changed to steam. 100 °C is the **boiling point** of water.

And when steam is cooled, the opposite changes take place:

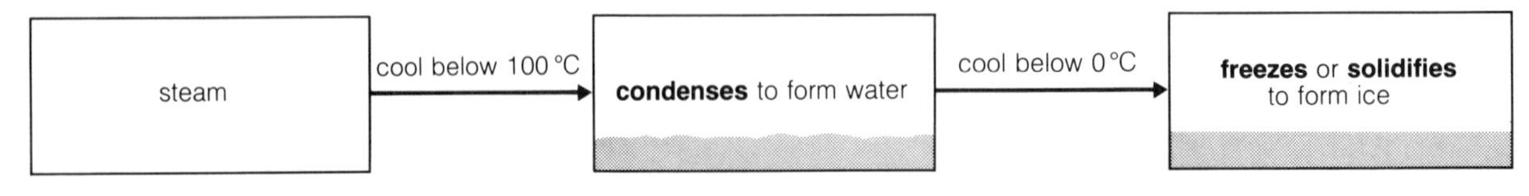

You can see that:
condensing is the opposite of evaporating;
freezing is the opposite of melting;
the freezing point of water is the same as the melting point of ice, 0 °C.

Other things can change state too

It is not only water that can change state – even iron and diamond can melt and boil. Some melting and boiling points are given in this table. Notice how different they can be:

Substance	Melting point/°C	Boiling point/°C
Oxygen	−219	−183
Ethanol	−15	78
Sodium	98	890
Sulphur	119	445
Iron	1540	2900
Diamond	3550	4832

In this play, the 'fog' was carbon dioxide, subliming from lumps on the stage floor.

A few substances change straight from solid to gas, without becoming liquid, when they are heated. This change is called **sublimation**. Carbon dioxide and iodine both sublime.

Using melting and boiling points

No two substances have the same pair of melting and boiling points. Besides, the melting and boiling points of a substance change if even a tiny amount of another substance is mixed with it. So melting and boiling points can give useful clues:

1 they can be used to **identify** a substance
2 they can be used to tell whether a substance is **pure**.
(A pure substance has nothing else mixed with it.)

For example:

This substance melts at 119 °C and boils at 445 °C. Can you identify it? (Look in the table.)

This water was taken from the sea. It freezes at about −2 °C and boils at about 101 °C. Is it pure?

A substance is not pure if it has other things mixed with it. These are called **impurities**. Sea water contains many impurities, mainly salt. **An impurity lowers the freezing point of a substance and raises its boiling point.**

Salt lowers the freezing point of water, which means it will also make ice melt. This truck spreads salt on icy roads to make them safe.

Questions

1 Write down two properties of a solid, two of a liquid and two of a gas.
2 What word means the same as freezing?
3 What word means the opposite of boiling?
4 Room temperature is taken as 20 °C. Which substance in the table above is:
  a a liquid at room temperature?
  b a gas at room temperature?
5 Which has the lower freezing point, oxygen or ethanol?
6 Give two ways in which melting points and boiling points can be useful.
7 What does *to identify a substance* mean?
8 What is a pure substance?
9 A sample of ethanol boils at 81 °C. What can you say about it?

1.3 The particles in solids, liquids and gases

You saw on page 9 that a substance can change from solid to liquid to gas. The individual *particles* of the substance are the same in each state. It is their *arrangement* that is different:

State	How the particles are arranged	Diagram of particles
Solid 	The particles in a solid are packed tightly in a fixed pattern. There are strong forces holding them together, so they cannot leave their positions. The only movements they make are tiny vibrations to and fro.	
Liquid 	The particles in a liquid can move about and slide past each other. They are still close together but are not in a fixed pattern. The forces that hold them together are weaker than in a solid.	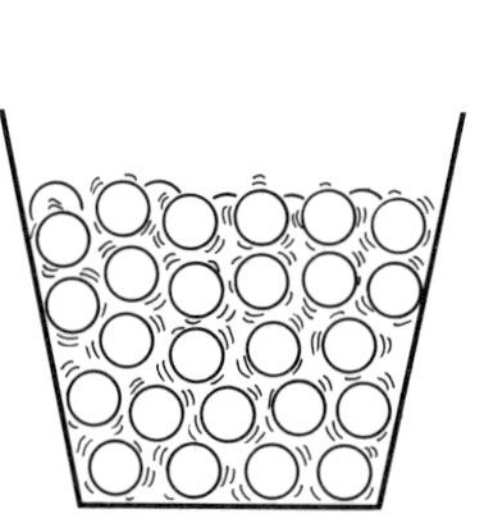
Gas 	The particles in a gas are far apart, and they move about very quickly. There are almost no forces holding them together. They collide with each other and bounce off in all directions.	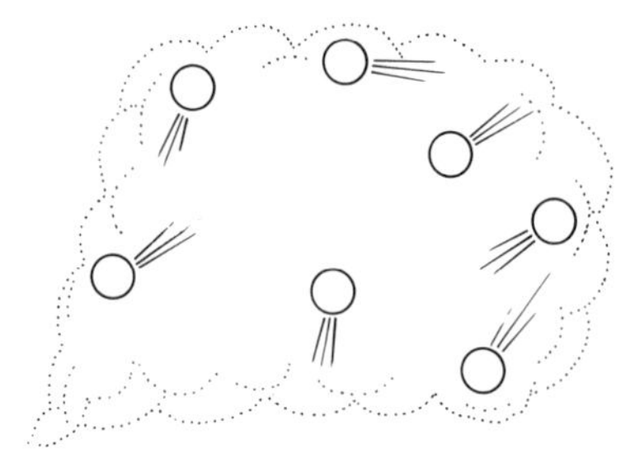

Changes of state

Melting When a solid is heated, its particles get more energy and vibrate more. This makes the solid **expand**. At the melting point the particles vibrate so much that they break away from their positions. The solid becomes a liquid.

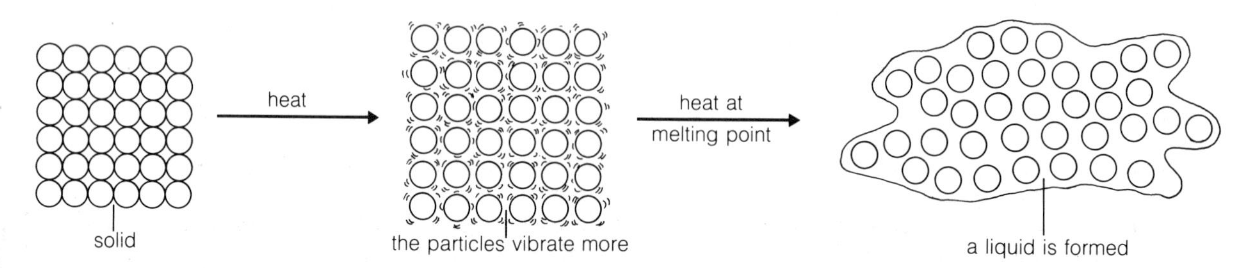

Boiling When a liquid is heated, its particles get more energy and move faster. They bump into each other more often and bounce further apart. This makes the liquid expand. At the boiling point, the particles get enough energy to overcome the forces holding them together. They break away from the liquid and form a gas.

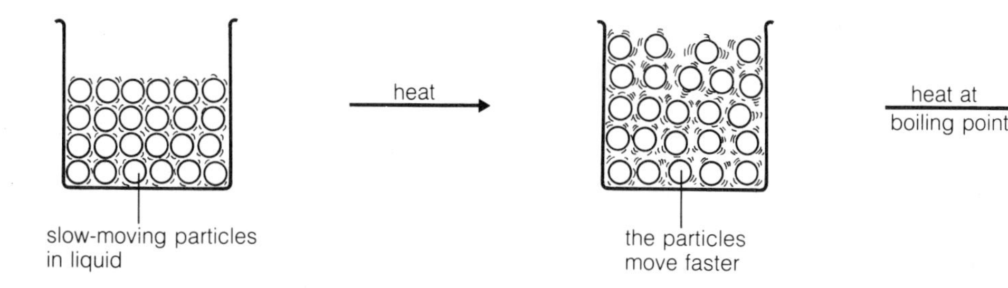

the particles get enough energy to escape

Evaporating Some particles in a liquid have more energy than others. Even when a liquid is well below boiling point, *some* particles have enough energy to escape and form a gas. This is called **evaporation**. It is why puddles of rain dry up in the sunshine.

Condensing and solidifying When a gas is cooled, the particles lose energy. They move more and more slowly. When they bump into each other, they do not have enough energy to bounce away again. They stay close together and a liquid forms. When the liquid is cooled, the particles slow down even more. Eventually they stop moving, except for tiny vibrations, and a solid forms.

Compressing a gas

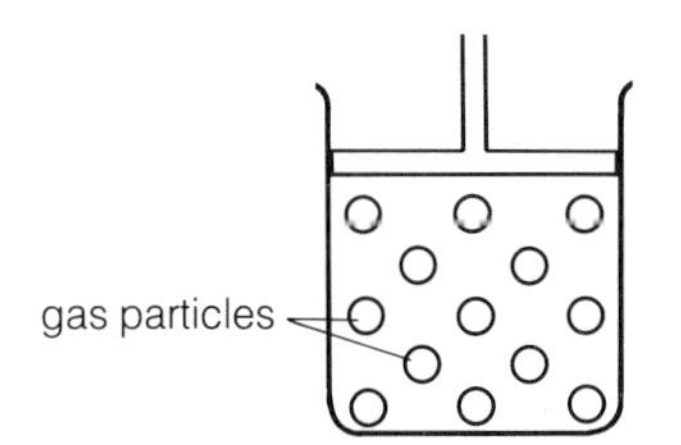

There is a lot of space between the particles in a gas. You can force the particles closer . . .

plunger pushed in

gas compressed into a smaller volume

. . . by pushing in the plunger. The gas gets squeezed or **compressed** into a smaller volume.

If enough force is applied to the plunger, the particles get so close together that the gas turns into a liquid. But liquids and solids can not be compressed because their particles are already close together.

This diver carries a compressed mixture of oxygen and helium so that he can breathe underwater.

Questions

1 Using the idea of particles, explain why:
   a it is easy to pour a liquid
   b a gas will completely fill any container
   c a solid expands when it is heated

2 Draw a diagram to show what happens to the particles in a liquid, when it boils.

3 Explain why a gas can be compressed into a smaller volume, but a solid can't.

1.4 A closer look at gases

When you blow up a balloon, you fill it with air particles moving at speed. The particles knock against the sides of the balloon and exert **pressure** on it. The pressure is what keeps the balloon inflated. In the same way, *all* gases exert pressure. The pressure depends on the **temperature** of the gas and the **volume** it fills, as you will see below.

When you blow air into a balloon the gas particles exert pressure on the balloon and make it inflate. The more you blow the greater the pressure.

How gas pressure changes with temperature

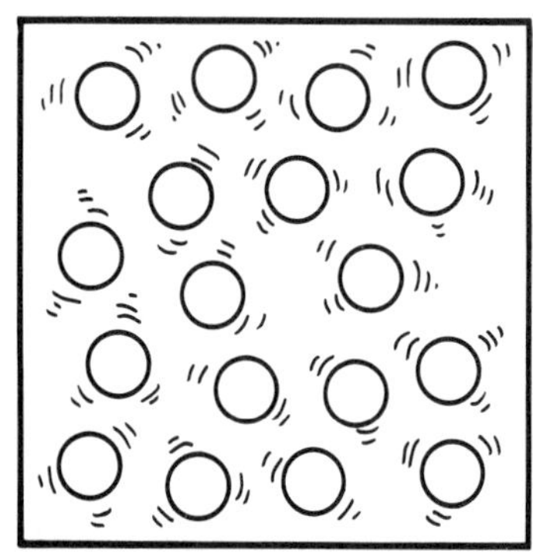

The particles in this gas are moving at speed. They knock against the walls of the container and exert pressure on them.

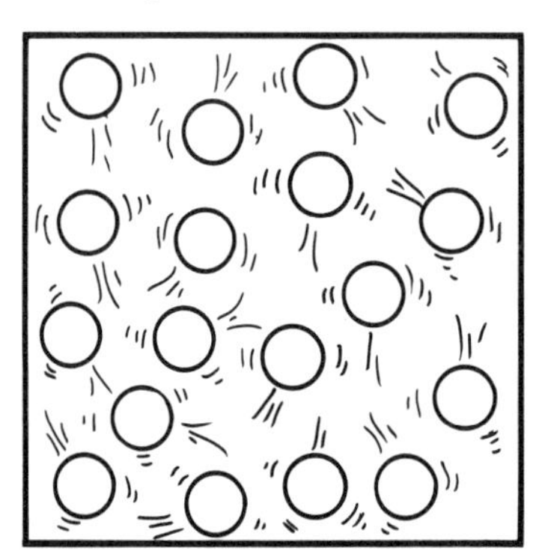

Now the gas is heated. The particles take in heat energy and move even faster. They strike the walls more often and with more force. So the gas pressure increases.

The same happens with all gases:
If the volume of a gas is constant, its pressure increases with temperature.

In a pressure cooker, water vapour is heated to well over 100°C. So it reaches very high pressure. It is dangerous to open a pressure cooker without first letting it cool.

How gas pressure changes with volume

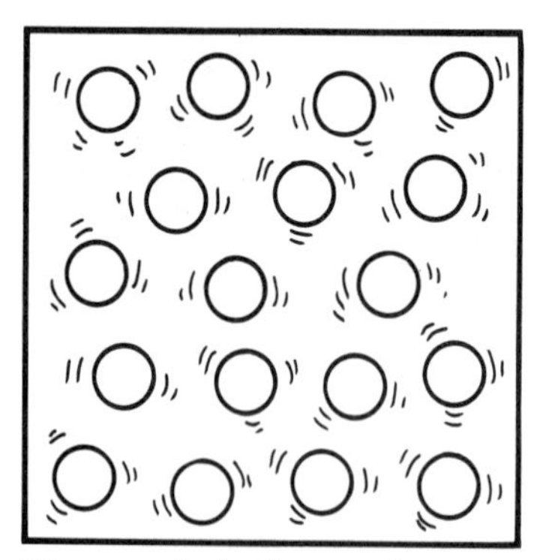

Here again is the gas from above. Its pressure is due to the particles colliding with the walls of the container.

This time the gas is squeezed into a smaller volume. The particles now hit the walls more often. So the gas pressure is greater.

The same thing is true for all gases:
When a gas is squeezed into a smaller volume, its pressure increases.

How gas volume changes with temperature

Now let's see what happens if the gas pressure is kept constant, but the temperature changed:

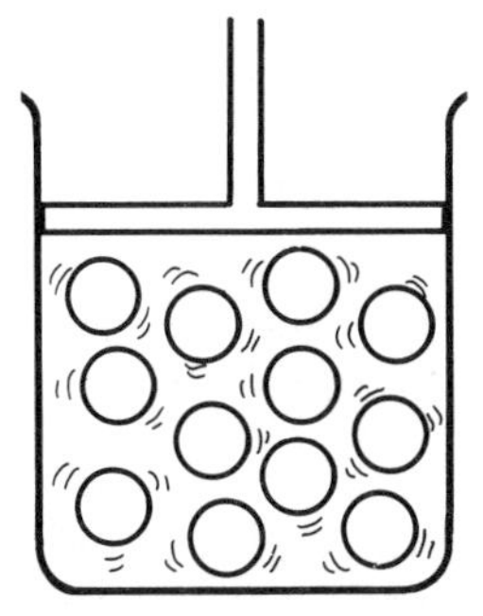

The plunger in this container can move freely in and out. When the gas is heated . . .

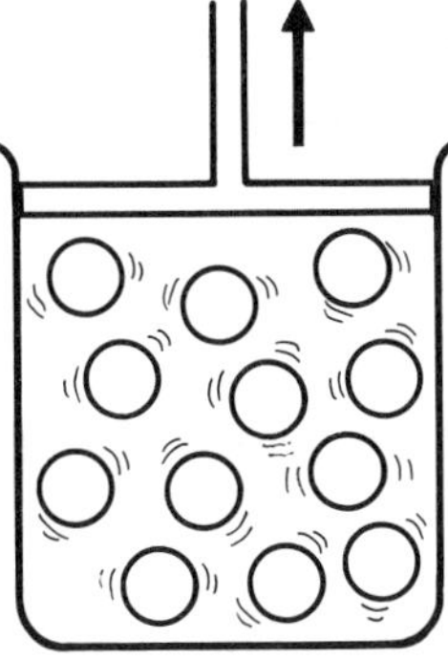

. . . the plunger moves out, so that the gas pressure stays the same. The gas now fills a larger volume.

This shows that:
If the pressure of a gas is constant, its volume increases with temperature.

The diffusion of gases

On page 7 you saw that gases **diffuse**. A particle of ammonia gas has about half the mass of a particle of hydrogen chloride gas. So will it diffuse faster? Let's see:

1. Cotton wool soaked in ammonia solution is put into one end of a long tube. It gives off ammonia gas.
2. At exactly the same time, cotton wool soaked in hydrochloric acid is put into the other end of the tube. It gives off hydrogen chloride gas.
3. The gases diffuse along the tube. White smoke forms where they meet.

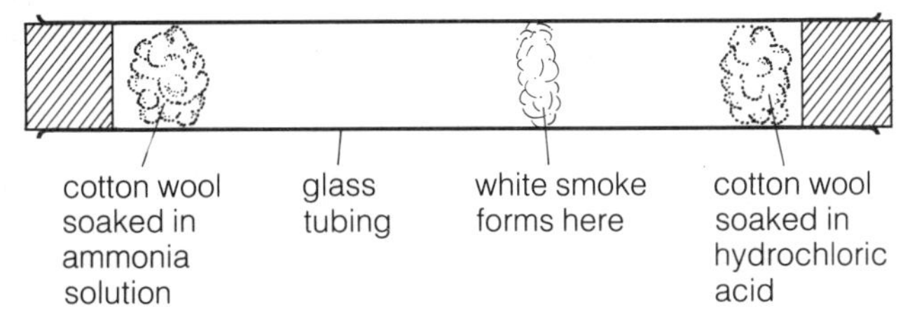

The white smoke forms closer to the right-hand end of the tube. So the ammonia particles have travelled further than the hydrogen chloride particles, in the same length of time.
The lighter the particles of a gas, the faster the gas will diffuse.

These heavenly farmyard smells! 'Smells' are caused by gas particles diffusing through the air. They dissolve in moisture in the lining of your nose. Then cells in the lining send a message to your brain.

Questions

1. What causes the pressure in a gas?
2. Why does a balloon burst if you keep on blowing?
3. A gas is in a sealed container. How do you think the pressure will change if the container is cooled? Explain your answer.
4. A gas flows from one container into a larger one. What happens to its pressure? Draw diagrams to explain.
5. Of all gases, hydrogen diffuses fastest. What can you tell from that?

1.5 Mixtures

Many things are not pure substances, they are **mixtures**. A mixture contains more than one substance. Here are some examples:

A mixture of water and carbon dioxide (liquid + gas)

A mixture of water, caffeine and other substances (liquid + solids)

A mixture of copper and nickel (solid + solid)

A mixture of water and ethanoic acid (liquid + liquid)

Solutions

Often it is hard to tell that something is a mixture. Sugar in water is one example: the mixture is clear and you cannot see the sugar. A mixture like this is called a **solution**.
The sugar is **soluble** in water, so it has **dissolved**. The sugar is called the **solute** and water is the **solvent**:

solute + solvent ⟶ solution

When sugar dissolves in water, its particles separate and spread out among the water particles. The separate particles are too small to be seen so the solution looks clear:

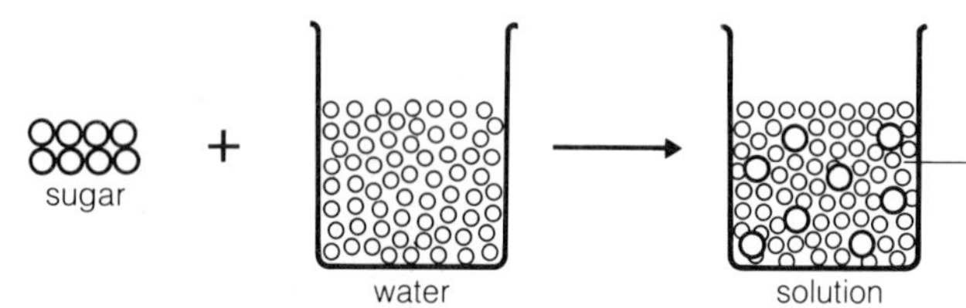

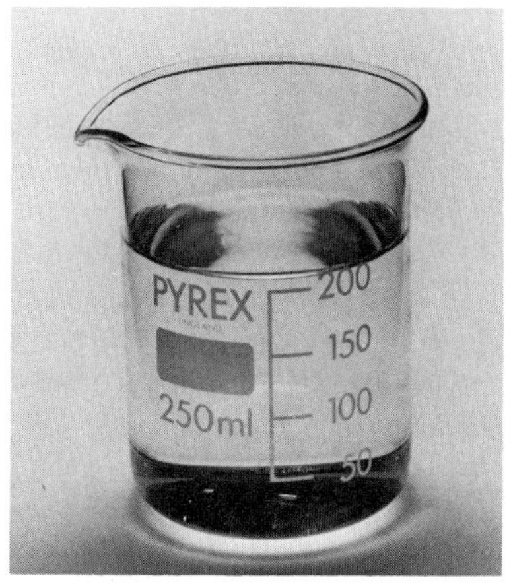

A mixture of sugar and water.

Suspensions

Sometimes it is easy to recognise a mixture. Look at the chalk in water on the right, for example. There are white specks in the water because the chalk has not dissolved – it is **insoluble** in water. A mixture like this is called a **suspension**.
In a suspension, the particles of solid do not all separate. Instead they stay in clusters that are large enough to be seen.

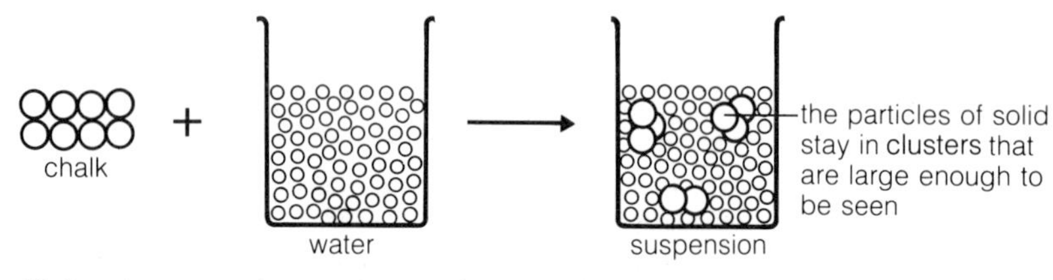

A mixture of chalk and water.

If the clusters of particles are heavy, they sink to the bottom and form a **sediment**.

Mixing liquids

When alcohol is added to water, the two liquids mix completely, forming a clear solution. They are said to be **miscible**. But when cooking oil is added to water, the two liquids do not mix properly. They are **immiscible**:

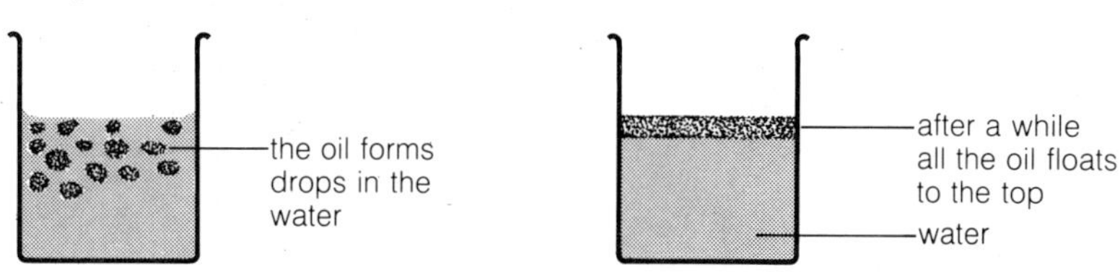

When cooking oil is *shaken* with water, the oil forms tiny drops in the water, and the mixture looks milky. This type of mixture is called an **emulsion**. Some common emulsions are:

milk – drops of butter fat in water
emulsion paint – drops of coloured oils in water
French dressing – drops of olive oil in vinegar

On shaking well, an emulsion of oil in vinegar forms.

Other solvents

When water is the solvent, the solution is called an **aqueous solution** (from *aqua*, the Latin word for water).
Water is the most common solvent, but many others are used in industry and about the house. They are needed to dissolve substances that are insoluble in water. Some examples are:

Solvent	It dissolves
White spirit	Gloss paint
Propanone	Grease, nail polish
Tetrachloromethane	Grease
Trichloroethane	The white substance in typists' correction fluid
Ethanol	Glues, printing inks, the scented substances used in perfumes and aftershaves

All the solvents above evaporate easily at room temperature – they are **volatile**. This means that glues, paints, and typists' correction fluid dry easily. Aftershaves feel cool because ethanol cools the skin when it evaporates.

Ethanol is used as a solvent for perfume. It is a volatile liquid. Why is that an advantage?

Questions

1 Write one sentence to explain each term:
suspension sediment aqueous solution
2 What is the difference between a solution and a suspension?
3 Explain what each of these words means:
insoluble immiscible
4 Is it a solution, suspension, or emulsion?
a lemonade **b** double cream
c toothpaste **d** shampoo
e chicken soup **f** white liquid bath cleaner
5 Name 3 solvents other than water, and give one use for each.

1.6 Solubility

Saturated solutions

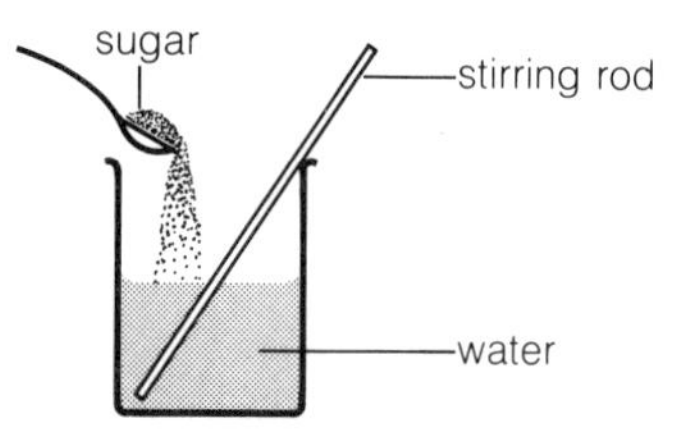

Sugar is soluble in cold water. Stirring helps it to dissolve . . .

sugar solution

but if you keep on adding sugar to the solution . . .

extra sugar sinks to bottom

eventually no more dissolves. The solution is **saturated**.

More sugar dissolves in hot water than in cold. Here is the same sugar solution at three different temperatures:

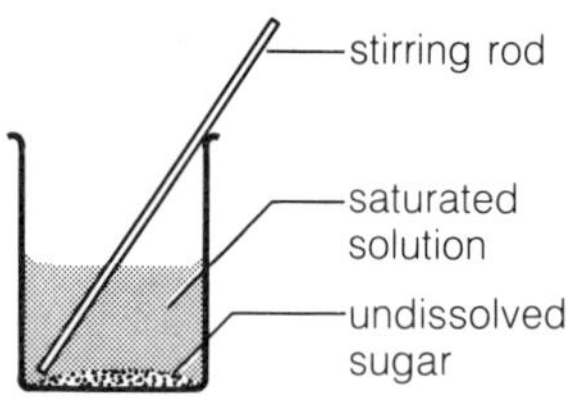

At 20 °C
This solution is saturated. You can tell by the undissolved crystals at the bottom.

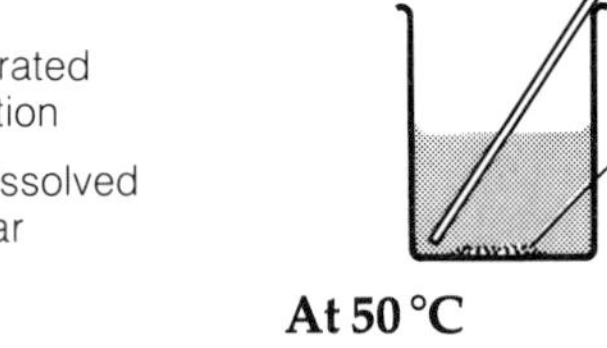

At 50 °C
Some more sugar has dissolved, but the solution is still saturated.

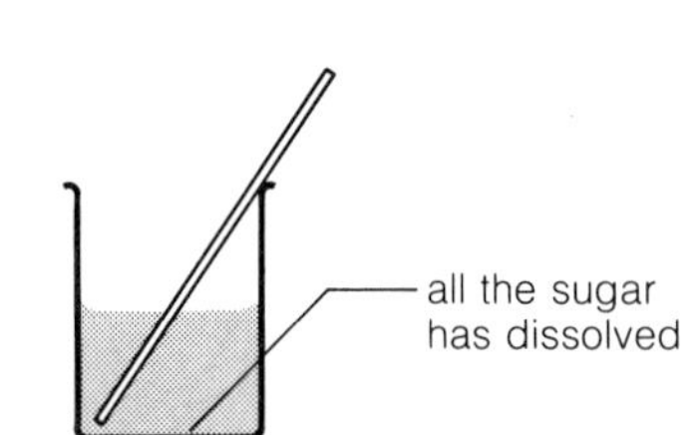

At 80 °C
Now the solution is *just* saturated. If extra sugar is added, it will not dissolve. It will sink to the bottom.

If the last solution above is heated to 100 °C, it will no longer be saturated. It will be able to dissolve more sugar. So you can see that saturation depends on temperature:
A saturated solution is one that can dissolve no more solute at a given temperature.

Solubility

Some solutes are more soluble in water than others. Compare these:

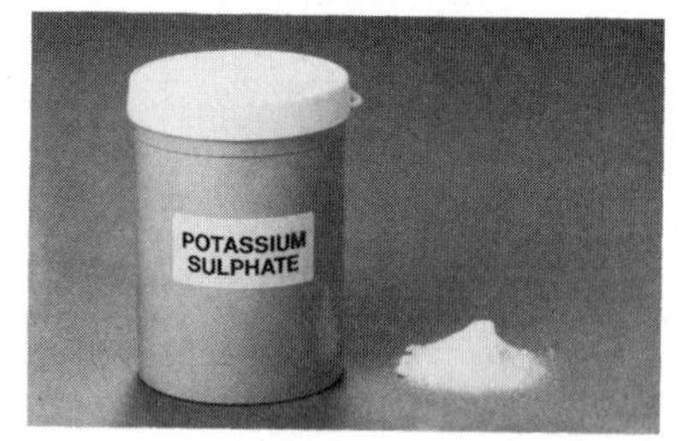

15 grams of this can dissolve in 100 grams of water at 50 °C.

39 grams of this can dissolve in 100 grams of water at 50 °C.

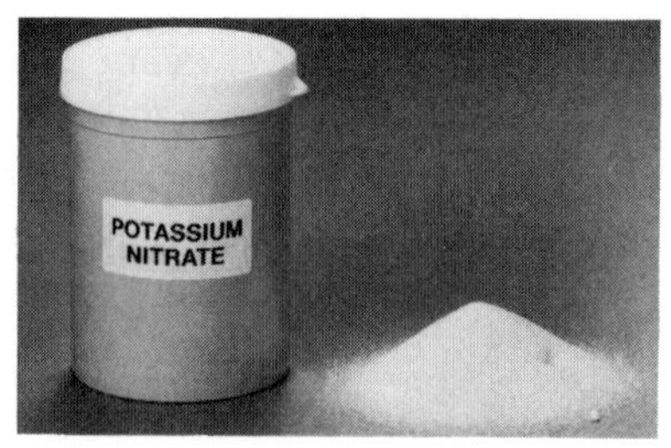

80 grams of this can dissolve in 100 grams of water at 50 °C.

For each solute, the amount shown gives a saturated solution at 50 °C, and is called the **solubility** of the solute.
The solubility of a solute in water, at a given temperature, is the maximum amount that can dissolve in 100 grams of water at that temperature.
So the solubility of sodium chloride in water at 50 °C is 39 grams.

Calculating solubility

Example 12 grams of potassium sulphate dissolve in 75 grams of water at 60 °C. What is its solubility in water at that temperature?

You need to find how much dissolves in *100 grams* of water:

in 75 grams of water, 12 grams dissolve, so

in 1 gram of water, $\frac{12}{75}$ grams dissolve, and

in 100 grams of water, $\frac{12}{75} \times 100$ grams dissolve. $\frac{12 \times 100}{75} = 16$

The solubility of potassium sulphate in water at 60 °C is **16 grams**.

Solubility curves

The table below shows that the solubility of copper(II) sulphate in water increases with temperature. The same is true for most solids.

Temperature/°C	0	10	20	30	40	50	60	70
Solubility of copper(II) sulphate/g	14	17	21	24	29	34	40	47

The table can be used to plot a graph like the one on the right. It is called a **solubility curve**. You can use it to find the solubility of copper(II) sulphate at any temperature from 0°C to 70°C. For example:
at 15°C its solubility is 19 grams
at 55°C its solubility is 37 grams.

The solubility curve for copper(II) sulphate

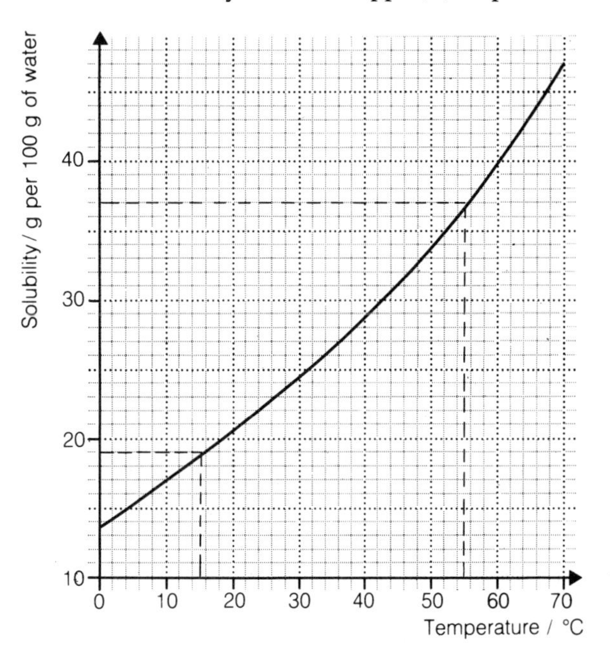

The solubility of gases

Unlike most solids, gases get *less* soluble in water as the temperature rises. For example:

Temperature/°C	0	20	40
Solubility of oxygen/g	0.007	0.004	0.003

Luckily such tiny amounts of dissolved oxygen are enough for fish and other water life.

Questions

1 What is a saturated solution?
2 The solubility of sodium chloride in water at 100 °C is 40 g. Explain what that means.
3 Find the solubility of X and Y in water at 60 °C, if at that temperature:
4 g of X dissolves in 25 g of water
21 g of Y dissolves in 50 g of water

4 Use the graph above to answer these:
a What is the solubility of copper(II) sulphate in water at: 46 °C? 68 °C?
b How much copper(II) sulphate will dissolve in 100 g of water at: 36 °C? 26 °C?
c What would you see when a saturated solution of copper(II) sulphate is cooled from 36 °C to 26 °C?

1.7 Separating mixtures (I)

Often, only one substance from a mixture is needed, so it has to be **separated** from the mixture. Below are some ways of doing this.

How to separate a solid from a liquid

By filtering Chalk can be separated from water by filtering the suspension through filter paper. The chalk gets trapped in the filter paper while the water passes through it. The chalk is called the **residue**. The water is called the **filtrate**.
Other suspensions can be separated in exactly the same way.

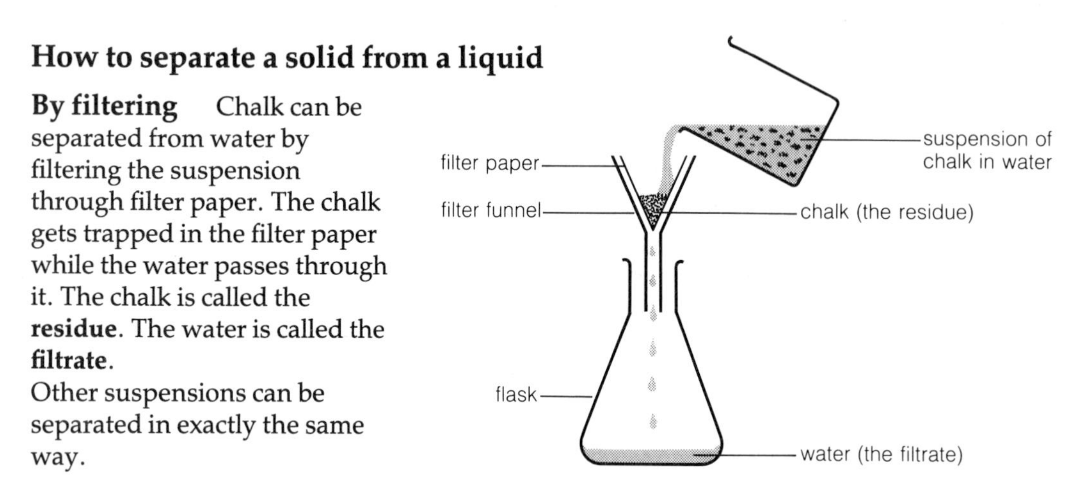

By centrifuging A centrifuge is used to separate *small* amounts of suspension. In a centrifuge, test-tubes of suspension are spun round very fast, so that the solid gets flung to the bottom:

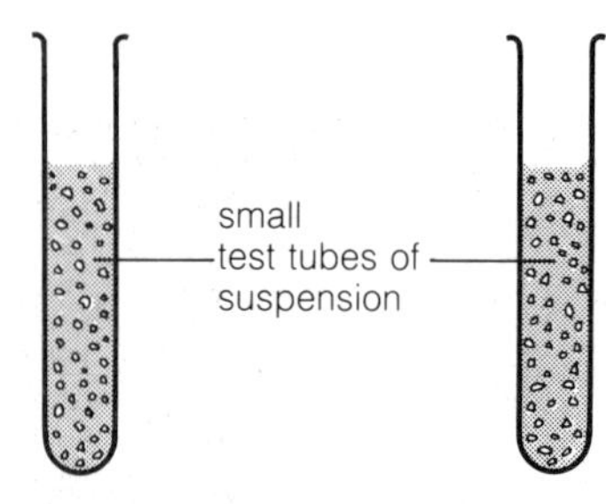

Before centrifuging, the solid is mixed all through the liquid.

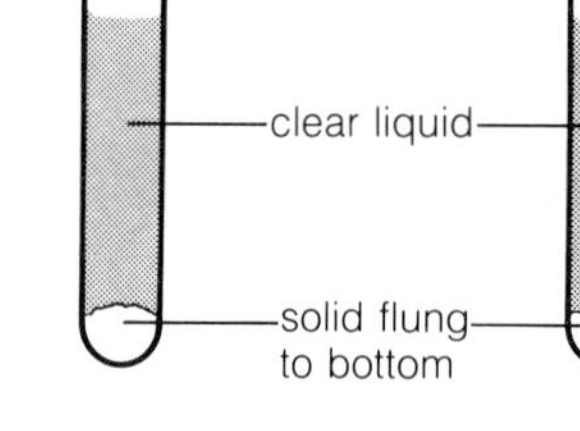

After centrifuging, all the solid has collected at the bottom.

A centrifuge.

The liquid can be **decanted** (poured out) from the test-tubes, or removed with a small pipette. The solid is left behind.

By evaporating the solvent If the mixture is a *solution*, the solid cannot be separated by filtering or centrifuging. This is because it is spread all through the solvent in tiny particles. Instead, the solution is heated so that the solvent evaporates, leaving the solid behind. Salt is obtained from its solution by this method:

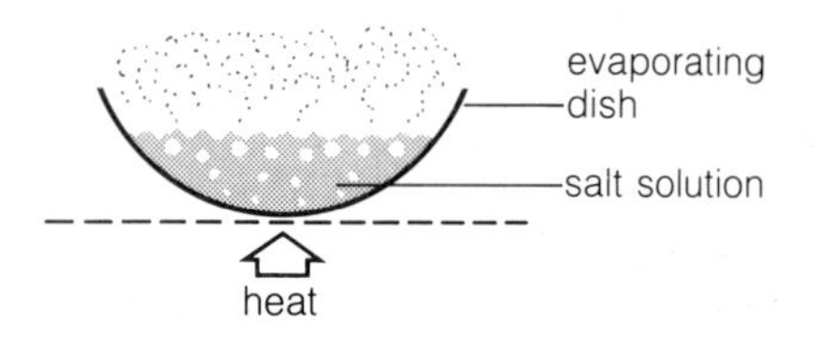

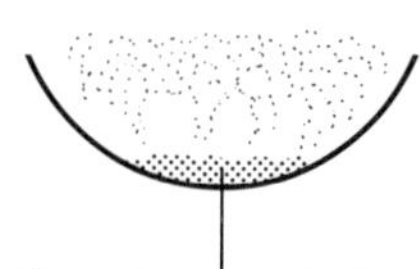

Evaporating the water from a salt solution.

By crystallising You can separate many solids from solution by letting them form crystals. Copper(II) sulphate is an example:

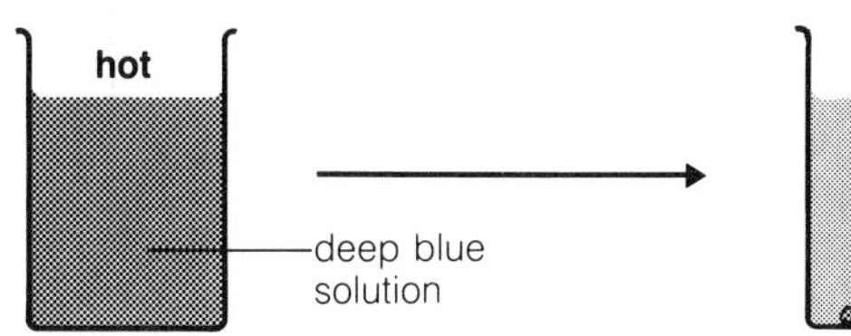

This is a saturated solution of copper(II) sulphate in water at 70 °C. If it is cooled to 20 °C . . .

crystals begin to appear, because the compound is *less soluble* at 20 °C than at 70 °C.

The process is called **crystallisation**. It is carried out like this:

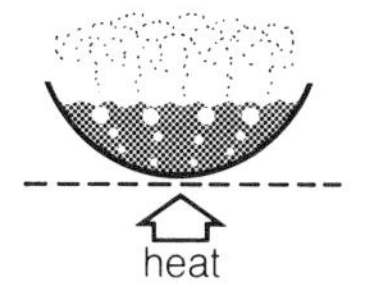

1 A solution of copper(II) sulphate is heated, to get rid of some water. As the water evaporates, the solution becomes more concentrated.

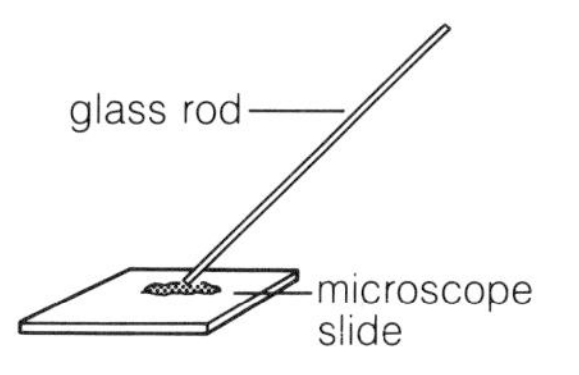

2 The solution can be checked to see if it is concentrated enough, by placing one drop on a microscope slide. Crystals should form quickly on the cool glass.

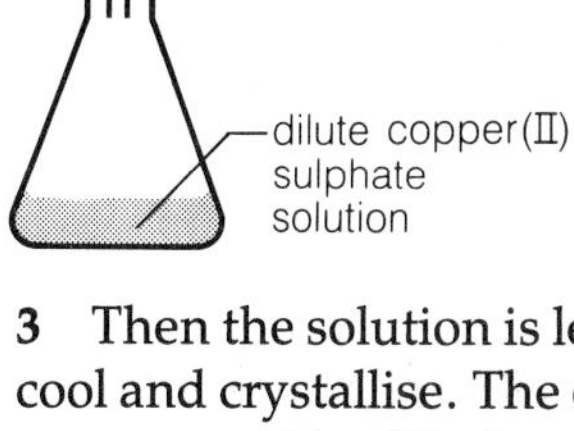

3 Then the solution is left to cool and crystallise. The crystals are removed by filtering, rinsed with water and dried with filter paper.

How to separate a mixture of two solids

By dissolving one of them A mixture of salt and sand can be separated like this:

1 Water is added to the mixture, and it is stirred. The salt dissolves.
2 The mixture is then filtered. The sand is trapped in the filter paper, but the salt solution passes through.
3 The sand is rinsed with water and dried in an oven.
4 The salt solution is evaporated to dryness.

This method works because the salt is soluble in water, and the sand is not. Water could *not* be used to separate salt and sugar, because it dissolves both of them. Ethanol could be used instead, because it dissolves sugar but not salt. Ethanol is inflammable, so it should be evaporated from the sugar solution over a water bath, as on the right.

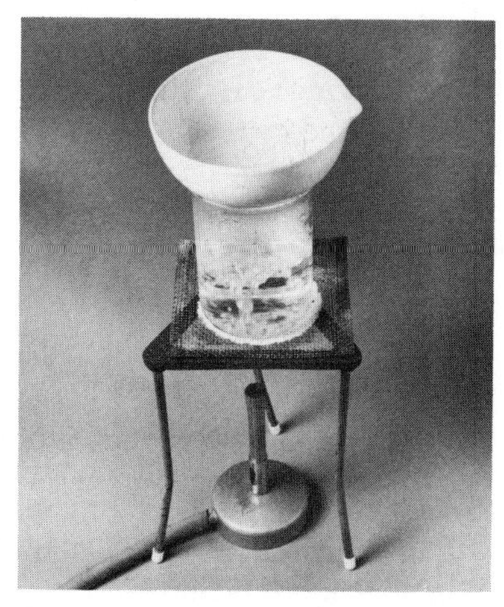

Evaporating ethanol from a sugar solution, over a water bath.

Questions

1 What is: a filtrate? a residue?
2 Describe two ways of separating the solid from the liquid in a suspension.
3 Sugar *cannot* be separated from sugar solution by filtering. Explain why.
4 What happens when a solution is evaporated?
5 Describe how you would crystallise potassium nitrate from its aqueous solution.
6 How would you separate salt and sugar? Mention any special safety precaution you would take.

1.8 Separating mixtures (II)

How to separate the solvent from a solution

By simple distillation This is a way of getting pure solvent out of a solution. The apparatus is shown on the right. It could be used to obtain pure water from salt water, for example. This is what happens:

1 The solution is heated in the flask. It boils, and steam rises into the condenser. The salt is left behind.
2 The condenser is cold, so the steam condenses to water in it.
3 The water drips into the beaker. It is completely pure. It is called **distilled water**.
This method could also be used to obtain pure water from sea water, or from ink.

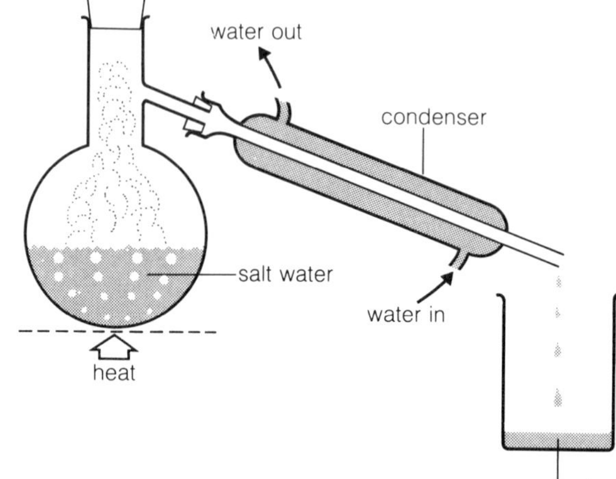

How to separate two liquids

Using a separating funnel If two liquids are **immiscible**, they can be separated with a separating funnel. For example, when a mixture of oil and water is poured into the funnel, the oil floats to the top, as shown on the right. When the tap is opened, the water runs out. The tap is closed again when all the water has gone.

oil
water
tap

By fractional distillation If two liquids are **miscible**, they must be separated by fractional distillation. The apparatus is shown below. It could be used to separate a mixture of ethanol and water, for example. These are the steps:

1 The mixture is heated. At about 78 °C, the ethanol begins to boil. Some water evaporates too, so a mixture of ethanol vapour and water vapour rises up the column.
2 The vapours condense on the glass beads in the column, making them hot.
3 When the beads reach about 78 °C, ethanol vapour no longer condenses on them. Only the water vapour does. The water drips back into the flask, while the ethanol vapour is forced into the condenser.
4 There it condenses. Liquid ethanol drips into the beaker.
5 Eventually, the thermometer reading rises above 78 °C. This is a sign that all the ethanol has been separated, so heating can be stopped.

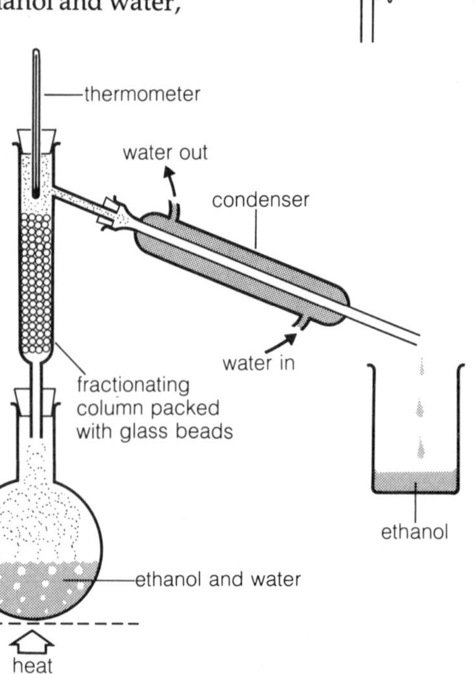

How to separate a mixture of coloured substances

Paper chromatography This method can be used to separate a mixture of coloured substances. For example, it will separate the coloured substances in black ink:

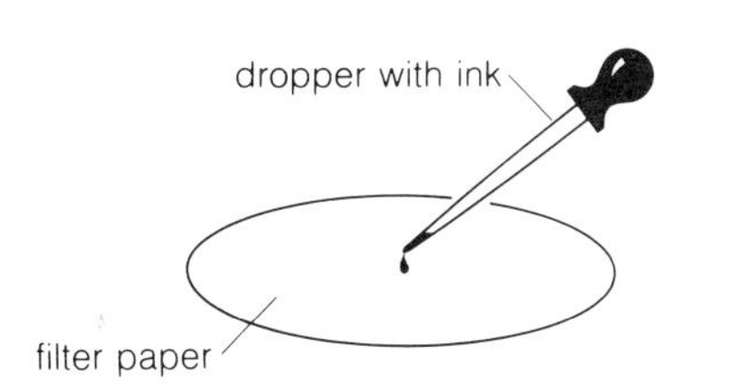

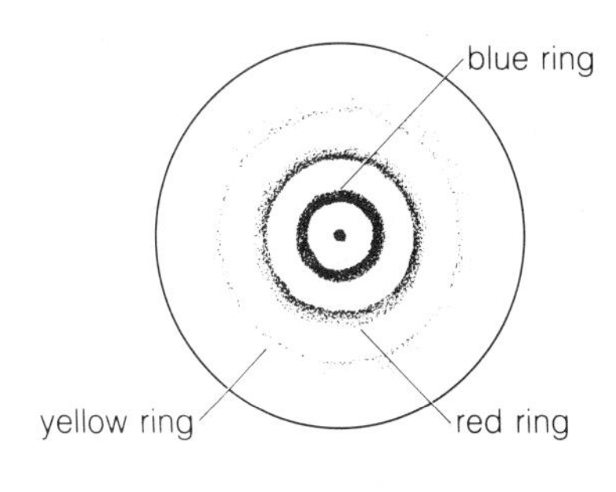

1 A small drop of black ink is placed at the centre of a piece of filter paper, and allowed to dry. Three or four more drops are added on the same spot.

2 Water is then dripped onto the spot, one drop at a time. The ink slowly spreads out into coloured rings.

3 Suppose there are three rings, yellow, red and blue. This shows that the ink contains three substances, coloured yellow, red and blue.

The filter paper with its coloured rings is called a **chromatogram**. In the example above, the outside ring is yellow. This shows that the yellow substance is the most soluble in water. When the water drips onto the ink spot, it dissolves the yellow substance the most easily, and carries it furthest away. Can you tell which substance is the least soluble?

Paper chromatography is often used to **identify** the substances in a mixture. For example, mixture X is thought to contain the substances A, B, C, and D, which are all soluble in propanone. The mixture could be checked like this:

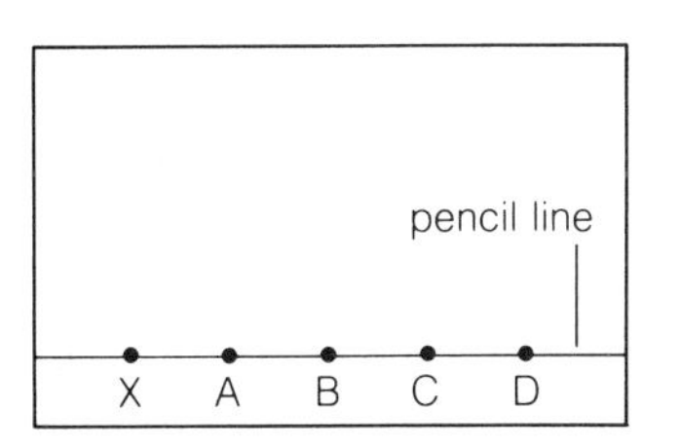

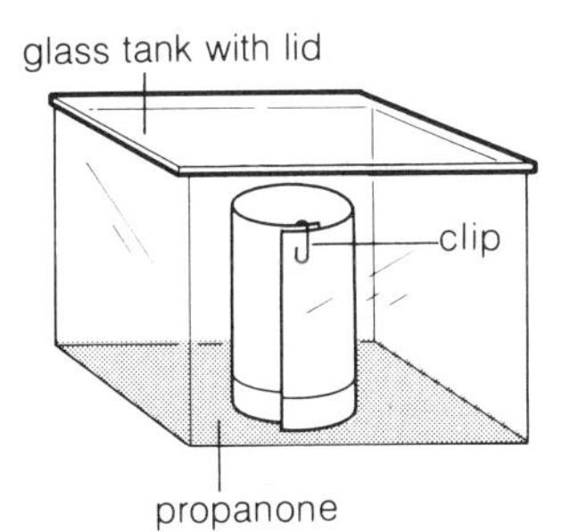

1 Some X, A, B, C, and D are dissolved in tiny amounts of propanone. A spot of each solution is placed on a line, on a sheet of filter paper, and labelled.

2 The paper is stood in a little propanone, in a covered glass tank. The solvent rises up the paper; when it has nearly reached the top, the chromatogram is taken out.

3 X has separated into three spots. Two of them are at the same height as A and B, so X must contain substances A and B. Does it also contain C and D?

Questions

1 How would you obtain pure water from ink? Draw the apparatus you would use, and explain how the method works.
2 Why are condensers so called? What is the reason for running cold water through them?
3 Water and turpentine are immiscible. How would you separate a mixture of the two?
4 Explain how fractional distillation works.
5 In the chromatogram above, how can you tell that X does not contain substance C?

Questions on Chapter 1

1 A large crystal of potassium manganate(VII) was placed in the bottom of a beaker of cold water and left for several hours.

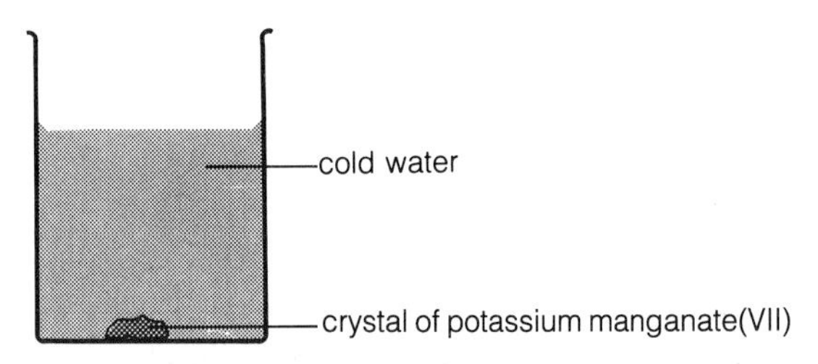

a Describe what would be seen after five minutes.
b Describe what would be seen after several hours.
c Explain your answers using the idea of particles.
d Name the TWO processes which have taken place during the experiment.

2 Describe the arrangement of particles in a solid, a liquid and a gas.

3 Draw diagrams to show what happens to the particles when:
a water freezes to ice
b steam condenses to water

4 The graph below is a heating curve for a pure substance. It shows how the temperature rises with time, when the solid is heated until it melts, and then the liquid is heated until it boils.

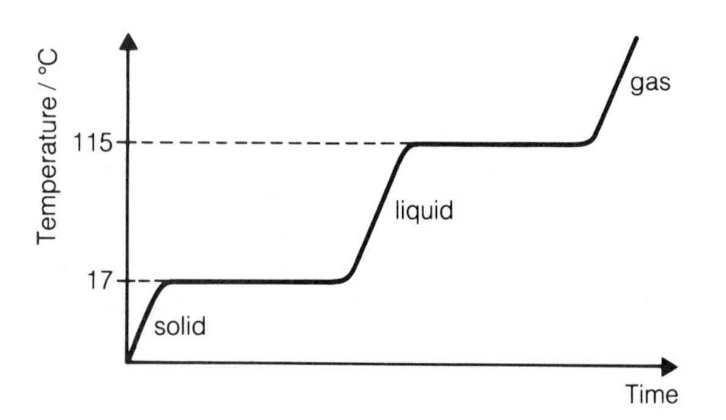

a What is the melting point of the substance?
b What is its boiling point?
c What happens to the temperature while the substance changes state?
d How can you tell that the substance is not water?

5 Sketch the heating curve for pure water, between − 10 °C and 110 °C. Mark in the temperatures at which water changes state, and its state for each sloping part of the graph.

6 A cooling curve shows how the temperature of a substance changes with time, as it is cooled from a gas to a solid. Below is the cooling curve for one substance:

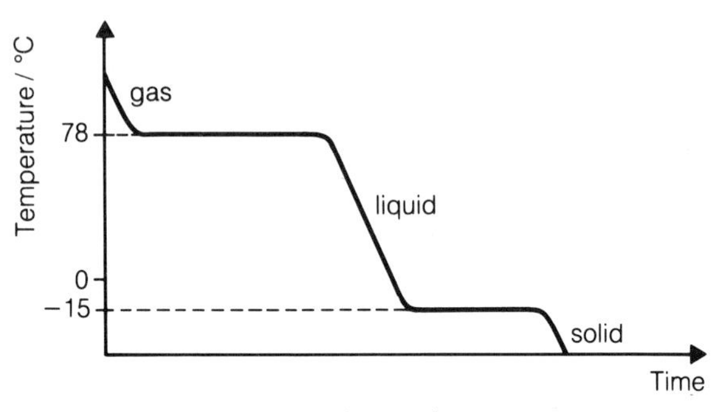

a What is the state of the substance at room temperature (20 °C)?
b Use the list of melting and boiling points on page 9 to identify the substance.

7 The graph shows the solubility curves for copper(II) sulphate (A) and sodium chloride (B), both plotted on the same axes.

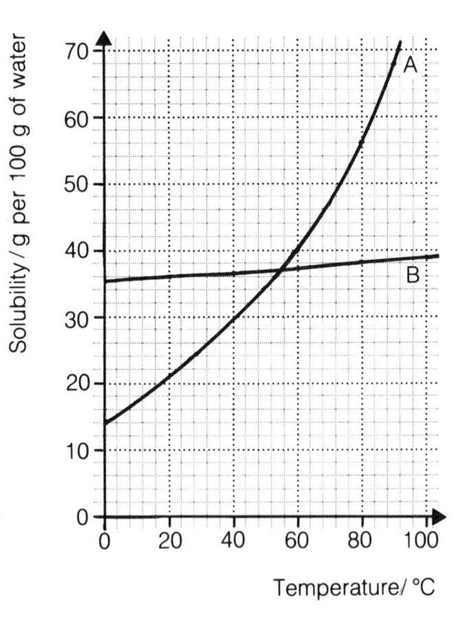

a What is the solubility of each substance at:
i 40 °C? ii 60 °C?
b Which substance is more soluble at:
i 30 °C? ii 70 °C?
c At what temperature do both substances have the same solubility?
d At what temperature would 45 g of copper(II) sulphate just dissolve in 100 g of water?
e Would it be possible to dissolve 50 g of sodium chloride in 100 g of water?

8 Look again at the solubility curves in question 7.

a If a saturated solution of copper(II) sulphate in 100 g of water is cooled from 80 °C to 20 °C, how much copper(II) sulphate will crystallise out of solution?

b If a saturated solution of sodium chloride in 100 g of water is cooled from 80 °C to 20 °C, how much sodium chloride will crystallise out of solution?

c To obtain sodium chloride from its solution in water, the solution must be evaporated to dryness, rather than left to crystallise. Why is this?

9 The graph below shows how the solubility of calcium hydroxide varies with temperature.

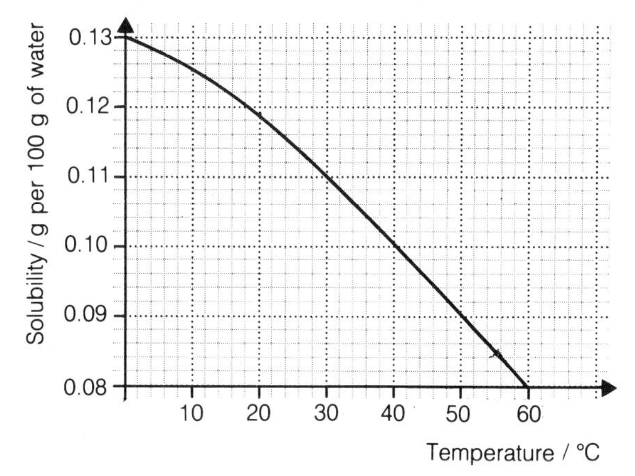

a What is unusual about the solubility of calcium hydroxide?

b What is its solubility at 50 °C?

c What would be seen if a saturated solution of calcium hydroxide was heated slowly to its boiling point?

10 The table below shows the solubilities of some *gases* in water at different temperatures.
The solubilities are in g per 100 g of water.

Gas	Temperature/ °C		
	0	**20**	**40**
Ammonia	90	53	31
Hydrogen chloride	82	72	63
Oxygen	0.007	0.004	0.003
Nitrogen	0.003	0.002	0.001

a What is the solubility in water at 40 °C of:
i ammonia? **ii** oxygen?

b Which two of the gases are only very slightly soluble in water?

c Do the gases become more soluble or less soluble as the temperature rises?

d How could the solubility of a gas in water be increased at a given temperature? (Think of fizzy drinks.)

11 Describe the relationship that exists between:

a gas volume and pressure

b gas volume and temperature

c gas pressure and temperature

12

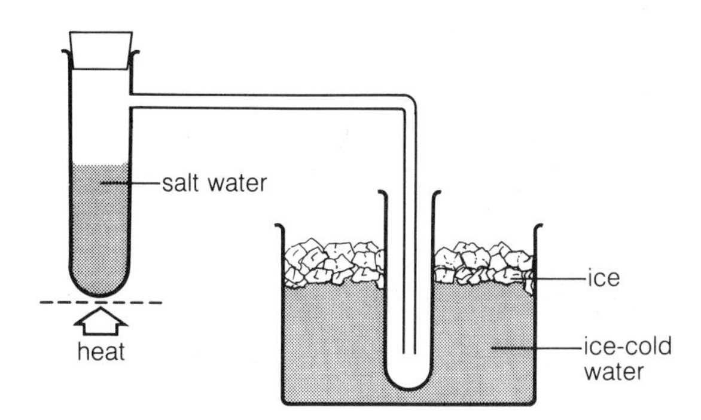

The apparatus above can be used to obtain pure water from salt water.

a What is the purpose of the ice-cold water?

b Why must the glass arm from the first tube reach far down into the second tube?

c Explain how the method works.

d What is this separation method called?

13 A mixture of salt and sugar has to be separated, using the solvent ethanol.

a Which of the two substances is soluble in ethanol?

b Draw a diagram to show how you would separate the salt.

c How could you obtain sugar crystals from the sugar solution, *without* losing the ethanol in the process?

d Draw a diagram of the apparatus for **c**.

14 Eight coloured substances were spotted on a piece of filter paper, which was then stood in a covered glass tank containing a little propanone. Three of the substances were the basic colours, red, blue and yellow. The others were dyes, labelled A, B, C, D, E. The resulting chromatogram is shown below:

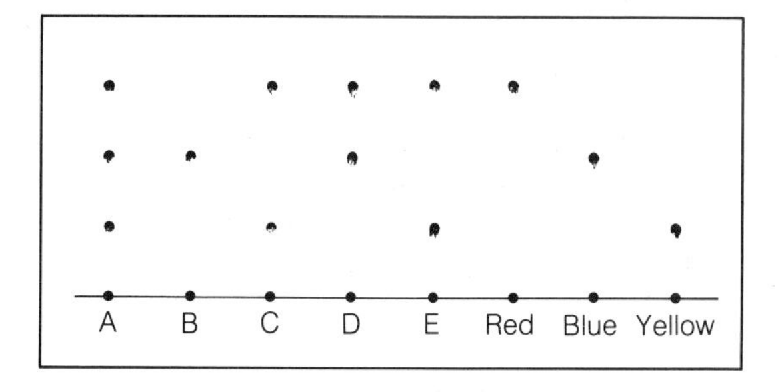

a Which dye contains only one basic colour?

b Which dye contains all three basic colours?

c Which basic colour is most soluble in propanone?

2.1 Atoms, elements and compounds

Atoms

This piece of sodium is made of billions of tiny particles called **sodium atoms**.

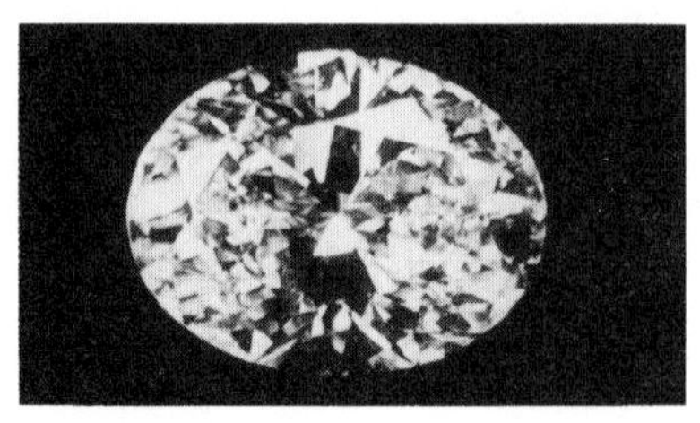

Diamond is a form of carbon. This diamond is made of billions of **carbon atoms**, which are different from sodium atoms.

Mercury is made of **mercury atoms**, which are different from both sodium atoms and carbon atoms.

Single atoms are far too small to be seen, even with the most powerful microscope. For example, about four billion sodium atoms would fit side-by-side on the full stop at the end of this sentence. However, in spite of their small size, scientists have managed to find out a great deal about atoms. They have found that every atom consists of a **nucleus**, and a cloud of particles called **electrons** that whizz non-stop round the nucleus.
The drawing on the right shows what a sodium atom might look like, greatly magnified.
(You can find out more about the nucleus and electrons on page 26.)

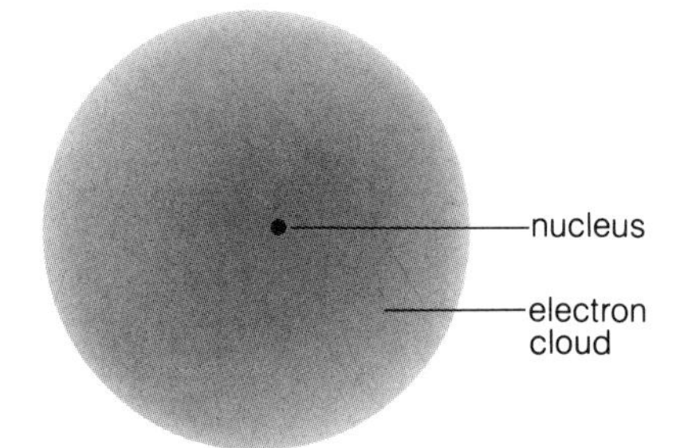

Elements

Sodium is made of sodium atoms only, so it is an **element**.
An element is a substance that is made of only one kind of atom.
Diamond (carbon) and mercury are also elements.
Altogether, 105 different elements are known. Of these, 90 have been obtained from the earth's crust and atmosphere, and 15 have been artificially made by scientists.
Every element has a name and a symbol. Here are some of them:

Element	Symbol	Element	Symbol
Aluminium	Al	Bromine	Br
Copper	Cu	Carbon	C
Iron	Fe	Chlorine	Cl
Lead	Pb	Hydrogen	H
Magnesium	Mg	Nitrogen	N
Mercury	Hg	Oxygen	O
Potassium	K	Phosphorus	P
Silver	Ag	Sulphur	S
Sodium	Na	Silicon	Si

It is easy to remember that the symbol for <u>al</u>uminium is Al, and for <u>c</u>arbon is C. But some symbols are harder to remember, because they are taken from the Latin names for the elements. For example: potassium has the symbol K, from its Latin name <u>k</u>alium. Sodium has the symbol Na, from its Latin name <u>na</u>trium.

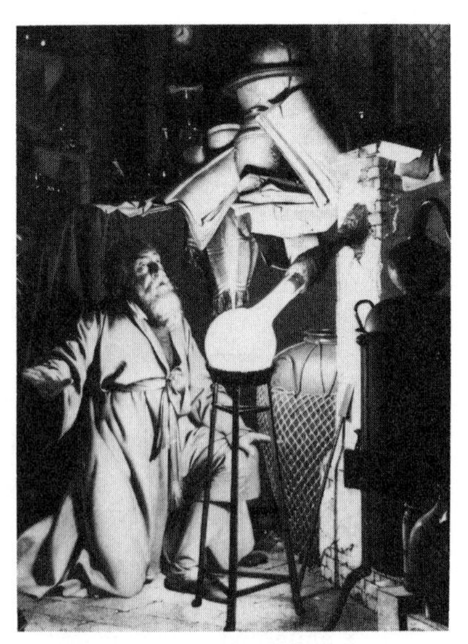

A few elements, such as copper, have been known for thousands of years. But most were discovered in the last 400 years. This shows Henry Brand, the German alchemist, on his discovery of phosphorus in 1659. He extracted it from urine, by accident, during his search for the elixir of life. To his amazement it glowed in the dark.

The metals The elements in the list on page 24 are in two columns for a good reason. The ones on the left are **metals**, while those on the right are **non-metals**.
Over 80 of the elements are metals. Although they all look different, they have many properties in common. Here are a few:

1 They allow electricity and heat to pass through them easily—they are good **conductors** of electricity and heat.
2 They are all solids at room temperature, except mercury, and most of them have high melting points.
3 Most of them can be hammered into different shapes (they are **malleable**) and drawn into wires (they are **ductile**).

The non-metals Only about one-fifth of the elements are non-metals. They are quite different from metals:

1 They are poor conductors of electricity and heat. (Carbon is an exception.)
2 They usually have low melting points (eleven of them are gases and one is a liquid at room temperature).
3 When solid non-metals are hammered, they break up—they are **brittle**.

Nerves of steel? 16 elements (including iron, but mainly oxygen, carbon, nitrogen, calcium and phosphorus) combine to make the hundreds of compounds in the human body.

Compounds

Elements can combine with each other to form **compounds**.
A compound contains atoms of different elements joined together.
Although there are only 105 elements, there are millions of compounds. This table shows three common ones:

Name of compound	Elements in it	How the atoms are joined up
Water	Hydrogen and oxygen	O; H H
Carbon dioxide	Carbon and oxygen	O C O
Ethanol	Carbon, hydrogen and oxygen	H H; H C C O H; H H

Symbols for compounds The symbol for a compound is called its **formula**. It is made up from the symbols of the elements. The formula for water is $\mathbf{H_2O}$ and for ethanol is $\mathbf{C_2H_5OH}$. Can you see why? Can you guess the formula for carbon dioxide? (Check your answer on page 60.) Note that the plural of **formula** is **formulae**.

Questions

1 What is an atom?
2 What is the centre part of an atom called?
3 Explain what an element is.
4 Explain what these words mean:
malleable ductile brittle
5 Write down three properties of non-metals.
6 What is: **a** a compound? **b** a formula?
7 What is H_2O? What does the $_2$ in it show?
8 A certain compound has the formula NaOH. What elements does it contain?

2.2 More about the atom

Protons, neutrons and electrons

On page 24 you saw that all atoms consist of a **nucleus** and a cloud of **electrons** that move round the nucleus. The nucleus is itself a cluster of two sorts of particles, **protons** and **neutrons**.
All the particles in an atom are very light. Their mass is measured in **atomic mass units**, rather than grams. Protons and electrons also have an **electric charge**:

Particle in atom	Mass	Charge
Proton ○	1 unit	Positive charge (+1)
Neutron ●	1 unit	None
Electron •	Almost nothing	Negative charge (−1)

The nucleus is very tiny compared with the rest of the atom. If the atom was the size of a football stadium, the nucleus (sitting on the centre spot) would be the size of a pea!

How the particles are arranged

The sodium atom is a good one to start with. It has **11** protons, **11** electrons and **12** neutrons. They are arranged like this:

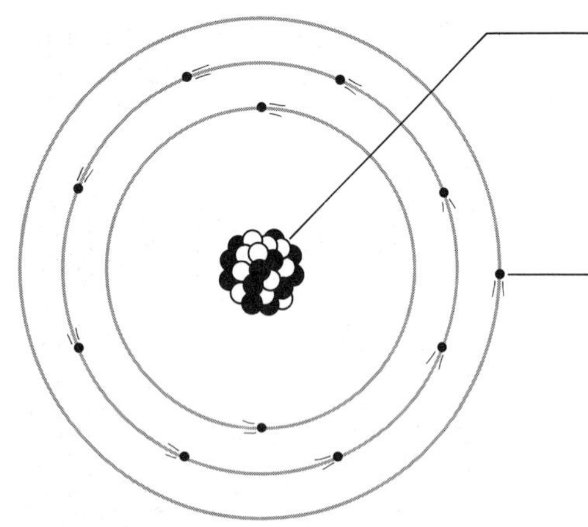

The different levels for the electrons are called **electron shells**. Each shell can hold only a limited number of electrons:

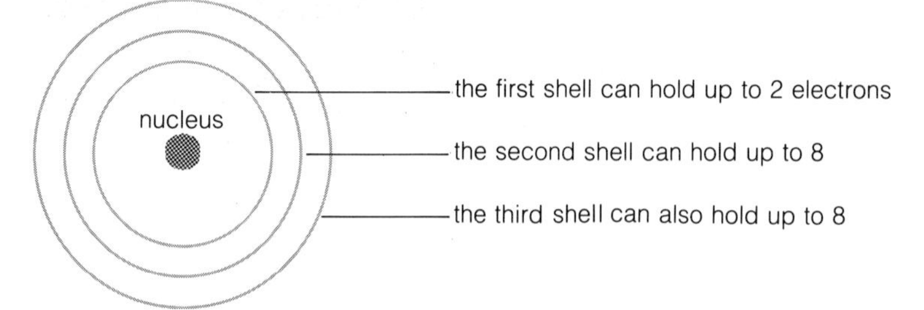

Notice how the electrons are arranged in the sodium atom:
2 in the first shell (it is full)
8 in the second shell (it is also full)
1 in the third shell (it is not full)
The atom is often written as **Na (2,8,1)**. Can you see why?
The (2,8,1) is its electron arrangement or **electronic configuration**.

Niels Bohr, a Danish scientist, was the first person to put forward the idea of electron shells. He died in 1962.

Atomic number and mass number

Atomic number Look again at the sodium atom on the opposite page. It has **11** protons. This fact could be used to identify it, because *only* a sodium atom has 11 protons. Every other sort of atom has a different number of protons.
The number of protons in an atom is therefore an important number. It is given a special name. It is called the **atomic number**.
The atomic number = the number of protons in an atom.
The atomic number of sodium is 11.

The sodium atom also has **11** electrons. So it has an equal number of protons and electrons. The same is true for every sort of atom:
Every atom has an equal number of protons and electrons.
Because of this, atoms have no overall charge. The charge on the electrons cancels the charge on the protons. You can check this for a sodium atom, on the right.

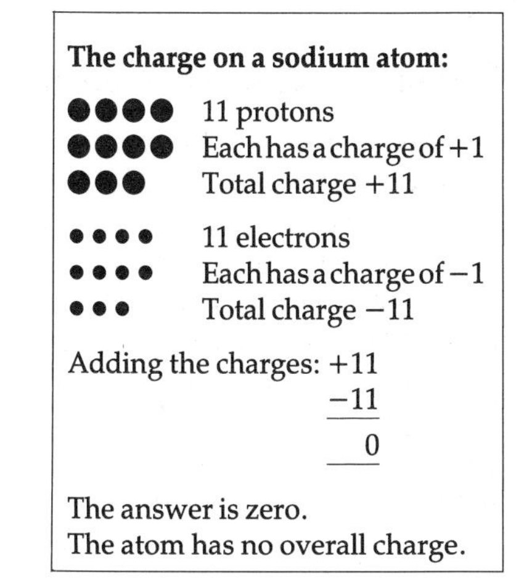

The charge on a sodium atom:

11 protons
Each has a charge of +1
Total charge +11

11 electrons
Each has a charge of −1
Total charge −11

Adding the charges: +11
−11
0

The answer is zero.
The atom has no overall charge.

Mass number The electrons in an atom have almost no mass. So the mass of an atom is nearly all due to its protons and neutrons. For this reason, the number of protons and neutrons in an atom is called its **mass number**.
The mass number = the number of protons + neutrons in an atom.
A sodium atom has 11 protons and 12 neutrons, so the mass number of sodium is 23.

Since the atomic number is the number of protons only, then:
mass number − atomic number = number of neutrons.
So, for a sodium atom, the number of neutrons = (23 − 11) = 12.

A hundred years ago, hardly anything was known about the atom. For example, the neutron was discovered only in 1932, by this British scientist Sir James Chadwick.

Shorthand for an atom

The sodium atom can be described in a short way, using:
the symbol for sodium (Na)
its atomic number (11)
its mass number (23)

The information is written as $^{23}_{11}\mathrm{Na}$.

From it you can tell that the sodium atom has 11 protons, 11 electrons and 12 neutrons (23 − 11 = 12). Chemists often describe atoms in this way. The information is always put in the same order:

$$^{\text{mass number}}_{\text{atomic number}}\text{Symbol}$$

Questions

1 Name the particles that make up the atom.
2 Which particle has:
  a a positive charge? b no charge?
  c almost no mass?
3 Draw a sketch of the sodium atom.
4 What does *electronic configuration* mean?
5 Write down the meaning of these terms:
  a atomic number b mass number
6 Name each of these atoms, and say how many protons, electrons and neutrons it has:
  $^{12}_{6}\mathrm{C}$ $^{16}_{8}\mathrm{O}$ $^{24}_{12}\mathrm{Mg}$ $^{27}_{13}\mathrm{Al}$ $^{64}_{29}\mathrm{Cu}$

2.3 Some different atoms

What makes two atoms different?

On page 26 you saw that sodium atoms have 11 protons. This is what makes them different from all other atoms. *Only* sodium atoms have 11 protons, and any atom with 11 protons *must* be a sodium atom.
In the same way, an atom with 6 protons must be a carbon atom, and an atom with 7 protons must be a nitrogen atom:
You can tell what an atom is from the number of protons it has.

The first twenty elements There are 105 elements altogether. Of these, hydrogen has the smallest atoms, with only 1 proton each. Helium atoms have 2 protons each, lithium atoms have 3 protons each, and so on up to hahnium atoms, which have 105 protons each. Below are the first twenty elements, arranged in order according to their number of protons:

Element	Symbol	Number of protons (atomic number)	Number of electrons	Number of neutrons	Number of protons + neutrons (mass number)
Hydrogen	H	1	1	0	1
Helium	He	2	2	2	4
Lithium	Li	3	3	4	7
Beryllium	Be	4	4	5	9
Boron	B	5	5	6	11
Carbon	C	6	6	6	12
Nitrogen	N	7	7	7	14
Oxygen	O	8	8	8	16
Fluorine	F	9	9	10	19
Neon	Ne	10	10	10	20
Sodium	Na	11	11	12	23
Magnesium	Mg	12	12	12	24
Aluminium	Al	13	13	14	27
Silicon	Si	14	14	14	28
Phosphorus	P	15	15	16	31
Sulphur	S	16	16	16	32
Chlorine	Cl	17	17	18	35
Argon	Ar	18	18	22	40
Potassium	K	19	19	20	39
Calcium	Ca	20	20	20	40

Drawing the different atoms

It is easy to draw the different atoms, if you remember these rules:

1. the protons and neutrons form the nucleus at the centre
2. the electrons are in electron shells around the nucleus
3. the first electron shell can hold a maximum of 2 electrons, the second can hold 8, and the third can also hold 8.

In the drawings below, **p** = proton, **e** = electron and **n** = neutron.

A hydrogen atom 1p, 1e	**A lithium atom** 3p, 3e, 4n	**A magnesium atom** 12p, 12e, 12n
e p	e e 3p 4n e	e e e e e e 12p 12n e e e e e e
Note that it has no neutrons. So its atomic number is 1 *and* its mass number is 1. It is described as $^{1}_{1}H$.	It has three electrons. The first shell can hold only two electrons, so a second shell is used. The atom is described as $^{7}_{3}Li$. Can you explain why?	The first two shells together hold only ten electrons. The remaining two electrons must go in a third shell. The atom is described as $^{24}_{12}Mg$. Can you explain why?

Isotopes

The atoms of an element are not always identical:

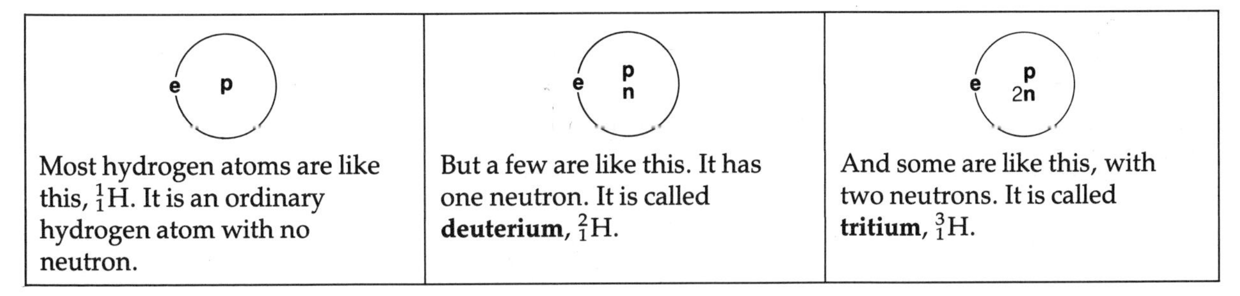

Most hydrogen atoms are like this, $^{1}_{1}H$. It is an ordinary hydrogen atom with no neutron.	But a few are like this. It has one neutron. It is called **deuterium,** $^{2}_{1}H$.	And some are like this, with two neutrons. It is called **tritium,** $^{3}_{1}H$.

All three atoms belong to the element hydrogen, *because they all have 1 proton*. They are called **isotopes** of hydrogen.
Isotopes are atoms of the same element, with the same number of protons but different numbers of neutrons.

Most elements have more than one isotope. For example carbon has three: $^{12}_{6}C$, $^{13}_{6}C$ and $^{14}_{6}C$. Chlorine has two: $^{35}_{17}Cl$ and $^{37}_{17}Cl$.

Questions

1. Which element has atoms with:
   a 5 protons? **b** 15 protons? **c** 20 protons?
2. A hydrogen atom can be described as $^{1}_{1}H$. Use the same method to describe the atoms of all the elements listed on page 28.
3. Draw a sketch of: **a** a lithium atom
   b a carbon atom **c** a neon atom
   Give the electronic configuration for each.
4. What is an isotope? Name the isotopes of hydrogen and write symbols for them.

2.4 The Periodic Table

Arranging the elements in groups

Two groups of elements are shown below. For each element, the electronic configuration of its atoms is given.
The three elements in each group have something in common:

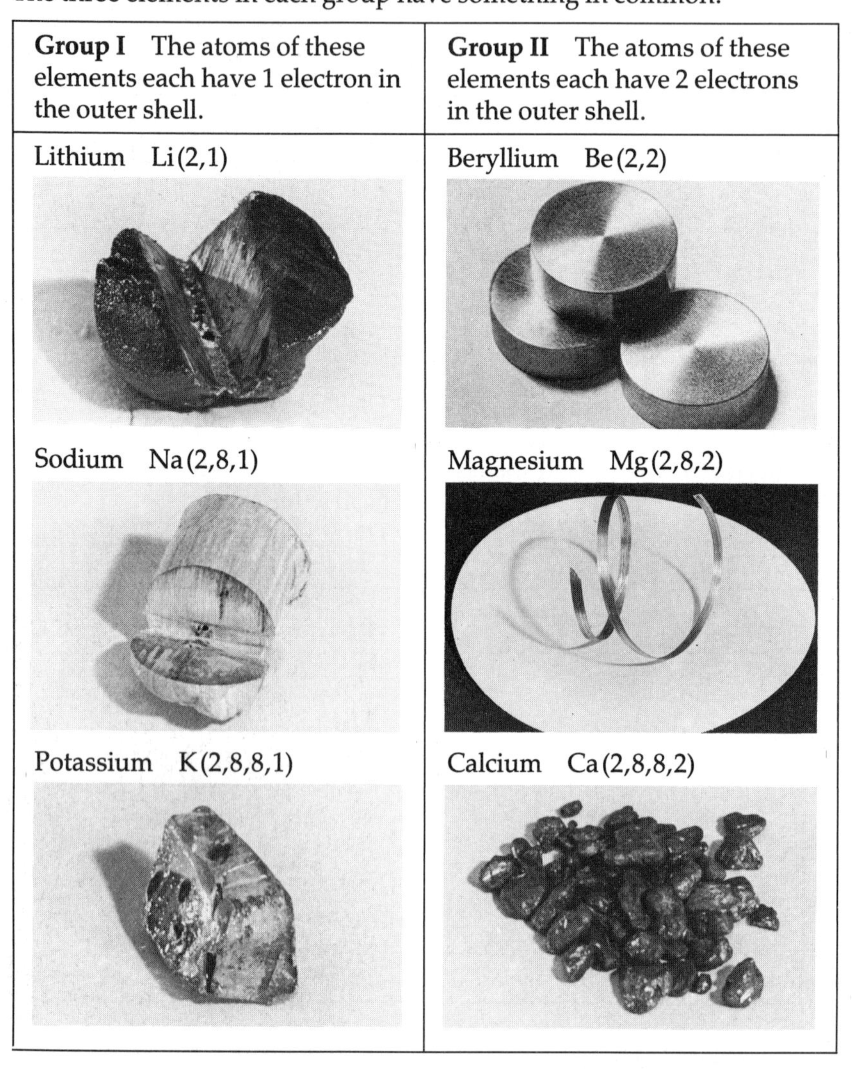

Group I The atoms of these elements each have 1 electron in the outer shell.	**Group II** The atoms of these elements each have 2 electrons in the outer shell.
Lithium Li(2,1)	Beryllium Be(2,2)
Sodium Na(2,8,1)	Magnesium Mg(2,8,2)
Potassium K(2,8,8,1)	Calcium Ca(2,8,8,2)

Using this idea, scientists have divided all the elements into groups. Here is how the arrangement works. First, the elements are listed in order of increasing atomic number. Hydrogen comes at the top, since its atomic number is 1. Next, the list is sub-divided in the following way:
The elements whose atoms have 1 outer-shell electron are picked out of the list, in order, and called **Group I**.
The elements whose atoms have 2 outer-shell electrons are picked out and called **Group II**.
Group III, **Group IV** and so on are formed in the same way.
Finally, the groups are arranged side-by-side to give the **Periodic Table**. This is shown briefly opposite, and in detail on page 202.

The Russian chemist Mendeleev, who drew up the first version of the Periodic Table in 1869.

The Periodic Table

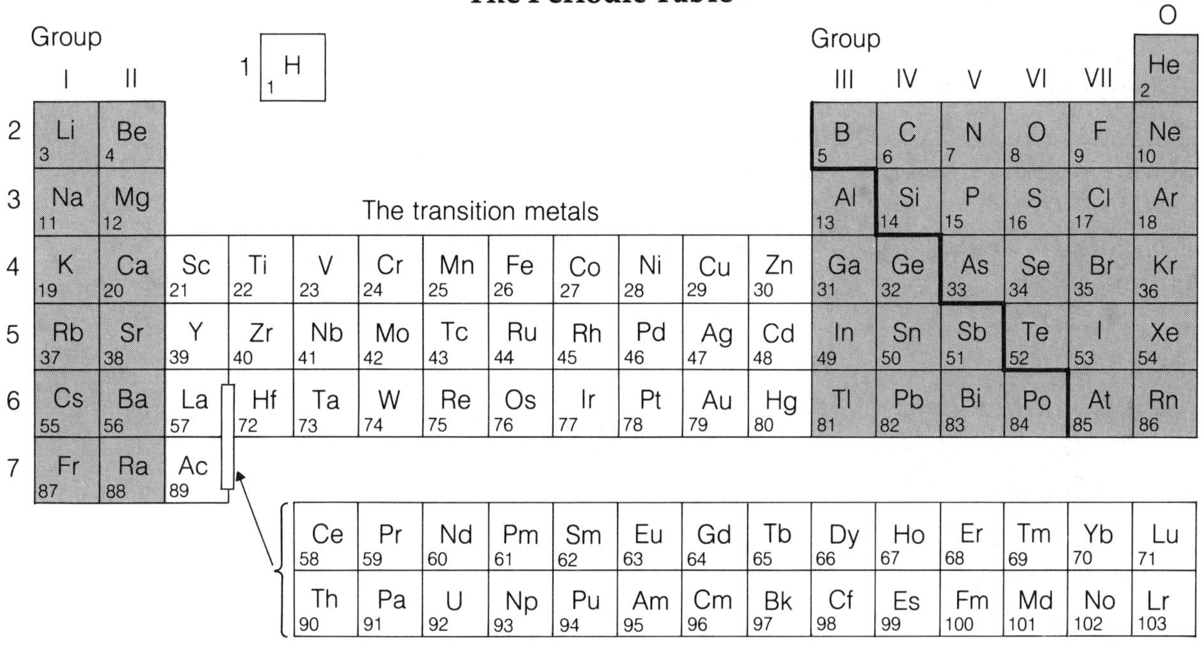

Period	Group I	Group II											Group III	Group IV	Group V	Group VI	Group VII	Group O
1									H 1									He 2
2	Li 3	Be 4											B 5	C 6	N 7	O 8	F 9	Ne 10
3	Na 11	Mg 12					The transition metals						Al 13	Si 14	P 15	S 16	Cl 17	Ar 18
4	K 19	Ca 20	Sc 21	Ti 22	V 23	Cr 24	Mn 25	Fe 26	Co 27	Ni 28	Cu 29	Zn 30	Ga 31	Ge 32	As 33	Se 34	Br 35	Kr 36
5	Rb 37	Sr 38	Y 39	Zr 40	Nb 41	Mo 42	Tc 43	Ru 44	Rh 45	Pd 46	Ag 47	Cd 48	In 49	Sn 50	Sb 51	Te 52	I 53	Xe 54
6	Cs 55	Ba 56	La 57	Hf 72	Ta 73	W 74	Re 75	Os 76	Ir 77	Pt 78	Au 79	Hg 80	Tl 81	Pb 82	Bi 83	Po 84	At 85	Rn 86
7	Fr 87	Ra 88	Ac 89															

Ce 58	Pr 59	Nd 60	Pm 61	Sm 62	Eu 63	Gd 64	Tb 65	Dy 66	Ho 67	Er 68	Tm 69	Yb 70	Lu 71
Th 90	Pa 91	U 92	Np 93	Pu 94	Am 95	Cm 96	Bk 97	Cf 98	Es 99	Fm 100	Md 101	No 102	Lr 103

The groups The table has eight groups of elements, plus a block of **transition metals**. The eight groups are shaded above. Look at Group IV. It contains the elements carbon (C), silicon (Si), germanium (Ge), tin (Sn), and lead (Pb). Their atoms each have 4 electrons in the outer shell. The atoms of Group V elements each have 5 electrons in the outer shell, and so on. Now look at the last group, Group O. Their atoms all have *full outer shells*.

Some of the groups have special names:
Group I is often called **the alkali metals**
Group II is **the alkaline earth metals**
Group VII is **the halogens**
Group O is **the noble gases.**

Look at the zig-zag line through the groups. It separates the **metals** from the **non-metals**. The metals are on the left.

The periods The horizontal rows in the table are called **periods**. Period 2 contains lithium (Li), beryllium (Be), boron (B), carbon (C), nitrogen (N), oxygen (O), fluorine (F), and neon (Ne).

The transition metals The atoms of these have more complicated electron arrangements. Note that the group contains many common metals, such as iron (Fe), nickel (Ni), and copper (Cu).

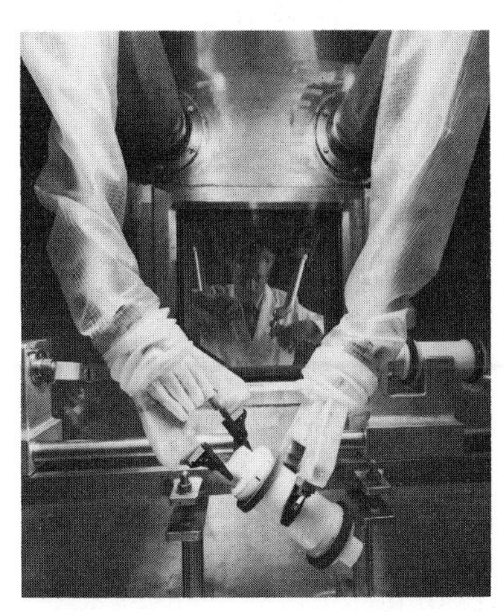

Some of the elements in the Periodic Table are dangerous to handle, because they are radioactive. Special safety equipment has to be used.

Questions

1 Explain why beryllium, magnesium and calcium are all in Group II of the Periodic Table.

2 Copy and complete:
   a In Group IV, the atoms have 4 . . .
   b In Group . . ., the atoms have 6 . . .
   c In Group O, the atoms have . . .

3 What are the rows in the Table called?

4 Use the larger table on page 202 to help you name the elements in: **a** Group V **b** Period 1 **c** Period 3

5 What is the special name for the elements in:
   a Group I? **b** Group VII **c** Group O?

6 Draw a large outline of the Periodic Table and mark in the names and symbols for the first twenty elements. Use the table on page 202 to help you.

2.5 Some groups from the Periodic Table

Families of elements

On the last two pages, you saw that the elements have been arranged in groups, in the Periodic Table.
A group of elements is sometimes called a **family**. This is because the elements in a group resemble each other. Sometimes they look alike, and usually they behave in the same way.
Below we look at some elements in different groups, to see how they resemble each other, and why.

All from the same family. Can you see the resemblance? Like humans, elements also belong to families, which are usually called **groups**. The elements in a group resemble each other.

Group I—the alkali metals

The first three elements of Group I are lithium, sodium and potassium. All three:

- are metals
- are so light that they float on water
- are silvery and shiny when freshly cut (but quickly tarnish)
- have low melting and boiling points, compared with other metals.

They also react in a similar way. For example:

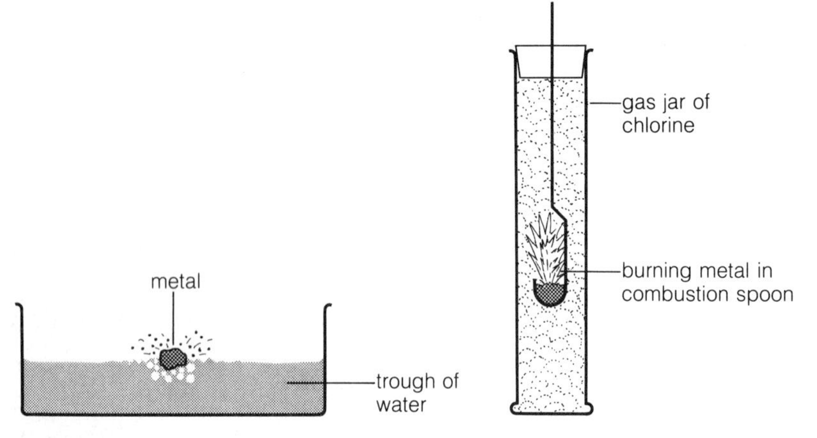

They react violently with water. Lithium floats, and a gas fizzes round it.
Sodium melts as it shoots across the water, and a gas fizzes fast.
Potassium melts immediately and the gas that forms catches fire.

They burn quickly in chlorine, with a bright flame.
Potassium burns the fastest, and sodium next. When they have burned away, all three leave a white solid behind.

So you can see these Group I metals have similar properties. Lithium is said to be the **least reactive** of them, because it reacts the most slowly. Potassium is the **most reactive** of the three.

Why they have similar properties In the elements of Group I, all the atoms have 1 electron in the outer shell. That is why the elements behave in a similar way.
Elements behave in a similar way if their atoms have the same number of outer-shell electrons.

A warning notice in a factory. Why should water not be used?

Group VII—the halogens

Chlorine, bromine and iodine are the first three halogens. All three:

are non-metals
are coloured (chlorine is a green gas, bromine is a red liquid, iodine is a black solid)
are poisonous

They too react in a similar way. For example, with iron:

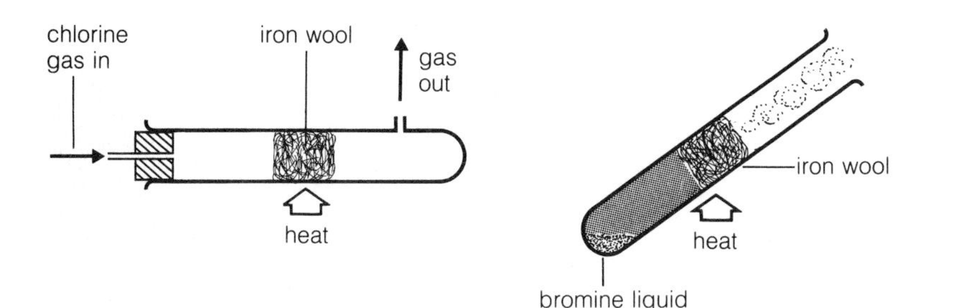

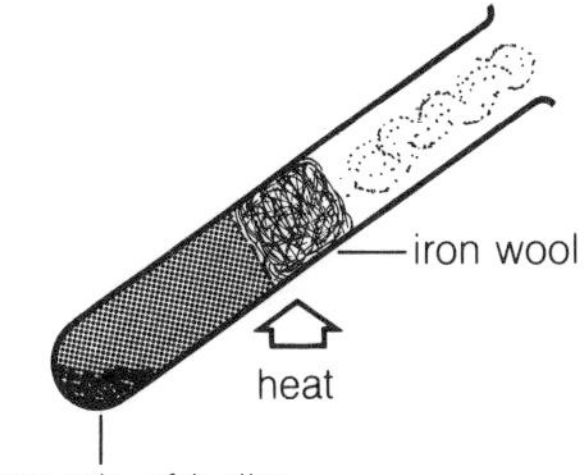

Hot iron wool glows brightly when chlorine passes over it. Brown smoke forms, and a brown solid is left behind.

The iron glows less brightly when bromine is used. Brown smoke and a brown solid are formed.

With iodine, the iron glows even less brightly. But once again, brown smoke and a brown solid are formed.

Chlorine is the most reactive of the halogens—it reacts the most easily with iron. Iodine is the least reactive of the three.

Why they have similar properties These elements have similar properties because their atoms all have 7 electrons in the outer shell.

Group O—the noble gases

The first three of these are helium, neon and argon. They:

are non-metals
are gases (they are all found in air)
are colourless

The striking thing about them is how *unreactive* they are: they normally will not react with anything. (In fact, until 1962 it was thought that all Group O elements were completely unreactive. Then scientists discovered that some of them could be forced to react weakly with some substances. Research is still going on.)

Why they have similar properties These elements have similar properties because their atoms all have full outer shells of electrons. That is why the elements are unreactive:

The noble gases are unreactive because their atoms have full outer shells of electrons.

Several of the noble gases are used in lighting. For example, xenon is used in lighthouse lamps, like this one. It gives a beautiful blue light.

Questions

1 Explain why lithium, sodium and potassium have similar properties.

2 The three experiments on this page should be carried out in a fume cupboard. Why?

3 Name three elements *not* mentioned on these two pages, which should all behave in a similar way.

4 The elements of Group O are unreactive. Explain what that means, in your own words.

2.6 Radioactivity

Chemical reactions involve the *electrons* in the outer shells of atoms. But some atoms behave in an unusual way that has nothing to do with their electrons, as you will see below.

Radioactive carbon atoms

Carbon has three isotopes, $^{12}_{6}C$, $^{13}_{6}C$ and $^{14}_{6}C$. They are often called carbon-12, carbon-13 and carbon-14:

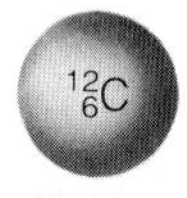

6 protons
6 electrons
6 neutrons

An atom of carbon-12. Most carbon atoms are like this.

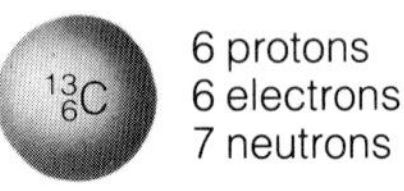

6 protons
6 electrons
7 neutrons

An atom of carbon-13. About 1% of carbon atoms are like this.

$^{14}_{6}C$

6 protons
6 electrons
8 neutrons

An atom of carbon-14. Only a few carbon atoms are like this.

The carbon-14 atom is the one that behaves strangely. Its nucleus is **unstable**, because of the extra neutrons. In time, every carbon-14 atom throws a particle out of its nucleus, and becomes a *nitrogen* atom. This process is called **decay**. Carbon-14 is said to be **radioactive**. When it decays, it gives out **radiation** from its nucleus.

Like carbon, many elements have radioactive isotopes or **radioisotopes**, that occur naturally. Other radioisotopes are made in laboratories or nuclear power stations. When radioisotopes decay they always give out radiation.

Radiation

There are three different types of radiation:

1 **Alpha particles.** These are made of two protons and two neutrons, like a helium nucleus. They shoot out of atoms at high speed, but slow down quickly in air. Paper and skin stop them.

2 **Beta particles.** These are fast electrons. They are formed by the breaking up of neutrons to give protons and electrons. They can travel 20 or 30 cm in air. They can pass through thin sheets of metal. They can also get through your skin.

3 **Gamma rays.** These are high energy rays. They can travel several metres in air, and through thick sheets of metal—except lead, which stops them fairly easily. They can pass deep into your body.

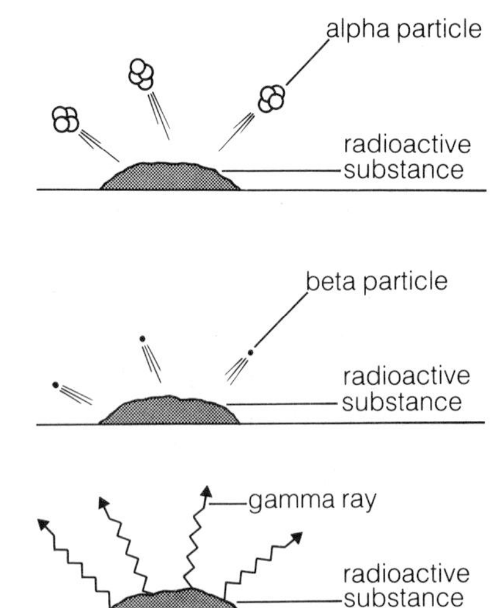

Radiation is dangerous It is dangerous to handle radioactive substances, or get too close to them without protection. Radiation in *tiny* amounts will not harm you. But small doses can lead to leukaemia and other cancers. Large doses cause radiation sickness—you vomit, and feel tired, you lose your appetite, your gums bleed, your hair falls out and you eventually die. Very high doses will burn skin black and cause death within minutes.
However, radiation can also be helpful, as you will soon see.

How radiation is detected

There are special instruments for detecting radiation. One of these is called a **Geiger counter**. The radiation triggers off electric signals in the counter. Then the signals are counted electronically.

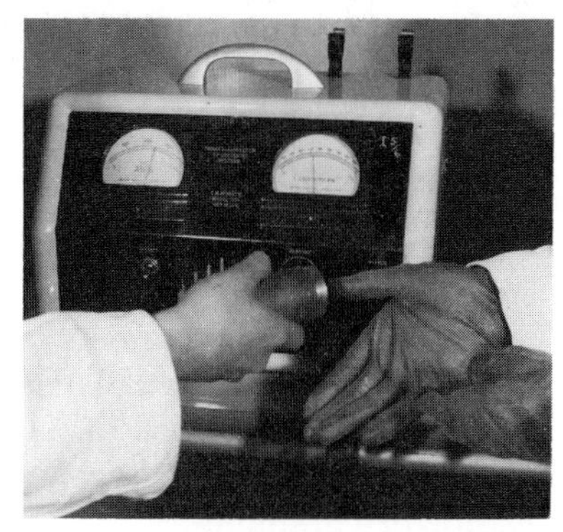

A Geiger counter being used to check rubber gloves for radioactivity.

Half-life

Carbon-14 atoms decay very slowly.
Suppose you start with 100 atoms of carbon-14.
After 5700 years, there will be half of them left (50).
After another 5700 years, half of the 50 will be left (25).
5700 years is called the **half-life** of carbon-14.
The half-life is the time it takes for half the radioactive atoms in a sample to decay.
Some radioactive atoms decay much faster than carbon-14.
The half-life of polonium-84 is only 3 minutes!

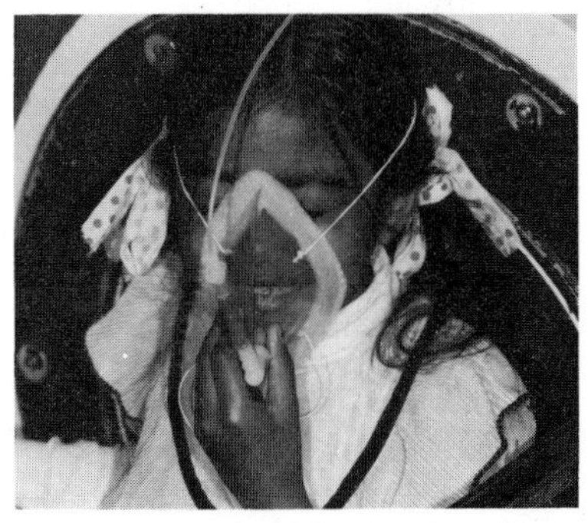

A lung check with krypton-81.

Making use of radiation

Tracers Oil is often carried long distances in buried pipes. Engineers can check the pipes for leaks by adding a radioactive substance to the oil. Some of it will leak out with the oil, into the ground. Then the leak can be located using a Geiger counter. Radioactive substances used in this way are called **tracers**.
Tracers are very useful in medicine. For example, the radioactive gas **krypton-81** is used for checking lungs. A small amount is breathed in. It decays in the lungs and the radiation can be 'watched' on a TV screen, where it shows up as bright spots. Dark patches show where the lungs are not working properly.

Cancer treatment Gamma rays are the most dangerous type of radiation, since they can pass deep into the body. They kill cells and can cause cancer. But a *weak* beam of gamma rays can *cure* cancer because it kills cancer cells more easily than healthy cells.
Cobalt-60 is the usual source of gamma rays for cancer treatment.

Carbon dating All living things contain carbon atoms. Plants get them from carbon dioxide in the air, then animals eat the plants. The carbon atoms always include some atoms of carbon-14.
When a plant or animal dies, it takes in no more carbon atoms. The carbon-14 atoms in it slowly decay, getting fewer as time goes on. By measuring the radiation from them, the age of the dead remains can be worked out.
This method is called **carbon dating**. It has been used for finding the ages of ancient bones, burial cloths, and wooden coffins, which were all once living things.

Carbon dating showed that this mummy from Chile was about 1100 years old.

Questions

1 Which isotope of carbon is radioactive?
Why is it radioactive, when the other two isotopes are not?

2 Name the different types of radiation, and describe each of them.

3 What is a Geiger counter?

4 The half-life of lead-214 is 27 minutes.
Explain what that means, in your own words.

5 Explain why carbon dating could be used to find the age of an ancient leather sandal.

2.7 Power from the nucleus

Nuclear fission

On page 34 you saw that a radioactive atom decays to form another atom, by giving out radiation.
Some large radioactive atoms can be made to behave in a different way. For example, an atom of uranium-235 splits into *two* separate atoms when it is struck by a neutron.
The splitting of an atom is called **nuclear fission**.
Fission of a uranium-235 atom produces a krypton atom, a barium atom and **3** neutrons, as well as gamma rays.

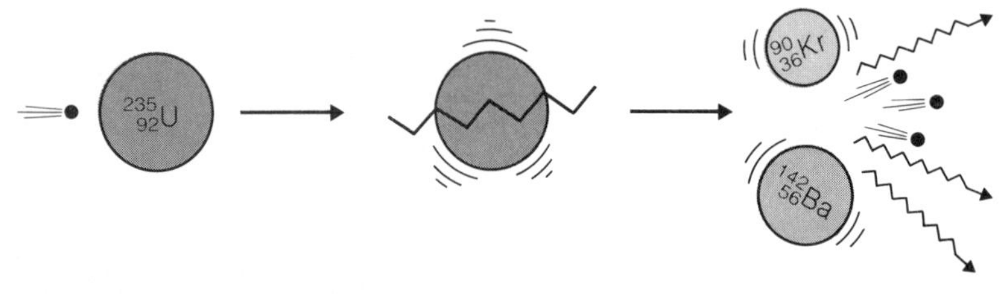

It also gives out a huge amount of heat energy.
The krypton and barium atoms are both radioactive, and in time decay.

A chain reaction

Look what happens when one atom in a group of uranium-235 atoms is struck by a neutron:

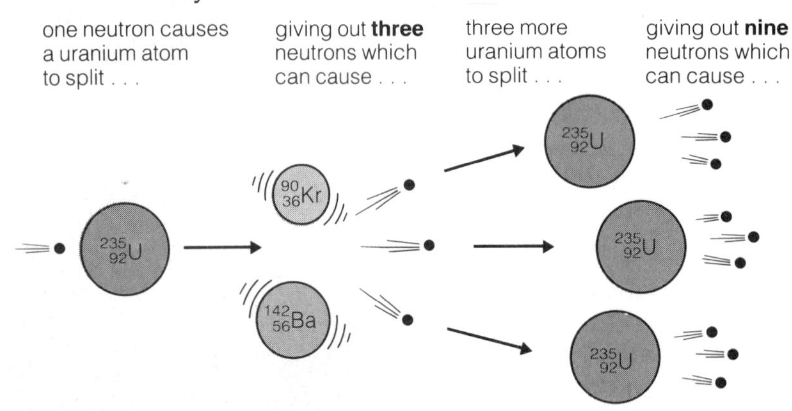

A reaction that 'grows' in this way is called a **chain reaction**.
This one is very fast. In only a fraction of a second, enormous amounts of heat and radiation are produced.

The atomic bomb

The chain reaction above is the basis of the **atomic bomb**.
In an atomic bomb, the chain reaction builds up until it gets out of control. Then the bomb explodes, giving out intense heat, a violent shock wave and a deadly burst of radiation. The earth is scorched and burned for miles around, the air is poisoned, and living things are killed.

An atomic bomb explodes.

The full effect of an atomic bomb is not felt in a day, or even a week. The products of the explosion are radioactive, and some of them have long half-lives. They settle to earth and slowly decay, causing radiation sickness, cancer and death.
The world's first atomic bomb was dropped on Hiroshima, in Japan, in August 1945. It probably contained about 10 kilograms of uranium. It killed 80 000 people within a day, and 60 000 more died within a year. Nowadays, some countries possess bombs that are 5000 times more powerful!

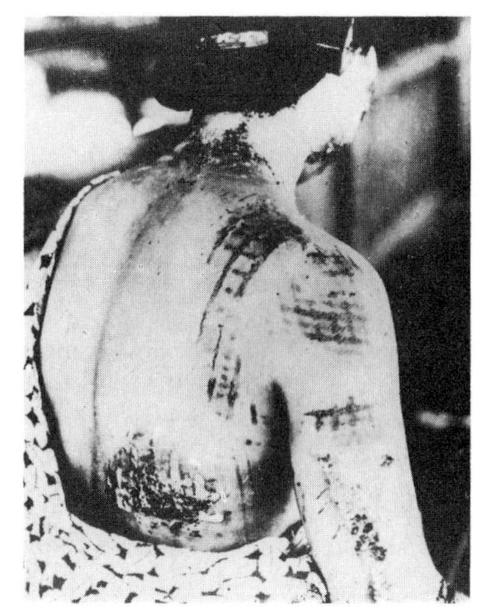
A survivor of Hiroshima. Her skin was burned through her checked dress.

Nuclear power

The chain reaction on page 36 also has a peaceful use. In some **nuclear power stations** it is used for creating electricity.
The chain reaction takes place in a tank called a **reactor**, and this time it is kept well under control. Boron can absorb neutrons, so rods of boron are placed over the reactor as **control rods**. They can be lowered into it to slow down the reaction, and raised again to speed it up.
This time, the heat from the reaction is the important thing. The heat is used to change water to steam. The steam drives turbines, and these drive generators. The result is electricity.

The boron control rods.

The nuclear power station at Oldbury in Gloucestershire.

The reactor is surrounded with thick steel and concrete, to make sure no radiation can escape.
The products of the fission are radioactive, as you saw before. They are stored in water for several months, until they partly decay. Then they are sealed into tanks and buried at sea or deep in the earth.
There are over 20 nuclear power stations in Britain, and they produce about $\frac{1}{5}$ of our electricity. Great care is taken to make sure they are not dangerous. Even so, many people would prefer to do without them, because they are afraid of leaks, or even explosions.

When the waste has become less radioactive, it is buried in concrete inside steel drums. These are then dumped at sea, or buried underground.

Questions

1 Draw a diagram to show what happens when a single atom of uranium-235 is struck by a neutron.
2 Make a list of all the things that are given out, during the fission of a uranium-235 atom. Which of them are harmful?
3 Explain why the chain reaction on the opposite page is able to build up very quickly.
4 Explain how the fission of uranium-235 is used to make electricity.
5 Why is boron useful, in a nuclear power station?

Questions on Chapter 2

1 Turn to page 28 and learn the names and symbols for the first twenty elements, in order. Then close the book and write them out.

2 **a** Give one difference between an element and a compound.
b The formulae of some compounds are given below. Write down the names of the elements they contain.
CO_2 $CaCl_2$ H_2S $PbCO_3$ KOH HgO

3 Hydrogen, deuterium and tritium are isotopes. Their structures are shown below.

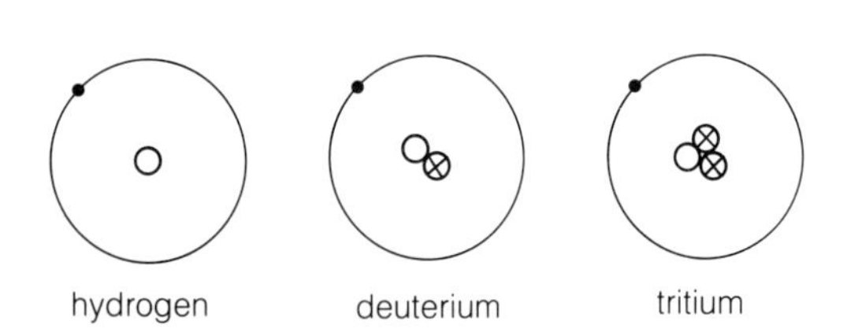

a Copy and complete the following key:
• represents ....
○ represents ....
⊗ represents ....
b What are the mass numbers of hydrogen, deuterium and tritium?
c Copy and complete this statement:
Isotopes of an element always contain the same number of .... and .... but different numbers of ....
d The average mass number of naturally-occurring hydrogen is 1.008. Which isotope is present in the highest proportion, in naturally-occurring hydrogen?

4 Copy and complete the following table for isotopes of some common elements:

Isotope	Name of element	Atomic number	Mass number	Number of p	e	n
$^{16}_{8}O$	oxygen	8	16	8	8	8
$^{18}_{8}O$						
$^{12}_{6}C$						
$^{13}_{6}C$						
$^{25}_{12}Mg$						
$^{26}_{12}Mg$						

5 For each of the six elements aluminium (Al), boron (B), nitrogen (N), oxygen (O), phosphorus (P), sulphur (S), write down:
a the period of the Periodic Table to which it belongs
b its group number in the Periodic Table
c its atomic number
d the number of electrons in one atom
e its electronic configuration
f the number of outer-shell electrons in one atom
Which of the above elements would you expect to have similar properties? Why?

6 The statements below are about metals and non-metals. Say whether each is true or false. (If false, give a reason.)
a All metals conduct electricity.
b All metals are solid at room temperature.
c Non-metals are good conductors of heat but poor conductors of electricity.
d Many non-metals are gases at room temperature.
e Most metals are brittle and break when hammered.
f Most non-metals are ductile.
g There are about four times as many metals as non-metals.

7 **a** Make a larger copy of this outline of the Periodic Table:

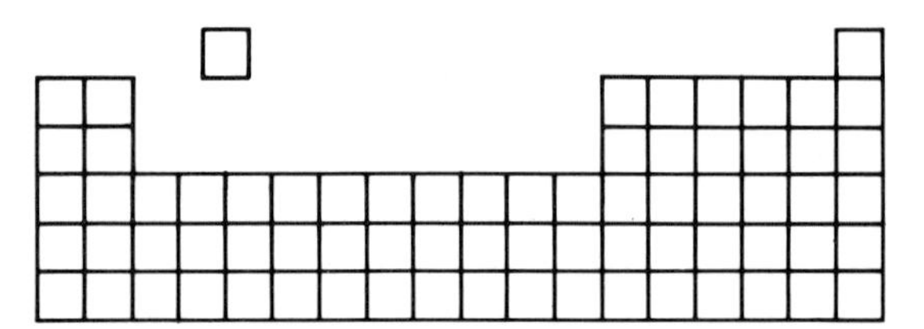

b Write in the group and period numbers.
c Draw a zig-zag line to show how the metals are separated from the non-metals, in the table.
d Now put the letters A to L in the correct places in the table, to fit these descriptions:
A the lightest element
B any noble gas
C the element with atomic number 5
D the element with 6 electrons in its atoms
E any element with 6 outer-shell electrons
F the most reactive alkali metal
G the least reactive alkali metal
H the most reactive halogen
I a Group III metal
J an alkaline earth metal
K a transition metal
L a Group V non-metal

8 This table gives data for some elements:

Name	Symbol	M.pt. °C	B.pt. °C	Electrical conductivity
Aluminium	Al	660	2450	good
Bromine	Br	−7	58	poor
Calcium	Ca	850	1490	good
Chlorine	Cl	−101	−35	poor
Copper	Cu	1083	2600	good
Helium	He	−270	−269	poor
Iron	Fe	1540	2900	good
Lead	Pb	327	1750	good
Magnesium	Mg	650	1110	good
Mercury	Hg	−39	357	good
Nitrogen	N	−210	−196	poor
Oxygen	O	−219	−183	poor
Phosphorus	P	44	280	poor
Potassium	K	64	760	good
Sodium	Na	98	890	good
Sulphur	S	119	445	poor
Tin	Sn	230	2600	good
Zinc	Zn	419	906	good

Use the table to answer these questions.

a What is the melting point of iron?

b Which element melts at −7 °C?

c Which element boils at 280 °C?

d Which element has a boiling point only 1 °C higher than its melting point?

e Over what temperature range is sulphur a liquid?

f Which element has:

i the highest melting point?

ii the lowest melting point?

g Divide the elements into two groups, one of metals and the other of non-metals.

h List the *metals* in order of increasing melting point.

i Which metal is a liquid at room temperature (20 °C)?

j Which non-metal is a liquid at room temperature?

k List the elements which are gases at room temperature. What do you notice about the elements in this list?

9 Before the development of the Periodic Table, a scientist called Döbreiner discovered sets of three elements with similar properties, which he called **triads**. Some of these triads are shown as members of the same group, in the modern Periodic Table. Complete the following triads by inserting the missing middle element:

chlorine (Cl), ..., iodine (I);
lithium (Li), ..., potassium (K)
calcium (Ca), ..., barium (Ba).

10 Many scientists contributed to the development of the modern Periodic Table. One of them was the Russian chemist Mendeleev. In 1869 he arranged the elements that were then known in a table very similar to the one in use today. He realized that gaps should be left for elements that had not yet been discovered, and even went so far as to predict the properties of several of these elements. Rubidium is an alkali metal that lies below potassium in Group I. Here is some data for Group I:

Element	Atomic number	M.pt. °C	B.pt. °C	Chemical reactivity
Lithium	3	180	1330	quite reactive
Sodium	11	98	890	reactive
Potassium	19	64	760	very reactive
Rubidium	37	?	?	?
Caesium	55	29	690	violently reactive

a Using your knowledge of the Periodic Table, predict the missing data for rubidium.

b In a rubidium atom:

i how many electron shells are there?

ii how many electrons are there?

iii how many outer-shell electrons are there?

11 **a** What is the *half-life* of a radioactive isotope?

b The following graph shows the decay of a radioactive isotope, over four minutes:

i What is the half-life of this isotope?

ii What could be used to detect the radiation?

iii How many atoms will be left after 5 minutes?

iv When the decay takes place in a sealed container, helium gas collects in the container. Name one type of radiation produced in the decay.

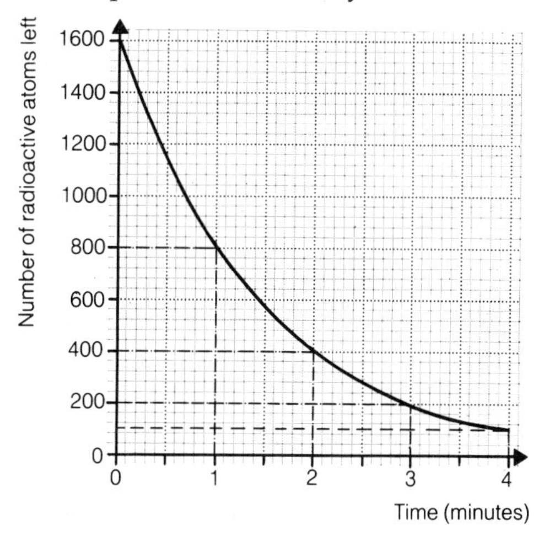

3.1 Why compounds are formed

Most elements form compounds

On page 32 you saw that sodium reacts with chlorine:

When sodium is heated, and placed in a jar of chlorine, it burns with a bright flame.

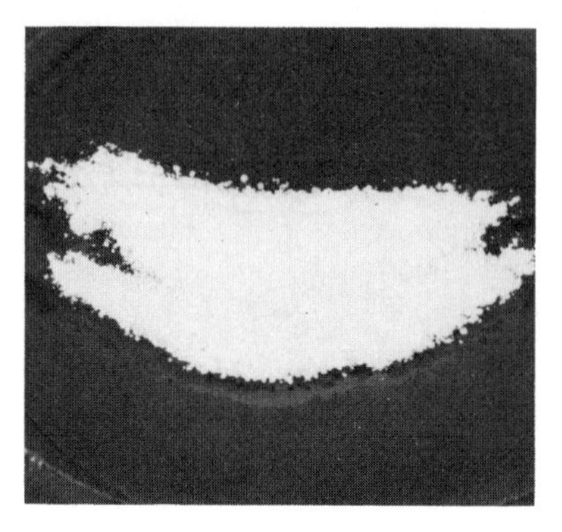

The result is a white solid, which has to be scraped from the sides of the jar.

The white solid is called **sodium chloride**. It is formed by atoms of sodium and chlorine joining together, so it is a **compound**.
The reaction can be described like this:

sodium + chlorine $\longrightarrow$ sodium chloride

The + means *reacts with*, and the $\longrightarrow$ means *to form*.
Most elements react to form compounds. For example:

lithium + chlorine $\longrightarrow$ lithium chloride
hydrogen + chlorine $\longrightarrow$ hydrogen chloride

The noble gases do not usually form compounds

The noble gases are different from other elements, because they do not usually form compounds, as you saw on page 33. For this reason, their atoms are described as **unreactive** or **stable**. They are stable because their outer electron shells are *full*:
A full outer shell makes an atom stable.

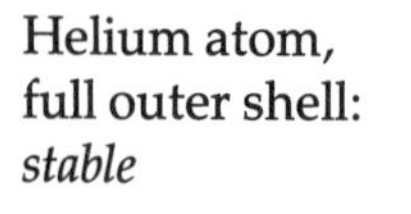

Helium atom, full outer shell: *stable*

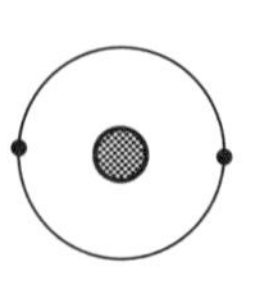

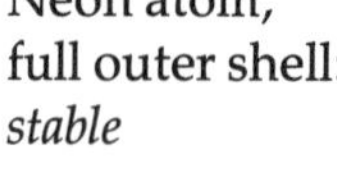

Neon atom, full outer shell: *stable*

Argon atom, full outer shell: *stable*

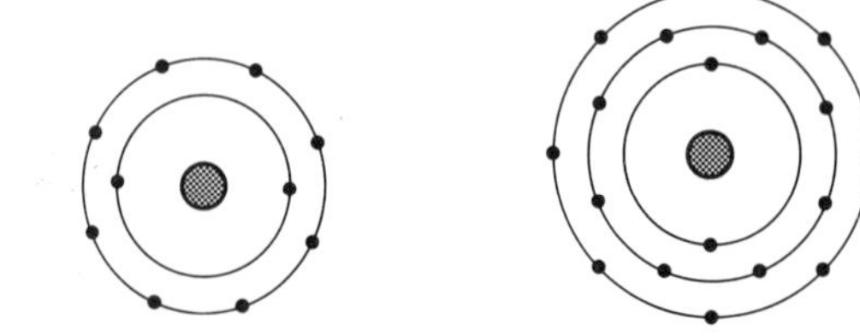

Only the noble gas atoms have full outer shells. The atoms of all other elements have incomplete outer shells. That is why they react:
By reacting with each other, atoms can obtain full outer shells and so become stable.

Since helium is unreactive, and also very light, it is an ideal gas for filling balloons like these.

Losing or gaining electrons

The atoms of some elements can obtain full shells by *losing* or *gaining* electrons, when they react with other atoms:

Losing electrons The sodium atom is a good example. It has just 1 electron in its outer shell. It can obtain a full outer shell by losing this electron to another atom. The result is a **sodium ion**:

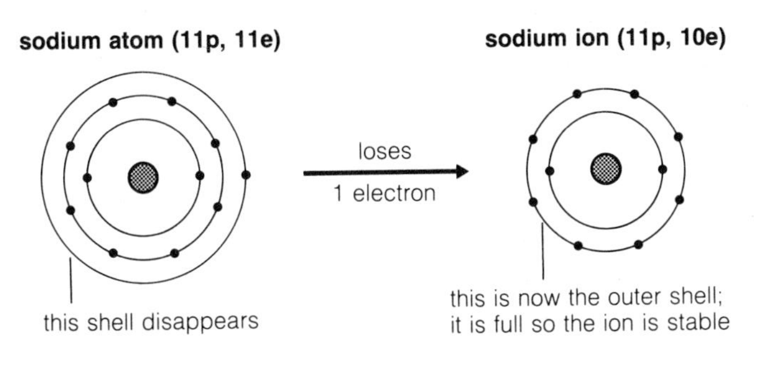

The charge on the sodium ion:

the charge on 11 protons is	+11
the charge on 10 electrons is	−10
total charge	+ 1

The sodium ion has 11 protons but only 10 electrons, so it has a charge of **+1**, as you can see from the box on the right above.

The symbol for sodium is Na, so the sodium ion is called $\mathbf{Na^+}$. The $^+$ means *1 positive charge*. Na^+ is a **positive ion**.

Gaining electrons A chlorine atom has 7 electrons in its outer shell. It can reach a full shell by accepting just 1 electron from another atom. It becomes a **chloride ion**:

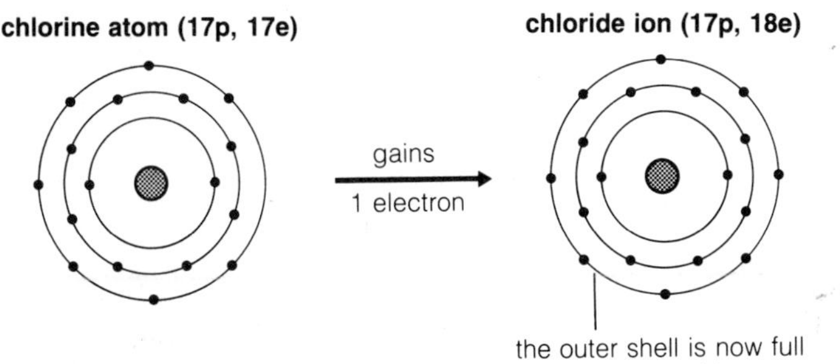

The charge on the chloride ion:

the charge on 17 protons is	+17
the charge on 18 electrons is	−18
total charge	− 1

The chloride ion has a charge of −1, so it is a **negative ion**. Its symbol is $\mathbf{Cl^-}$.

Ions Any atom becomes an ion if it loses or gains electrons. **An ion is a charged particle. It is charged because it contains an unequal number of protons and electrons.**

Questions

1 What is another word for *unreactive*?
2 Why are the noble gas atoms unreactive?
3 Explain why all other atoms are reactive.
4 Draw a diagram to show how a sodium atom obtains a full outer shell.
5 Explain why the sodium ion has a charge of +1.
6 Draw a diagram to show how a chlorine atom can obtain a full outer shell.
7 Write down the symbol for a chloride ion. Why does this ion have a charge of −1?
8 Explain what an ion is, in your own words.
9 Why do noble gas atoms *not* form ions?

3.2 The ionic bond

The reaction between sodium and chlorine

As you saw on the last page, a sodium atom can lose one electron, and a chlorine atom can gain one, to obtain full outer shells. So, when a sodium atom and a chlorine atom react together, the sodium atom loses its electron *to the chlorine atom*, and two ions are formed. Here, sodium electrons are shown as • and chlorine electrons as ×, but remember that all electrons are exactly the same:

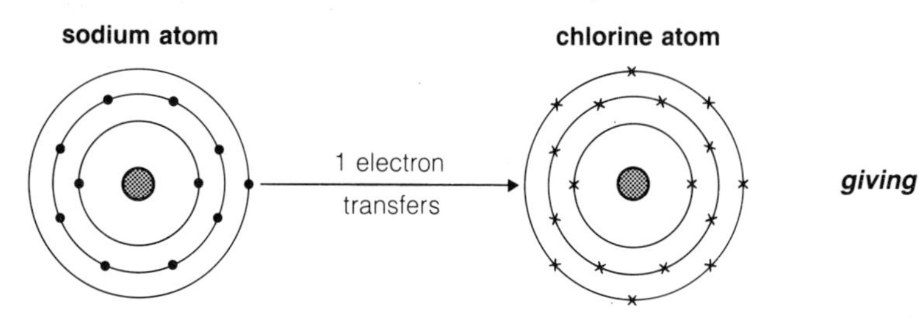

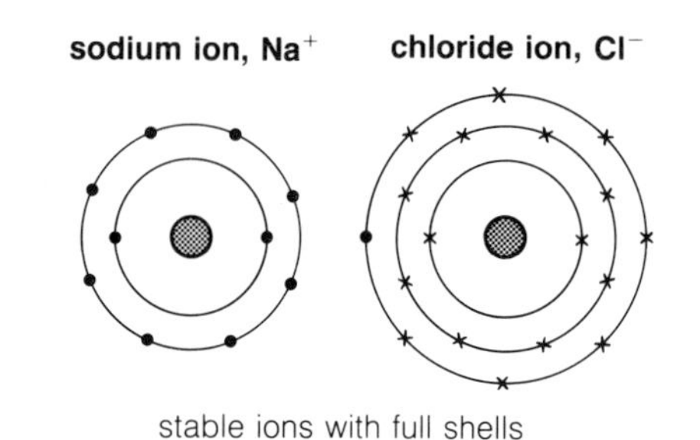

stable ions with full shells

The two ions have opposite charges, so they attract each other. The force of attraction between them is strong. It is called an **ionic bond**, or sometimes an **electrovalent bond**.

How solid sodium chloride is formed

When sodium reacts with chlorine, billions of sodium and chloride ions form, and are attracted to each other. But the ions do not stay in pairs. Instead, they cluster together, so that each ion is surrounded by six ions of opposite charge. They are held together by strong ionic bonds. (Each ion forms six bonds.)
The pattern grows until a giant structure of ions is formed. It contains equal numbers of sodium and chloride ions. This giant structure is the compound **sodium chloride**, or **salt**.

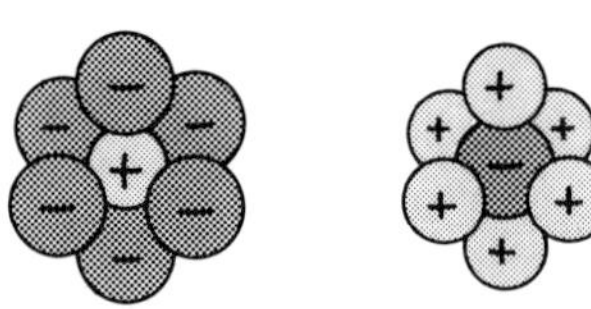

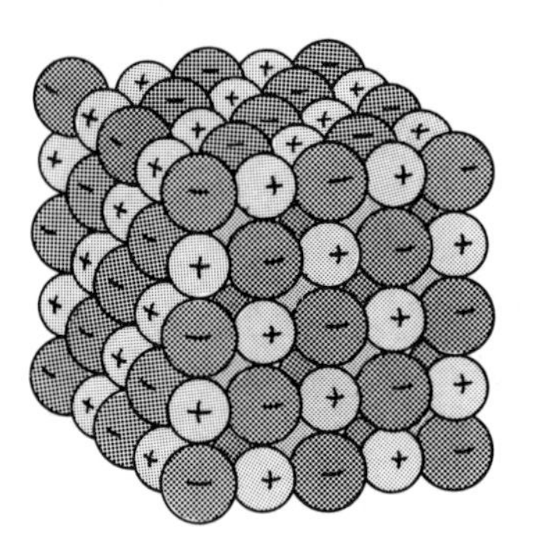

Because sodium chloride is made of ions, it is called an **ionic compound**. It contains one Na^+ ion for each Cl^- ion, so its formula is **NaCl**.
The charges in the structure add up to zero:

the charge on each sodium ion is	+1
the charge on each chloride ion is	−1
total charge	0

The compound therefore has no overall charge.

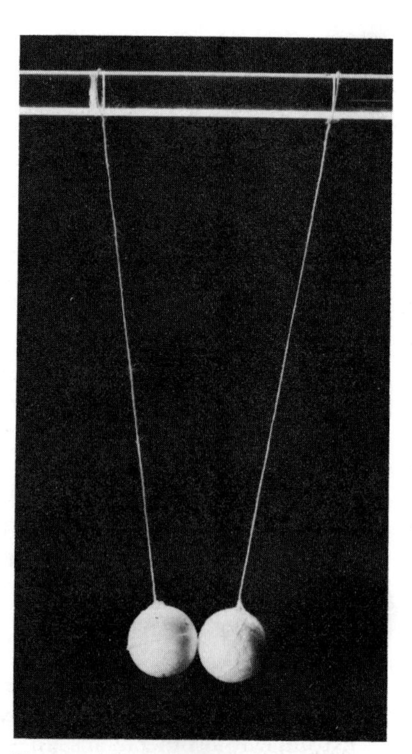

These polystyrene spheres have been given opposite charges, so they are attracted to each other. The same happens with ions of opposite charge.

Other ionic compounds

Sodium is a **metal**, and chlorine is a **non-metal**. They react together to form an **ionic compound**. Other metals can also react with non-metals to form ionic compounds. Below are two more examples.

Magnesium and oxygen A magnesium atom has 2 outer electrons and an oxygen atom has 6. Magnesium burns fiercely in oxygen. During the reaction, each magnesium atom loses its 2 outer electrons to an oxygen atom. Magnesium ions and oxide ions are formed:

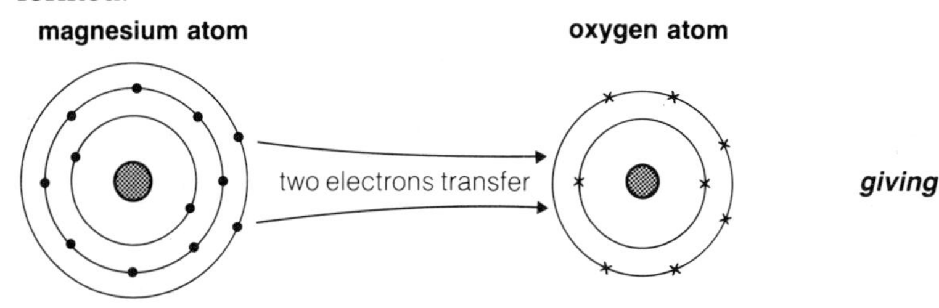

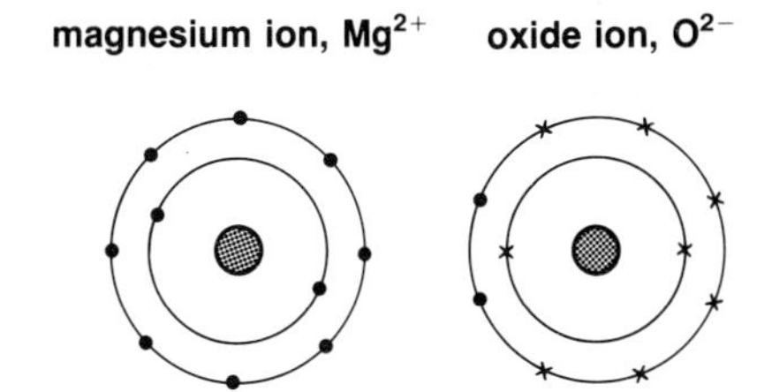

The ions attract each other, because of their opposite charges. Like the ions on the last page, they group together into a giant ionic structure. The resulting compound is called **magnesium oxide**. Magnesium oxide contains one magnesium ion for each oxide ion, so its formula is **MgO**. The compound has no overall charge:

the charge on each magnesium ion is	2+
the charge on each oxide ion is	2−
total charge	0

Magnesium and chlorine To obtain full outer shells, a magnesium atom must lose 2 electrons, and a chlorine atom must gain 1 electron. So when magnesium burns in chlorine, each magnesium atom reacts with *two* chlorine atoms, to form **magnesium chloride**:

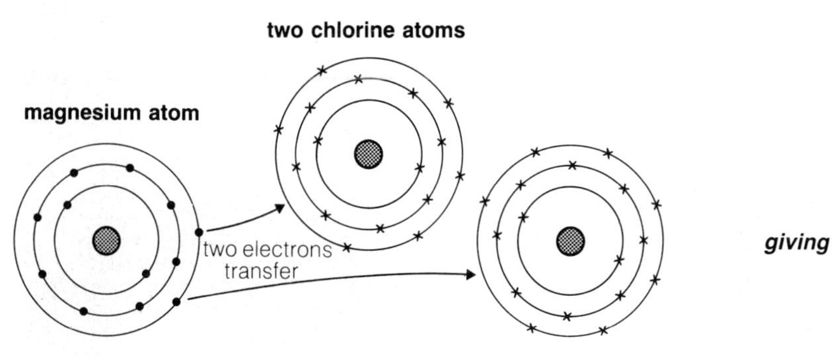

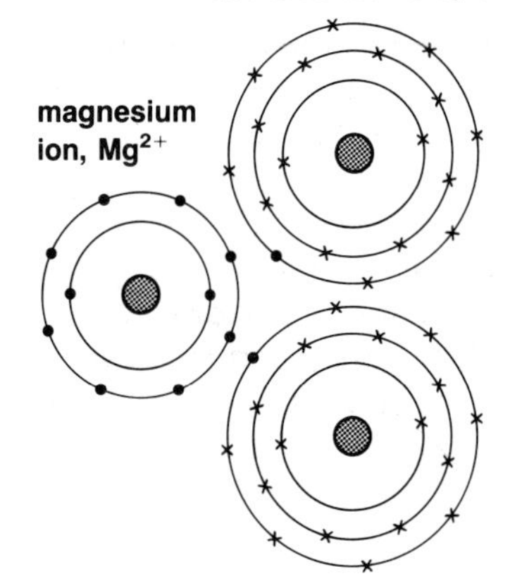

The ions form a giant ionic structure, with *two* chloride ions for each magnesium ion. The formula of magnesium chloride is therefore **$MgCl_2$**. The compound has no overall charge. Can you explain why?

Questions

1 Draw a diagram to show what happens when a sodium atom reacts with a chlorine atom.

2 What is an ionic bond? What is the other name for it?

3 Sketch the structure of sodium chloride, and explain why its formula is NaCl.

4 Explain why:

a a magnesium ion has a charge of 2+

b the ions in magnesium oxide stay together

c magnesium chloride has no overall charge

d the formula of magnesium chloride is $MgCl_2$.

3.3 Some other ions

Ions of the first twenty elements

Not every element forms ions, during reactions. In fact, out of the first twenty elements in the Periodic Table, only eleven easily form ions. These ions are given below, with their names. The shading shows non-metals:

Group I	II		III	IV	V	VI	VII	O
		H^+ hydrogen						none
Li^+ lithium	Be^{2+} beryllium		none	none	none	O^{2-} oxide	F^- fluoride	none
Na^+ sodium	Mg^{2+} magnesium		Al^{3+} aluminium	none	none	S^{2-} sulphide	Cl^- chloride	none
K^+ potassium	Ca^{2+} calcium	transition metals						

Note that hydrogen and the metals form **positive ions**, which have the same names as the atoms. The non-metals form **negative ions** and their names end in *-ide*.
The elements in Groups IV and V do not usually form ions, because their atoms would have to gain or lose several electrons, and that takes too much energy. The elements in Group O do not form ions because their atoms already have full shells.

Bath salts contain Na^+ and CO_3^{2-} ions; Epsom salts contain Mg^{2+} and SO_4^{2-} ions. Can you give the names and formulae of the three main compounds illustrated here?

The names and formulae of their compounds

The names To name an ionic compound, you just put the names of the ions together, with the positive one first:

Ions in compound	Name of compound
K^+ and F^-	Potassium fluoride
Ca^{2+} and S^{2-}	Calcium sulphide

The formulae The formulae of ionic compounds can be worked out by these steps:

1. Write down the name of the ionic compound.
2. Write down the symbols for its ions.
3. The compound must have no overall charage, so **balance** the ions, until the positive and negative charges add up to zero.
4. Write down the formula without the charges.

Example 1

1. Lithium fluoride.
2. The ions are Li^+ and F^-.
3. One Li^+ is needed for every F^-, to make the total charge zero.
4. The formula is LiF.

Example 2

1. Sodium sulphide.
2. The ions are Na^+ and S^{2-}.
3. Two Na^+ ions are needed for every S^{2-} ion, to make the total charge zero: $Na^+ \, Na^+ \, S^{2-}$.
4. The formula is Na_2S. (What does the $_2$ show?)

Transition metal ions

Some transition metals form only one type of ion:

silver forms only Ag^+ ions

zinc forms only Zn^{2+} ions

but most of them can form more than one type. For example, copper and iron can each form two:

Ion	Name	Example of compound
Cu^+	copper(I) ion	copper(I) oxide, Cu_2O
Cu^{2+}	copper(II) ion	copper(II) oxide, CuO
Fe^{2+}	iron(II) ion	iron(II) chloride, $FeCl_2$
Fe^{3+}	iron(III) ion	iron(III) chloride, $FeCl_3$

The (II) in a name shows that the ion has a charge of 2+. What do the (I) and (III) show?

Compound ions

So far, all the ions have been formed from single atoms. But ions can also be formed from groups of joined atoms. These are called **compound ions**, and the most common ones are shown on the right.
Remember, each is just one ion, even though it contains more than one atom.
The formulae of their compounds can be worked out as before. Some examples are shown below.

NH_4^+, the ammonium ion

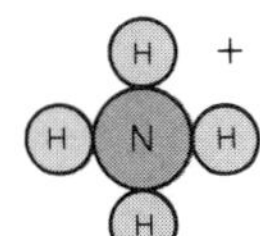

OH^-, the hydroxide ion

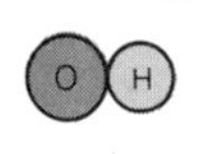

NO_3^-, the nitrate ion

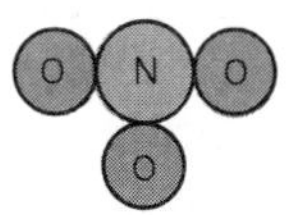

SO_4^{2-}, the sulphate ion

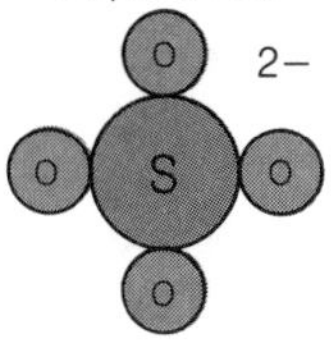

2–

CO_3^{2-}, the carbonate ion

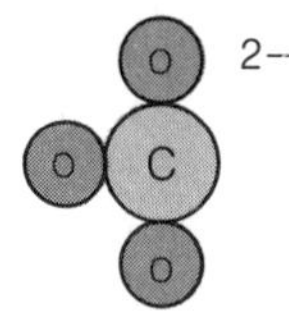

HCO_3^-, the hydrogen carbonate ion

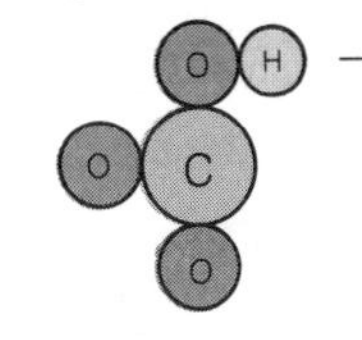

Example 3

1. Sodium carbonate.
2. The ions are Na^+ and CO_3^{2-}.
3. Two Na^+ are needed, to balance the charge on one CO_3^{2-}.
4. The formula is Na_2CO_3.

Example 4

1. Calcium nitrate.
2. The ions are Ca^{2+} and NO_3^-.
3. Two NO_3^- are needed, to balance the charge on one Ca^{2+}.
4. The formula is $Ca(NO_3)_2$. Note that a bracket is put round the NO_3, before the $_2$ is put in.

Questions

1. Explain why a calcium ion has a charge of 2+.
2. Why is the charge on an aluminium ion 3+?
3. Write down the symbols for the ions in:
   a potassium chloride **b** calcium sulphide
   c lithium sulphide **d** magnesium fluoride
4. Now work out the formula for each compound in question 3.
5. Work out the formula for each compound:
   a copper(II) chloride **b** iron(III) oxide
6. Write a name for each compound: $CuCl$, FeS, Na_2SO_4, $Mg(NO_3)_2$, NH_4NO_3, $Ca(HCO_3)_2$
7. Work out the formula for:
   a sodium sulphate **b** potassium hydroxide
   c silver nitrate **d** ammonium nitrate

3.4 The covalent bond

Sharing electrons

When two non-metal atoms react together, *both of them need to gain electrons*, to reach full shells. They can manage this only by sharing electrons between them. Atoms can share only their outer electrons, so just the outer electrons are shown in the diagrams below.

Hydrogen A hydrogen atom has only one electron. Its shell can hold two electrons, so is not full. When two hydrogen atoms get close enough, their shells overlap and then they can share electrons:

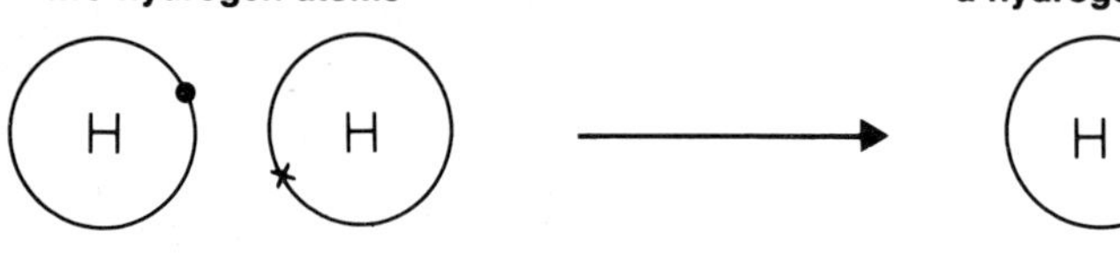

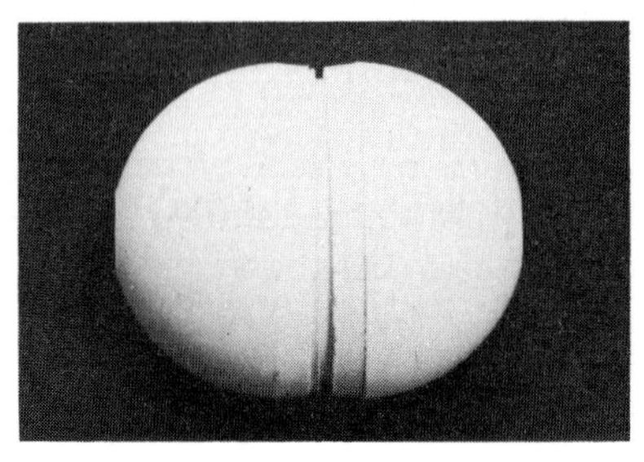

A model of the hydrogen molecule. The molecule can be shown as H–H. The line represents a single bond.

Because the atoms share electrons, there is a strong force of attraction between them, holding them together. This force is called a **covalent bond**. The bonded atoms form a **molecule**.
A molecule is a small group of atoms which are held together by covalent bonds.
Hydrogen gas is made up of hydrogen molecules, and for this reason it is called a **molecular** substance. Its formula is $\mathbf{H_2}$.
Several other non-metals are also molecular. For example:

chlorine, Cl_2 iodine, I_2 oxygen, O_2
nitrogen, N_2 sulphur, S_8 phosphorus, P_4

In H_2, the $_2$ tells you the number of hydrogen atoms in each molecule.
Because it has two atoms in each molecule, hydrogen is described as **diatomic**.

Chlorine A chlorine atom needs a share in one more electron, to obtain a full shell. So two chlorine atoms bond covalently like this:

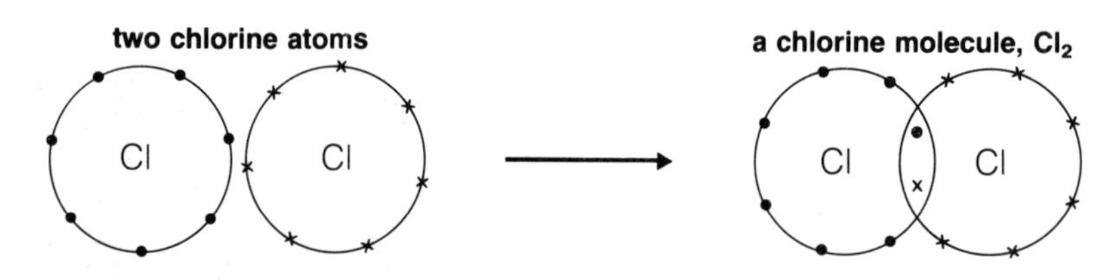

A model of the chlorine molecule.

Oxygen The formula for oxygen is O_2, so each molecule must contain two atoms. Each oxygen atom has only six outer electrons; it needs a share in two more to reach a full shell:

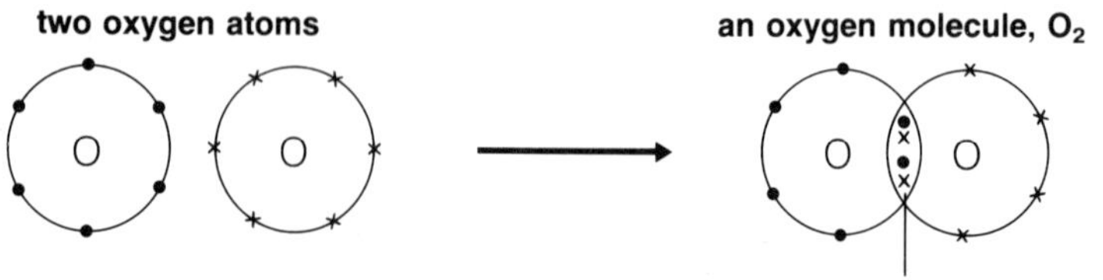

A model of the oxygen molecule. The molecule can be shown as O=O. The lines represent a double bond.

Since the oxygen atoms share two pairs of electrons, the bond between them is called a **double covalent bond**, or just a **double bond**.

Covalent compounds

On the opposite page you saw that several non-metal elements exist as molecules. A huge number of compounds also exist as molecules. In a molecular compound, atoms of *different* elements share electrons with each other. These compounds are often called **covalent compounds** because of the covalent bonds in them. Water, ammonia and methane are all covalent compounds.

Water The formula of water is H_2O. In each molecule, an oxygen atom shares electrons with two hydrogen atoms, and they all reach full shells:

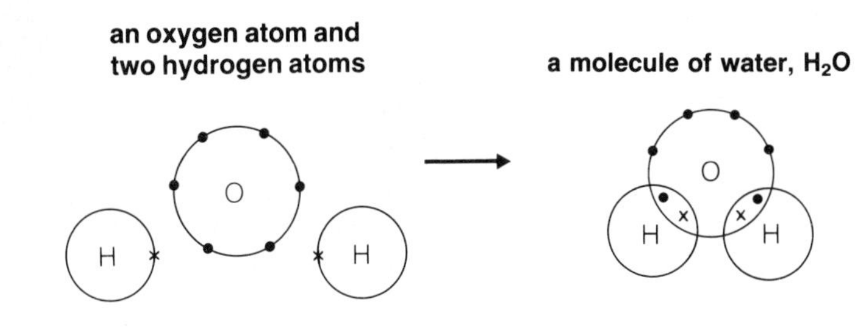

A model of the water molecule.

Ammonia Its formula is NH_3. Each nitrogen atom shares electrons with three hydrogen atoms, and they all reach full shells:

a nitrogen atom and three hydrogen atoms

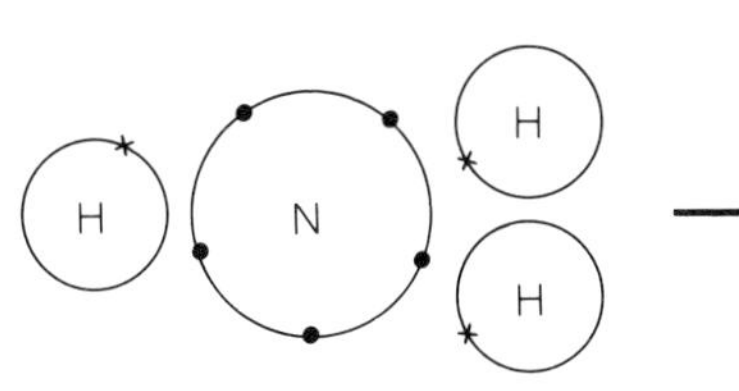

a molecule of ammonia, NH_3

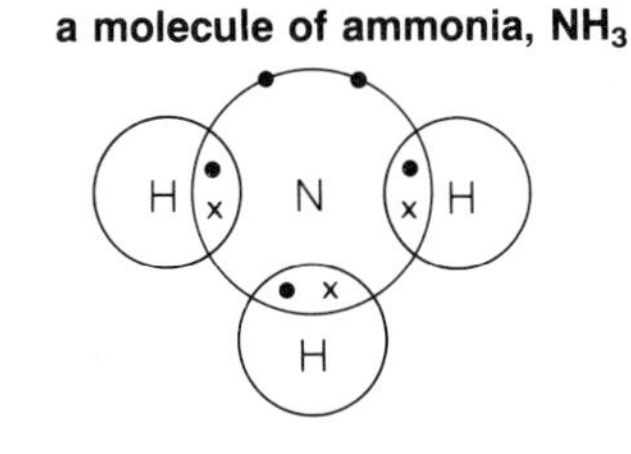

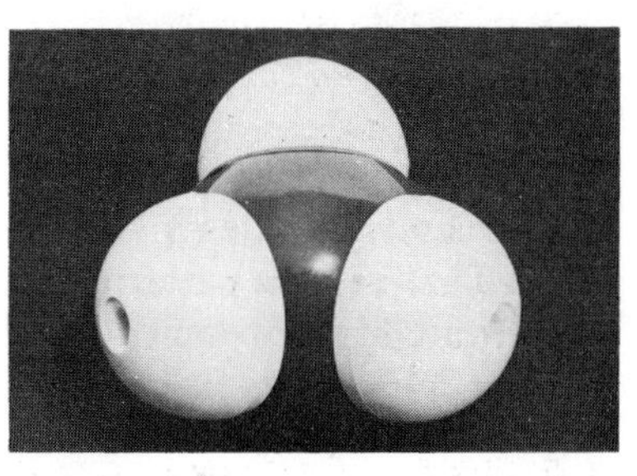

A model of the ammonia molecule.

Methane Its formula is CH_4. Each carbon atom shares electrons with four hydrogen atoms, and they all obtain full shells:

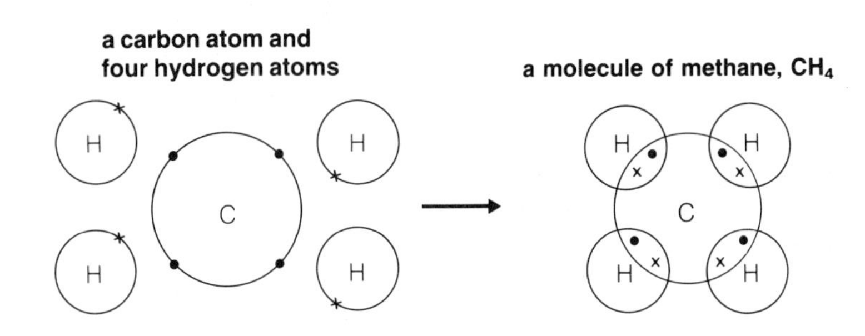

A model of the methane molecule.

Questions

1 What is the name of the bond between atoms that share electrons?
2 What is a molecule?
3 Give five examples of molecular elements.
4 Draw a diagram to show the bonding in:
   a chlorine **b** oxygen
5 The bond between two oxygen atoms is called a double bond. Why?
6 Show the bonding in a molecule of:
   a water **b** methane
7 Hydrogen chloride (HCl) is also molecular. Draw a diagram to show the bonding in it.

3.5 Ionic and molecular solids

On page 10 you saw that solids are made of particles packed closely together, in a regular pattern. If the particles are **ions**, the solids are called **ionic solids**. If they are **molecules**, the solids are called **molecular** or **covalent** solids. The two types of solids have quite different properties, as you will see below.

Some ionic solids you can find in the laboratory:

sodium chloride
sodium hydroxide
copper(II) sulphate
iron(II) chloride
silver nitrate

Ionic solids

One of the most common ionic solids is sodium chloride—it is ordinary table salt. You already know quite a lot about its structure:

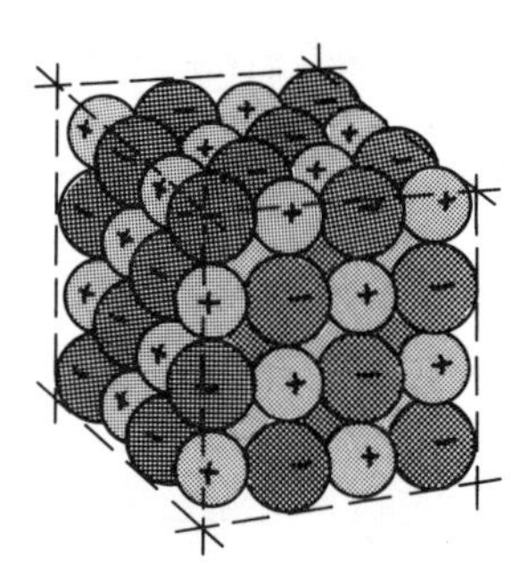

Sodium chloride is made of sodium and chloride ions, packed in a regular pattern. This arrangement is called a **lattice**. The ions are held together by strong **ionic bonds**.

The pattern repeats millions of times. The result is a piece of solid with straight edges and flat faces, called a **crystal**. Above is a crystal of sodium chloride, magnified 35 times.

The crystals look white and shiny. A box of table salt contains millions of them.

Sodium chloride is typical of ionic solids. In *all* ionic solids, the ions are packed in a regular pattern, and held together by strong ionic bonds. This means that all ionic solids are **crystalline.** The crystals may be quite large, or so small that you would need a microscope to see them properly.

Their properties Ionic solids have these properties:

1 They have high melting and boiling points. For example—

sodium chloride melts at 808 °C and boils at 1465 °C
sodium hydroxide melts at 319 °C and boils at 1390 °C.

The reason is that the ionic bonds are very strong, so it takes a lot of heat energy to break up the lattice and form a liquid. That explains why all ionic substances are solid at room temperature.

2 They are usually soluble in water.
3 They are insoluble in many other solvents, for example in tetrachloromethane and petrol.
4 They do not conduct electricity, when solid. Electricity is a stream of moving charges. Although the ions *are* charged, they cannot move, because the ionic bonds hold them firmly in position. (You will learn more about this in Chapter 6.)
5 They do conduct electricity when they are melted, or dissolved. This is because the ions are then free to move.

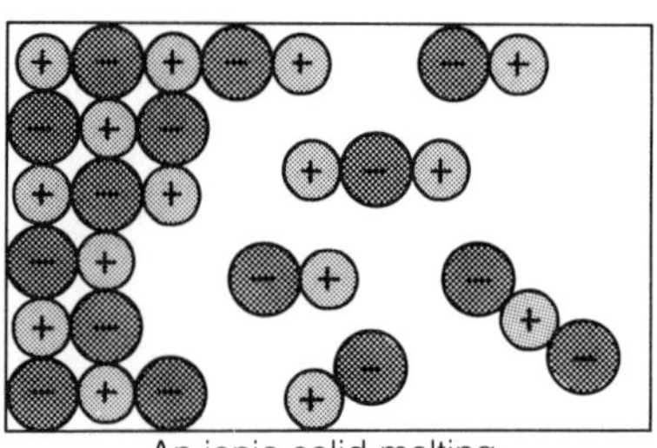

An ionic solid melting

Molecular solids

Iodine is a good example of a molecular solid. Each iodine molecule contains two atoms, held together by a strong covalent bond.

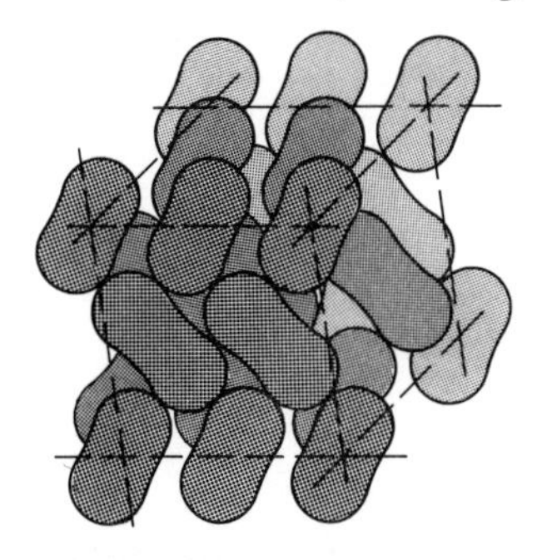

Here ⬭ represents an iodine molecule. The molecules are packed in a regular pattern, and held together by weak forces.

The pattern repeats millions of times, and the result is a crystal. Above is a single iodine crystal, magnified 15 times.

Iodine crystals are grey-black and shiny. This jar contains millions of them.

Iodine is typical of molecular solids. In *all* molecular solids, the molecules are held together in a regular pattern; so the solids are crystalline. The forces that hold the molecules together are weak.

Their properties Molecular solids have these properties:

1 They have low melting points and boiling points—much lower than ionic solids do. In fact many molecular substances melt, and even boil, *below* room temperature, so are liquids or gases at room temperature. Here are some examples:

Substance	Melting point/ °C	Boiling point/ °C
oxygen	−219	−183
chlorine	−101	−35
water	0	100
naphthalene	80	218

The reason for these low values is that the molecules are held together by only weak forces, so not much heat energy is needed to separate them.

2 Unlike ionic solids, molecular solids are usually insoluble in water.

3 Unlike ionic solids, they are soluble in tetrachloromethane and petrol.

4 They do not conduct electricity. Molecules are not charged, so molecular substances cannot conduct, even when melted.

Some molecular substances and their state at room temperature:

Solids iodine, sulphur, naphthalene

Liquids bromine, water, ethanol

Gases oxygen, nitrogen, carbon dioxide

Questions

1 What is:
a an ionic solid? **b** a molecular solid?
Give three examples of each.

2 What is another name for molecular solids?

3 Explain how a crystal of sodium chloride is formed. Can you think of a reason why its faces are flat?

4 Why do ionic solids have high melting points?

5 List four properties of covalent solids.

6 Explain why many molecular substances are gases or liquids at room temperature, and give four examples.

7 You can buy solid air-fresheners in shops. Do you think these substances are ionic or covalent? Why?

3.6 The metals and carbon

The last two pages dealt with ionic and molecular solids. Now we come to some other types of solid: the metals and carbon.

The metals

In a metal, the atoms are packed tightly together in a regular pattern. The tight packing causes outer electrons to get separated from their atoms, and the result is a lattice of ions in a sea of electrons. Copper is a good example:

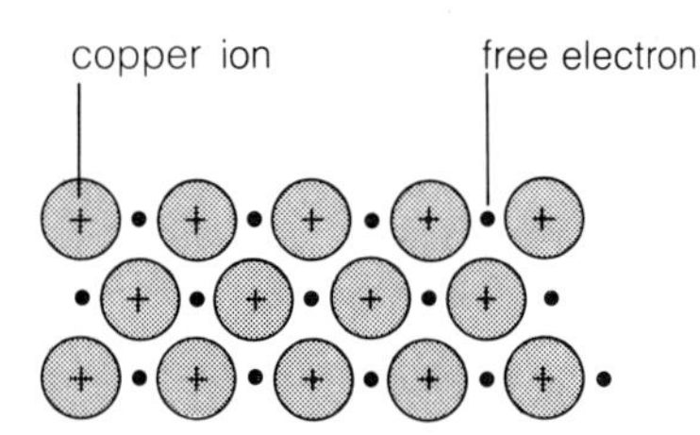

The copper ions are held together by their attraction to the electrons between them. (Opposite charges attract.)

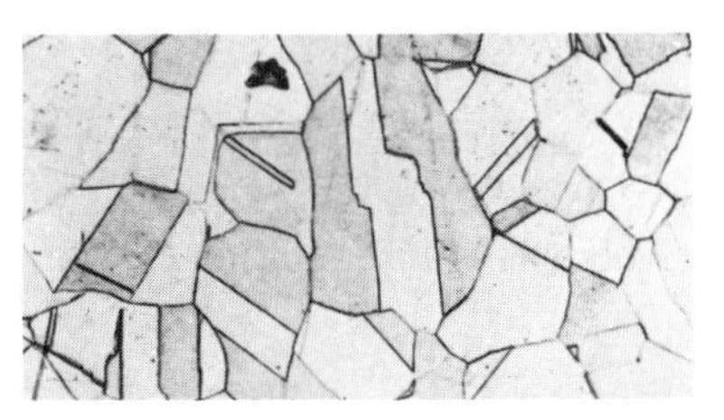

The regular arrangement of ions results in a crystal of copper. This picture shows copper magnified 100 times. The dark lines are the crystal edges.

A piece of copper wire contains millions of tiny crystals joined together.

Their properties Metals have these properties:

1 They are good conductors of electricity, because the free electrons can move through the lattice, carrying charge.
2 They are good conductors of heat. This is because the free electrons can take in heat energy, which makes them move faster. They pass some of it to their neighbours, during collisions. In this way the heat energy gets transferred all through the lattice.
3 Most metals can be hammered into different shapes (they are **malleable**) or drawn into wires (they are **ductile**). This is because the layers of ions can slide over each other.
4 They usually have high melting points, because it takes a lot of heat energy to break up the lattice. For example, copper melts at 1083 °C. There are some exceptions: the Group I metals have quite low melting points (sodium melts at 98 °C) and mercury is a liquid at room temperature.

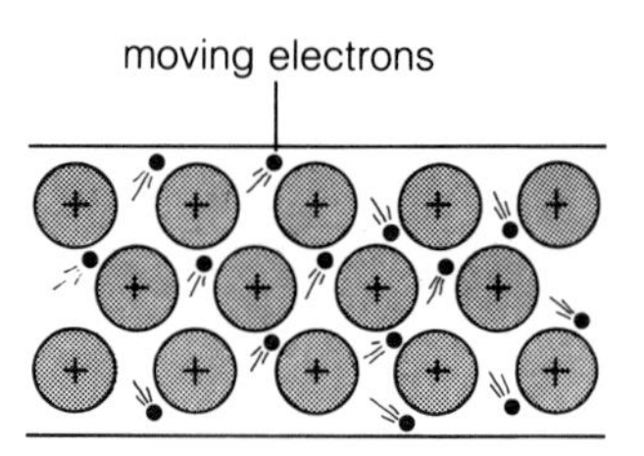

Carbon

Carbon occurs in two solid forms, **diamond** and **graphite**. These are very different, even though they both contain only carbon atoms. Pure diamond is a hard, colourless solid, that sparkles in the light, while graphite is a dark grey, greasy solid with a dull shine.
When an element has more than one form, it shows **allotropy**. The different forms are called **allotropes**. Diamond and graphite are therefore the allotropes of carbon.

Diamond and graphite, the allotropes of carbon.

Diamond A diamond is in fact a giant structure of carbon atoms:

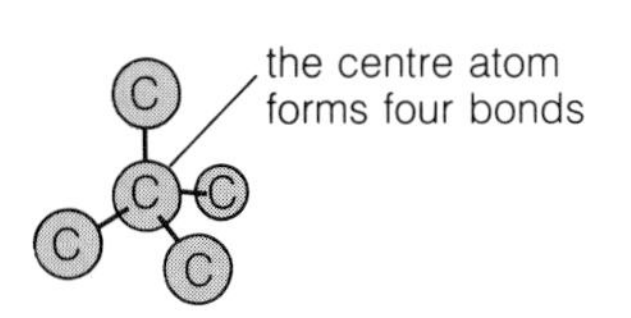

A carbon atom forms covalent bonds to *four* others, as shown above. Each outer atom then bonds to three more, and so on.

Eventually millions of carbon atoms are bonded together, in a giant covalent structure. This is part of it.

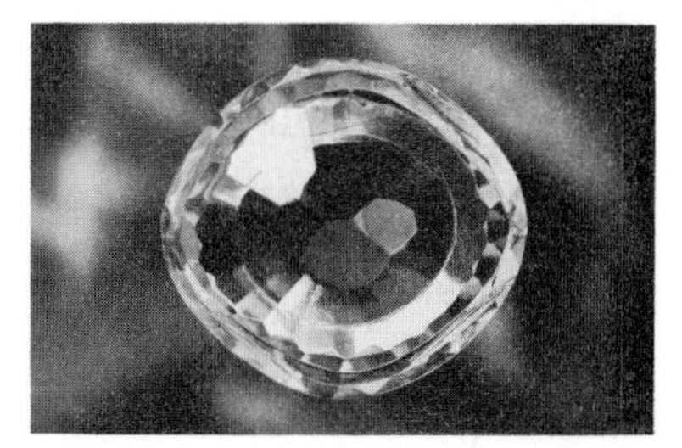

The result is a single crystal of diamond. The one shown here is the Koh-I-Noor, the largest in the world.

Diamond has these properties:

1 It is very hard—the hardest substance known. This is because each atom is held in place by four strong bonds. For the same reason, it has a very high melting point (3550 °C).
2 It cannot conduct electricity, because there are no ions or free electrons in it to carry charge.

There are several *compounds* which have giant covalent structures like that of diamond. **Silica** or **sand** is one of them. Like diamond, it is hard, has a high melting point, and does not conduct electricity.

Graphite Graphite is made of flat sheets of carbon atoms:

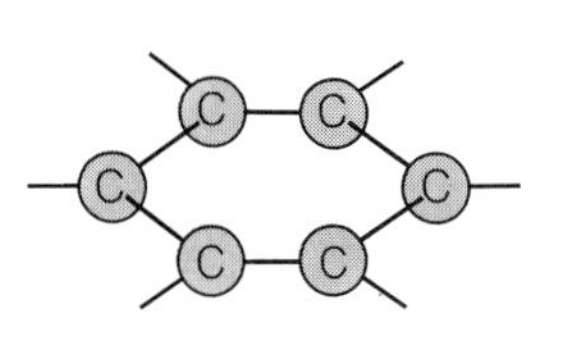

Each carbon atom forms covalent bonds to *three* others. This gives rings of six atoms, that join to make flat sheets.

weak forces

The sheets of atoms lie on top of each other, held together by weak forces.

This is a side view of a piece of graphite, magnified 5 million times! The dark lines are the sheets of atoms.

Graphite has these properties:

1 It is soft and slippery. This is because the sheets of atoms can slide over each other easily.
2 It is a good conductor of electricity. This is because each atom has four outer electrons, but forms only three bonds. The fourth electron is free to move through the graphite, carrying charge.

Questions

1 Describe the structure of a metal.
2 Copy and complete: Metals are .... because the ions are arranged in a .... pattern.
3 Why can metals conduct electricity? Would you expect a molten metal to conduct? Why?
4 What are allotropes? Give an example.
5 Draw a diagram to show how the carbon atoms bond in: **a** diamond **b** graphite
6 Why are diamonds so hard?
7 Why is graphite soft and slippery?

Questions on Chapter 3

1 The table below shows the structure of several particles:

Particle	Electrons	Protons	Neutrons
A	12	12	12
B	12	12	14
C	10	12	12
D	10	8	8
E	9	9	10

a Which three particles are neutral atoms?
b Which particle is a negative ion? What is the charge on this ion?
c Which particle is a positive ion? What is the charge on this ion?
d Which two particles are isotopes?
e Use the table on page 28 to identify the particles A to E.

2 This question is about the ionic bond formed between the metal lithium (atomic number 3) and the non-metal fluorine (atomic number 9).
a How many electrons are there in a lithium atom? Draw a diagram to show its electron structure. (You can show the nucleus as a dark circle at the centre.)
b How does a metal atom obtain a full outer shell of electrons?
c Draw the structure of a lithium ion, and write a symbol for the ion.
d How many electrons are there in a fluorine atom? Draw a diagram to show its electron structure.
e How does a non-metal atom become a negative ion?
f Draw the structure of a fluoride ion, and write a symbol for the ion.
g Draw a diagram to show what happens when a lithium atom reacts with a fluorine atom.
h Draw the arrangement of ions in the compound that forms when lithium and fluorine react together.
i Write a name and a formula for the compound in part **h**.

3 Na^+ O_2 Al CH_4 N I^-
a From the list above, select
i two atoms
ii two molecules
iii two ions
b What do the following symbols represent?
i Na^+
ii I^-
c Name the compound made up from Na^+ and I^- ions, and write a formula for it.

4 **a** The electronic configuration of a neon atom is (2,8). What is special about the outer shell of a neon atom?
b The electronic configuration of a calcium atom is (2,8,8,2). What must happen to a calcium atom for it to achieve noble gas structure?
c Draw a diagram of an oxygen atom, showing its eight protons (p), eight neutrons (n), and eight electrons (e).
d What happens to the outer shell electrons of a calcium atom, when it reacts with an oxygen atom?
e Name the compound that is formed when calcium and oxygen react together. What type of bonding does it contain?
f Write a formula for the compound in **e**.

5 **a** Write down the formula for each of the following:
i a nitrate ion
ii a sulphate ion
iii a carbonate ion
iv a hydroxide ion
b The metal strontium forms ions with the symbol Sr^{2+}. Write down the formula for each of the following:
i strontium oxide
ii strontium chloride
iii strontium nitrate
iv strontium sulphate

6 A molecule of a certain gas can be represented by the diagram on the right.

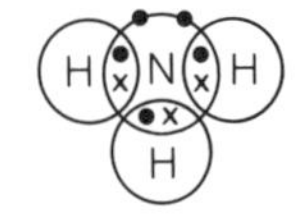

a What is the gas? What is its formula?
b What type of bonding holds the atoms together?
c Name another compound with this type of bonding.
d What do the symbols • and x represent?

7 Draw diagrams to show how the electrons are shared in the following molecules:
a fluorine, F_2
b water, H_2O
c methane, CH_4
d trichloromethane, $CHCl_3$
e oxygen, O_2
f hydrogen sulphide, H_2S
Draw the shapes of molecules **a**, **b** and **e**.

8 **a** An oxygen molecule is represented as O=O. What does the double line mean? How many electrons from each atom take part in bonding?
b A molecule of carbon dioxide (CO_2) can be drawn as O=C=O. Draw a diagram to show how the electrons are shared in the molecule.

9 Nitrogen is in Group V of the Periodic Table, and its atomic number is 7. It exists as molecules, each containing two atoms.

a Write the formula for nitrogen.

b What type of bonding would you expect between the two atoms?

c How many electrons does a nitrogen atom have in its outer shell?

d How many electrons must each atom obtain a share of, in order to gain a full shell of 8 electrons?

e Draw a diagram to show the bonding in a nitrogen molecule. You need show only the outer-shell electrons. (It may help you to look back at the bonding in an oxygen molecule, on page 46.)

f The bond in a nitrogen molecule is called a **triple bond**. Can you explain why?

g Nitrogen (N_2), oxygen (O_2), chlorine (Cl_2) and hydrogen (H_2) all exist as diatomic molecules. What does *diatomic* mean?

10 Arrange the following substances in three groups, according to their structure:
Group A—giant ionic structure
Group B—giant covalent structure
Group C—molecular structure

methane	sulphur	sodium chloride
silica	oxygen	naphthalene
iodine	graphite	water
diamond	ethanol	copper(II) sulphate
ice	ammonia	potassium hydroxide
nitrogen	bromine	hydrogen chloride
phosphorus	steam	sulphur dioxide

11 This table gives information about some properties of certain substances A to G.

Substance	M.pt. °C	Electrical conductivity		Solubility in water
		solid	liquid	
A	−112	poor	poor	insoluble
B	680	poor	good	soluble
C	−70	poor	poor	insoluble
D	1495	good	good	insoluble
E	610	poor	good	soluble
F	1610	poor	poor	insoluble
G	660	good	good	insoluble

a Which of the substances are metals? Give reasons for your choice.

b Which of the substances are ionic compounds? Give reasons for your choice.

c Two of the substances have very low melting points, compared with the rest. Explain why these could *not* be ionic compounds.

d Two of the substances are molecular. Which ones are they?

e Which substance is a giant covalent structure?

f Which substance would you expect to be very hard?

g Which substances would you expect to be soluble in tetrachloromethane?

12 Silicon lies directly below carbon in Group IV of the Periodic Table. The table below lists the melting and boiling points for silicon, carbon (diamond), and their oxides.

Substance	Symbol or formula	M.pt. °C	B.pt. °C
Carbon	C	3730	4530
Silicon	Si	1410	2400
Carbon dioxide	CO_2	sublimes at −78 °C	
Silicon dioxide	SiO_2	1610	2230

a In which state are the two *elements*, at room temperature (20 °C)?

b Is the structure of carbon (diamond) giant covalent or molecular?

c What type of structure would you expect silicon to have? Give reasons.

d In which state are the two oxides, at room temperature?

e What type of structure does carbon dioxide have?

f Does silicon dioxide have the same structure as carbon dioxide? What is the evidence for your answer?

13 Hydrogen bromide is a compound of the two elements hydrogen and bromine. It has a melting point of −87°C and a boiling point of −67°C. Bromine is one of the halogens (Group VII of the Periodic Table.)

a Is hydrogen bromide a solid, a liquid or a gas at room temperature (20°C)?

b Is it made of molecules, or does it have a giant structure? How can you tell?

c What type of bond is formed between the hydrogen and bromine atoms in hydrogen bromide? Show this on a diagram.

d Write a formula for hydrogen bromide.

e Name two other compounds that would have bonding similar to hydrogen bromide.

f Write formulae for these two compounds.

14 a Use the structures of diamond and graphite to explain why:

i graphite is used for the 'lead' in pencils

ii diamonds are used in cutting tools

b Give two reasons why:

i copper is used in electrical wiring

ii steel is used for domestic radiators

c Ethanol is used as the solvent in perfume and aftershave, because it evaporates easily. What does that tell you about the bonding in it?

4.1 The masses of atoms

Relative atomic mass

A single atom weighs hardly anything. For example, the mass of a single hydrogen atom is only about 0.000 000 000 000 000 000 000 002 grams. Very small numbers like that are awkward to use, so scientists had to find a simpler way to express the mass of an atom. Here is what they did:
First, they chose a carbon atom to be the standard atom.

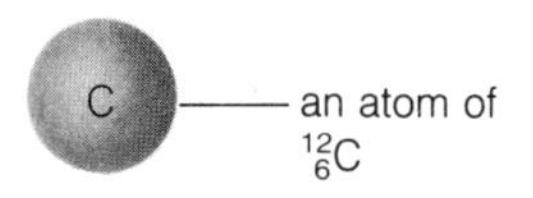

The mass spectrometer was invented by a British scientist called Aston, in 1919.

Next, they fixed its mass as exactly 12 units. (It has 6 protons and 6 neutrons, and they ignored the electrons.)
Then, they compared all the other atoms with this standard atom, using a machine called a mass spectrometer, and found values for their masses, like this:

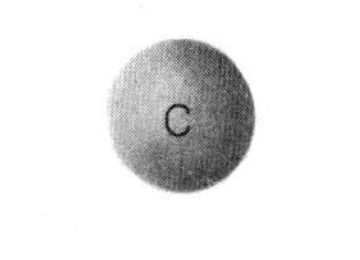		H
This is the standard atom, $^{12}_{6}C$. Its mass is exactly 12.	This magnesium atom is twice as heavy as the standard atom, so its mass must be 24.	This hydrogen atom is $\frac{1}{12}$ as heavy as the standard atom, so its mass must be 1.

The mass of an atom found by comparing it with the $^{12}_{6}C$ atom is called its **relative atomic mass**, or **RAM** for short.
So the RAM of hydrogen is 1 and the RAM of magnesium is 24.

RAM's and isotopes Not all atoms of an element are exactly the same. For example, when scientists examined chlorine in the mass spectrometer, they found there were two types of chlorine atom:

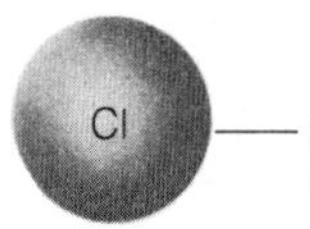

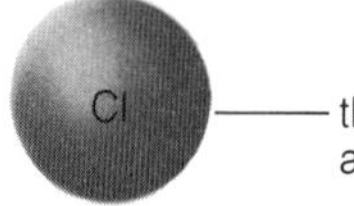

These atoms are the **isotopes** of chlorine. (They have different masses because one has two neutrons more than the other.)
It was found that out of every four chlorine atoms, three have a mass of 35 and one has a mass of 37; using this information, the *average* mass of a chlorine atom was calculated to be 35.5.
Most elements have more than one isotope, and these have to be taken into account when finding RAM's:
The RAM of an element is the average mass of its isotopes compared with an atom of $^{12}_{6}C$.
For most elements, the RAM's work out very close to whole numbers. They are usually rounded off to whole numbers, to make calculations easier.

To calculate the average mass of a chlorine atom, first find the total mass of four atoms:

$$3 \times 35 = 105$$
$$1 \times 37 = 37$$
$$142$$

The average mass $= \frac{142}{4} = 35.5$

The RAM's of some common elements Here is a list of them:

Element	Symbol	RAM	Element	Symbol	RAM
Hydrogen	H	1	Chlorine	Cl	35.5
Carbon	C	12	Potassium	K	39
Nitrogen	N	14	Calcium	Ca	40
Oxygen	O	16	Iron	Fe	56
Sodium	Na	23	Copper	Cu	64
Magnesium	Mg	24	Zinc	Zn	65
Sulphur	S	32	Iodine	I	127

Finding the mass of an ion:

Mass of sodium atom = 23 so
mass of sodium ion = 23

since a sodium ion is just a sodium atom minus an electron, and an electron has hardly any mass.
An ion has the same mass as the atom from which it is made.

Formula mass

Using a list of RAM's, it is easy to work out the mass of any molecule or group of ions. Check these examples using the information above:

Hydrogen gas is made of molecules. Each molecule contains 2 hydrogen atoms, so its mass is 2. ($2 \times 1 = 2$)

The formula of water is H_2O. A water molecule contains 2 hydrogen atoms and 1 oxygen atom, so its mass is 18. ($2 \times 1 + 16 = 18$)

Sodium chloride (NaCl) contains a sodium ion for every chloride ion. The mass of a 'unit' of sodium chloride is 58.5. ($23 + 35.5 = 58.5$)

The mass of a substance found in this way is called its **formula mass**, because it can be obtained by adding up the masses of the atoms in the formula. If the substance is made of molecules, its mass can also be called the **relative molecular mass**, or **RMM**.
So the RMM of hydrogen is 2 and the RMM of water is 18.
Here are some more examples:

Substance	Formula	Atoms in formula	RAM of atoms	Formula mass
Nitrogen	N_2	2N	N = 14	$2 \times 14 =$ **28**
Ammonia	NH_3	1N 3H	N = 14 H = 1	$1 \times 14 = 14$ $3 \times 1 = 3$ Total = **17**
Magnesium nitrate	$Mg(NO_3)_2$	1Mg 2N 6O	Mg = 24 N = 14 O = 16	$1 \times 24 = 24$ $2 \times 14 = 28$ $6 \times 16 = 96$ Total = **148**

Questions

1 What is the relative atomic mass of an element?
2 What is the RAM of the iodide ion, I^-?
3 Show that the formula mass for chlorine (Cl_2) is 71.
4 What is the RMM of butane, C_4H_{10}?
5 Work out the formula mass of:
a oxygen, O_2 **b** iodine, I_2
c methane, CH_4 **d** ethanol, C_2H_5OH
e ammonium sulphate $(NH_4)_2SO_4$

4.2 The mole

What is a mole?

On the last two pages you read about relative atomic masses and formula masses. These are not just boring numbers—they are very important things for a chemist to know:

If you work out the RAM or formula mass of a substance, and then weigh out that number of grams of the substance, you can say how many atoms or molecules it contains.

This is very useful, since single atoms and molecules are far too small to be counted.

For example, the RAM of carbon is 12. The photograph on the right shows 12 grams of carbon. The heap contains 602 000 000 000 000 000 000 000 carbon atoms.

This huge number of atoms is called a **mole** of atoms.
It is called **Avogadro's number**, after Avogadro, an Italian scientist who lived over a hundred years ago.
It is usually written in a shorter way as $\mathbf{6.02 \times 10^{23}}$.
(The 10^{23} shows that you must move the decimal point 23 places to the right to get the full number.)
Now look at these:

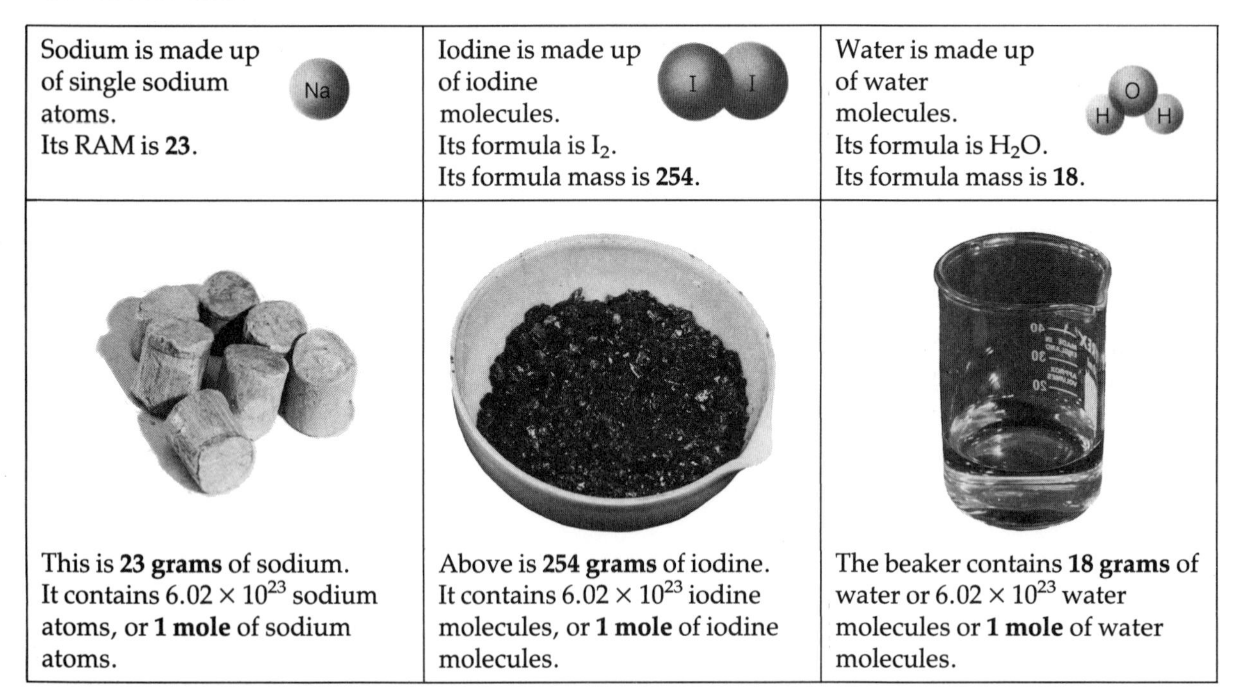

Sodium is made up of single sodium atoms. Its RAM is **23**.	Iodine is made up of iodine molecules. Its formula is I_2. Its formula mass is **254**.	Water is made up of water molecules. Its formula is H_2O. Its formula mass is **18**.
This is **23 grams** of sodium. It contains 6.02×10^{23} sodium atoms, or **1 mole** of sodium atoms.	Above is **254 grams** of iodine. It contains 6.02×10^{23} iodine molecules, or **1 mole** of iodine molecules.	The beaker contains **18 grams** of water or 6.02×10^{23} water molecules or **1 mole** of water molecules.

From these examples you should see that:
One mole of a substance is 6.02×10^{23} particles of the substance. It is obtained by weighing out the RAM or formula mass, in grams.

Finding the mass of a mole

You can find the mass of 1 mole of any substance, by these steps:

1 Write down the symbol or formula of the substance.
2 Find out its RAM or formula mass.
3 Express that mass in grams.

This table shows more examples:

Substance	Symbol or formula	RAM's	Formula mass	Mass of 1 mole
Helium	He	He = 4	4	**4 grams**
Oxygen	O_2	O = 16	$2 \times 16 = 32$	**32 grams**
Ethanol	C_2H_5OH	C = 12 H = 1 O = 16	$2 \times 12 = 24$ $6 \times 1 = 6$ $1 \times 16 = 16$ 46	**46 grams**

Some calculations on the mole

Example 1 Calculate the mass of:
(a) 0.5 moles of bromine atoms
(b) 0.5 moles of bromine molecules

(a) The RAM of bromine is 80, so 1 mole of bromine *atoms* has a mass of 80 grams. Therefore 0.5 moles of bromine *atoms* have a mass of 0.5×80 grams, or 40 grams.
(b) A bromine *molecule* contains 2 atoms, so its formula mass is 160. Therefore 0.5 moles of bromine *molecules* have a mass of 0.5×160 grams, or 80 grams.

Formula mass or RMM = 160

So, to find the mass of a given number of moles:
mass = mass of 1 mole × number of moles

Example 2 How many moles of oxygen molecules are there in 64 grams of oxygen, O_2?
The formula mass of oxygen gas is 32, so 32 grams of it is 1 mole.
Therefore 64 grams is $\frac{64}{32}$ moles, or 2 moles.

Formula mass or RMM = 32

So, to find the number of moles in a given mass:

$$\textbf{number of moles} = \frac{\textbf{mass}}{\textbf{mass of 1 mole}}$$

Questions

1 How many atoms are in 1 mole of atoms?
2 How many molecules are in 1 mole of molecules?
3 What name is given to the number 6.02×10^{23}?
4 Find the mass of 1 mole of:
 a hydrogen atoms **b** iodine atoms
 c chlorine atoms **d** chlorine molecules
5 Find the mass of 2 moles of:
 a oxygen atoms **b** oxygen molecules
6 Find the mass of 3 moles of ethanol, C_2H_5OH.
7 How many moles of molecules are there in:
 a 18 grams of hydrogen, H_2?
 b 54 grams of water?

4.3 Compounds and solutions

The percentage of an element in a compound

Methane is a compound of carbon and hydrogen, with the formula CH_4. The mass of a carbon atom is 12, and the mass of each hydrogen atom is 1, so its formula mass is 16.

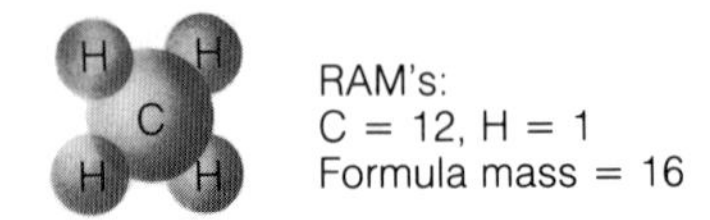

You can find what fraction of the total mass is carbon, and what fraction is hydrogen, like this:

$$\text{Mass of carbon as fraction of total mass} = \frac{\text{mass of carbon}}{\text{total mass}} = \frac{12}{16} \text{ or } \frac{3}{4}$$

$$\text{Mass of hydrogen as fraction of total mass} = \frac{\text{mass of hydrogen}}{\text{total mass}} = \frac{4}{16} \text{ or } \frac{1}{4}$$

These fractions are usually written as percentages. To change a fraction to a percentage, you just multiply it by 100:

$\frac{3}{4} \times 100 = \frac{300}{4} = 75$ per cent or 75% $\frac{1}{4} \times 100 = \frac{100}{4} = 25\%$

So 75% of the mass of methane is carbon, and 25% is hydrogen. We say that the **percentage composition** of methane is 75% carbon, 25% hydrogen.

The famous scientist John Dalton and a pupil, collecting methane from a pond. Methane also occurs in oil wells and natural gas wells. No matter where it comes from, its composition is always the same.

Calculations on composition

Here is how to calculate the percentage of an element in a compound:

1 Write down the formula of the compound.
2 Using a list of RAM's, work out its formula mass.
3 Write the mass of the element you want, as a fraction of the total.
4 Multiply the fraction by 100, to give a %.

Example Fertilisers contain nitrogen, which plants need to help them grow. One important fertiliser is ammonium nitrate, NH_4NO_3. What is the percentage of nitrogen in it?

The formula mass of the compound is 80, as shown on the right. The element we are interested in is nitrogen.

Mass of nitrogen in the formula $= 28$

Mass of nitrogen as fraction of total $= \frac{28}{80}$

Mass of nitrogen as percentage of total $= \frac{28}{80} \times 100 = \mathbf{35\%}$

Finding the formula mass for ammonium nitrate:

The RAM's are: N = 14, H = 1, O = 16.
The formula contains 2 N, 4 H and 3 O, so the formula mass is:

$2N = 2 \times 14 = 28$
$4H = 4 \times 1 = 4$
$3O = 3 \times 16 = \underline{48}$
Total $= \underline{\underline{80}}$

The concentration of a solution

Below are three solutions of copper(II) sulphate in water.
Each has the same volume, 1 dm^3. (**1 dm^3 is 1 litre**, or **1000 cm^3**.)

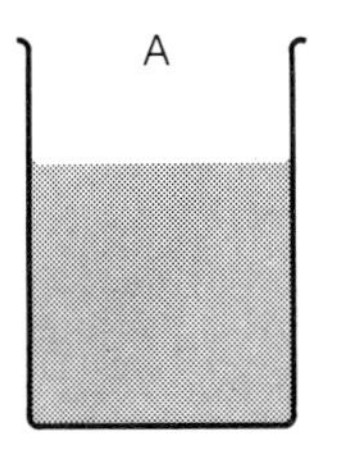

Only 2.5 grams of copper(II) sulphate were dissolved to make this solution. It is **dilute**.

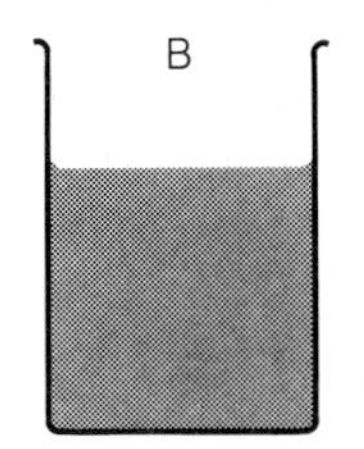

This one contains 25 grams of the compound. It is more **concentrated** than A.

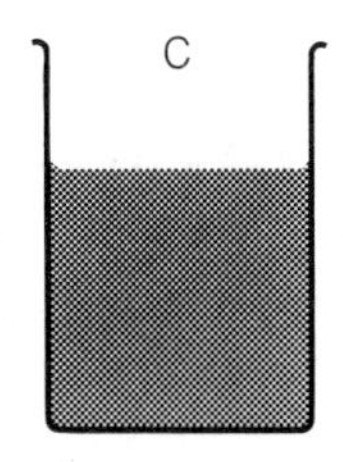

This one contains 250 grams of the compound, so it is more concentrated than either A or B.

Each solution above has a different **concentration.**
The concentration of a solution is the amount of solute, in grams or moles, that are dissolved in 1 dm^3 of solution.

The concentrations of the copper(II) sulphate solutions in grams are:
A — 2.5 grams per dm^3
B — 25 grams per dm^3
C — 250 grams per dm^3

Finding the concentration in moles

The formula mass of copper(II) sulphate must first be worked out. The answer is 250, as shown on the right. Therefore 1 mole of the compound has a mass of 250 grams.

> The formula of copper(II) sulphate is $CuSO_4.5H_2O$. The formula contains 1 Cu, 1 S, 9 O, and 10 H, so the formula mass is:
>
> | | | | | |
> |---|---|---|---|---|
> | 1 Cu | = | 1 × 64 | = | 64 |
> | 1 S | = | 1 × 32 | = | 32 |
> | 9 O | = | 9 × 16 | = | 144 |
> | 10 H | = | 10 × 1 | = | 10 |
> | | | Total | = | 250 |

Solution A It has 2.5 grams of the compound in 1 dm^3 of solution.

$$2.5 \text{ grams} = \frac{2.5}{250} \text{ moles} = 0.01 \text{ mole}$$

so its concentration is **0.01 mole per dm^3**.
Mole per dm^3 is often shortened just to M, so the concentration of solution A can be written as **0.01 M**.

Solution C It has 250 grams of the compound in 1 dm^3 of solution.
250 grams = 1 mole
so its concentration is **1 mole per dm^3**, or **1 M** for short.
A solution that contains 1 mole of solute per dm^3 is often called a **molar solution**, so C is a molar solution.

Questions

1. What percentage of sulphur is there in sulphur dioxide, SO_2? The RAM's are: S = 32, O = 16. What percentage of oxygen is there?
2. Find the percentage of hydrogen and oxygen in ammonium nitrate, NH_4NO_3.
3. Find the percentage of hydrogen in water, H_2O.
4. Work out the concentration of solution B above in moles per dm^3.
5. The formula mass of sodium hydroxide is 40. How many grams of sodium hydroxide are there:
   a in 1 litre of a molar solution?
   b in 500 cm^3 of a 1 M solution?

4.4 Formulae of compounds

What a formula tells you

The formula of carbon dioxide is CO_2. Some molecules of it are shown on the right. You can see that:

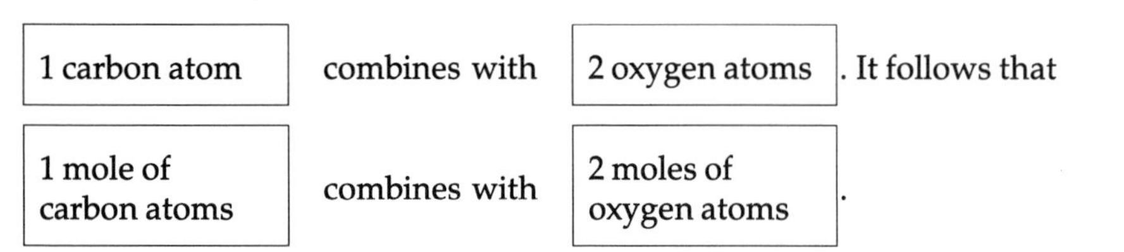

RAM's: C = 12, O = 16

Moles can be changed to grams, using RAM's. So we can write:

12 g of carbon | combines with | 32 g of oxygen

This means that 6 g of carbon combines with 16 g of oxygen, and so on. Therefore, from the formula of a compound, you can tell:
(i) how many moles of the different atoms combine
(ii) how many grams of the different elements combine
Now look at the formula of ammonia, NH_3. It shows that:

1 mole of nitrogen atoms (14 grams) | combines with | 3 moles of hydrogen atoms (3 grams)

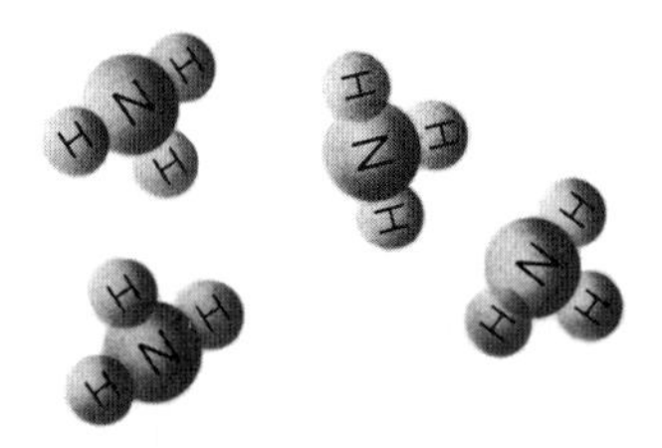

RAM's: H = 1, N = 14

Do you agree? Check the masses using the RAM's on the right.

Finding formulas from masses

From the formula of a compound, you can tell what masses of the elements will combine. But you can also do things the other way round. Starting with the masses that combine, you can work out the formula of the compound. These are the steps:

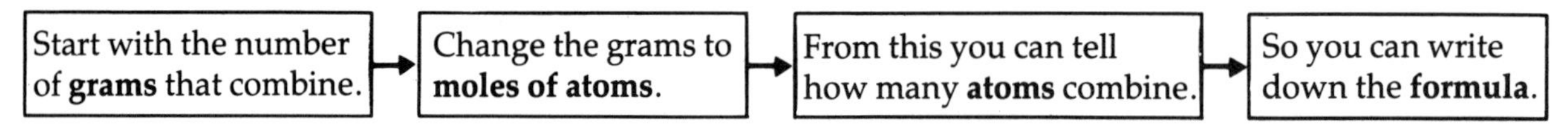

A formula found by this method is called an **empirical formula**.

Example 1 32 grams of sulphur combine with 32 grams of oxygen, to form the compound sulphur dioxide. What is its formula?

First, change the masses to moles of atoms.
The RAM of sulphur is 32, and the RAM of oxygen is 16, so:
$\frac{32}{32}$ moles of sulphur atoms combine with $\frac{32}{16}$ moles of oxygen atoms, or
1 mole of sulphur atoms combine with 2 moles of oxygen atoms, so
1 sulphur atom combines with 2 oxygen atoms.
The formula of sulphur dioxide is therefore SO_2.

Remember, to change masses to moles:

$$\textbf{no. of moles} = \frac{\textbf{mass}}{\textbf{mass of 1 mole}}$$

Example 2 20 grams of calcium react completely with 19 grams of fluorine, to form calcium fluoride. What is its formula?

The RAM's are: Ca = 40, F = 19. Changing masses to moles:

$\frac{20}{40}$ moles of calcium atoms react with $\frac{19}{19}$ moles of fluorine atoms, or
0.5 moles of calcium atoms react with 1 mole of fluorine atoms, so
1 mole of calcium atoms react with 2 moles of fluorine atoms.
The formula of calcium fluoride is therefore **CaF_2**.

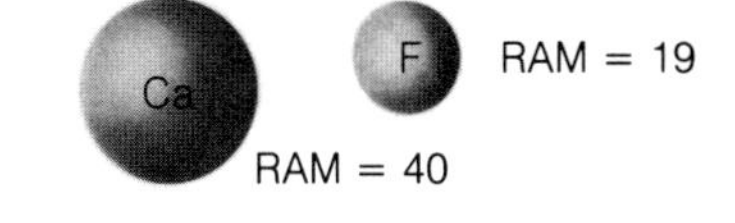

Finding masses by experiment

To work out a formula, you first need to know what masses of the elements combine. This must be found out by experiment.
For example, magnesium combines with oxygen like this:

magnesium + oxygen ⟶ magnesium oxide

and the masses that combine can be found like this:

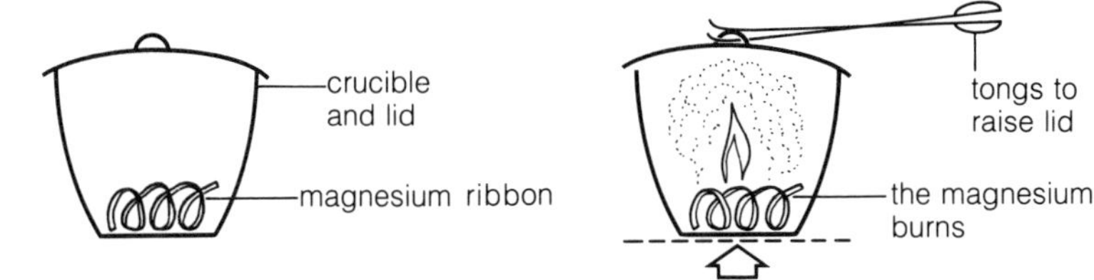

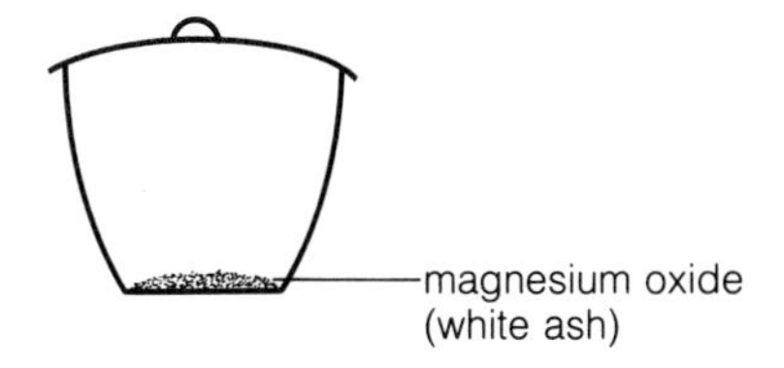

First, a crucible and lid is weighed. Next a coil of magnesium ribbon is added, and the new mass found.

Then the crucible is heated, and the lid carefully raised a little, at intervals, to let in oxygen. The magnesium burns brightly.

When burning is complete, the crucible is allowed to cool, with its lid still on.
Then it is weighed again.

The results Here are some sample results, and the calculation:

Mass of crucible + lid = 25.2 g
Mass of crucible + lid + magnesium = 27.6 g
Mass of crucible + lid + magnesium oxide = 29.2 g

Mass of magnesium = 27.6 g − 25.2 g = 2.4 g
Mass of magnesium oxide = 29.2 g − 25.2 g = 4.0 g
Mass of oxygen therefore = 4.0 g − 2.4 g = 1.6 g

So 2.4 g of magnesium combines with 1.6 g of oxygen. The formula of magnesium oxide can now be found.
The RAM's are: Mg = 24, O = 16. Changing the masses to moles:

$\frac{2.4}{24}$ moles of magnesium atoms combine with $\frac{1.6}{16}$ moles of oxygen atoms
0.1 moles of magnesium atoms combine with 0.1 moles of oxygen atoms
1 mole of magnesium atoms combines with 1 mole of oxygen atoms.
The formula of magnesium oxide is therefore **MgO**.

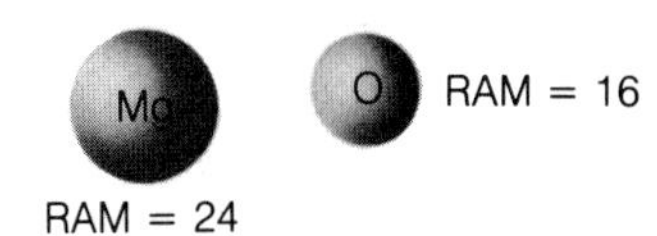

Questions

1 **a** How many moles of carbon atoms combine with 4 moles of hydrogen atoms, to form methane, CH_4?
b How many grams of hydrogen combine with 12 grams of carbon, to form methane?

2 How would you change *grams* to *moles of atoms*?

3 To form iron(II) sulphide, 56 g of iron combine with 32 g of sulphur. Find its formula.
(The RAM's are: Fe = 56, S = 32.)

4 In hydrogen sulphide, there is 1 g of hydrogen for every 16 g of sulphur. Find its formula.

Questions on Chapter 4

Relative atomic masses are given on page 202. Use the approximate values given in the table.

1 How many grams are there in:
a 1 mole of copper atoms?
b 1.5 moles of sulphur atoms?
c 2 moles of magnesium atoms?
d 5 moles of carbon atoms?
e 10 moles of chlorine atoms?
f 0.1 mole of nitrogen atoms?
g 0.2 moles of neon atoms?
h 0.6 moles of hydrogen atoms?
i 1.5 moles of oxygen atoms?

2 How many grams are there in:
a 1 mole of hydrogen molecules, H_2?
b 2 moles of hydrogen molecules, H_2?
c 1 mole of oxygen molecules, O_2?
d 0.5 moles of chlorine molecules, Cl_2?
e 2 moles of phosphorus molecules, P_4?
f 4 moles of sulphur molecules, S_8?

3 Find how many moles of atoms there are, in:
a 32 g of sulphur
b 48 g of magnesium
c 23 g of sodium
d 14 g of lithium
e 1.4 g of lithium
f 3.1 g of phosphorus
g 6.4 g of oxygen
h 5.4 g of aluminium
i 2 g of hydrogen
j 0.6 g of carbon

4 For each pair, decide which of the two substances contains the greater number of atoms.
a 80 g of sulphur, 80 g of calcium
b 80 g of sulphur, 80 g of oxygen
c 1 mole of sulphur atoms, 8 moles of chlorine atoms
d 1 mole of sulphur atoms, 1 mole of oxygen molecules
e 4 moles of sulphur atoms, $\frac{1}{8}$ mole of sulphur molecules (S_8)

5 For each pair, decide which of the two solutions contains the greater number of moles of solute.
a 1 litre of 1 M sodium chloride (NaCl), 1 litre of 2 M sodium chloride
b 500 cm^3 of 1 M sodium chloride, 1 litre of 1 M sodium chloride
c 1 litre of 0.1 M sodium chloride, 100 cm^3 of 2 M sodium chloride
d 1 litre of 1 M sodium chloride, 1 litre of 1 M sodium hydroxide (NaOH)
e 20 cm^3 of 0.5 M sodium chloride, 40 cm^3 of 1 M sodium chloride

6 How many grams are there in:
a 1 mole of water, H_2O?
b 5 moles of water?
c 1 mole of anhydrous copper(II) sulphate, $CuSO_4$?
d 1 mole of hydrated copper(II) sulphate, $CuSO_4.5H_2O$?
e 2 moles of ammonia, NH_3?
f 0.5 moles of ammonium carbonate, $(NH_4)_2CO_3$?
g 0.3 moles of calcium carbonate, $CaCO_3$?
h $\frac{1}{5}$ moles of magnesium oxide, MgO?
i 0.1 moles of sodium thiosulphate, $Na_2S_2O_3$
j 2 moles of iron(III)chloride, $FeCl_3$

7 1 mole of sodium carbonate (Na_2CO_3) contains 2 moles of sodium atoms, 1 mole of carbon atoms and 3 moles of oxygen atoms.
In the same way, write down the number of moles of each atom present in 1 mole of:
a lead oxide, Pb_3O_4
b ammonium nitrate, NH_4NO_3
c calcium hydroxide, $Ca(OH)_2$
d dinitrogen tetroxide, N_2O_4
e ethanol, C_2H_5OH
f ethanoic acid, CH_3COOH
g hydrated iron(II) sulphate, $FeSO_4.7H_2O$
h iron(III) ammonium sulphate, $NH_4Fe(SO_4)_2$

8 The formula of calcium oxide is CaO. The RAM's are: Ca = 40, O = 16.
Complete the following statements:
a 1 mole of Ca (...... g) and 1 mole of O (...... g) combine to form ...... mole of CaO (...... g).
b 4.0 g of calcium and ...... g of oxygen combine to form ...... g of calcium oxide.
c When 0.4 g of calcium reacts with oxygen, the increase in mass is ...... g.
d If 6 moles of CaO were decomposed to calcium and oxygen, ... moles of Ca and ...... moles of O_2 would be obtained.
e The percentage by mass of calcium in calcium oxide is ......%.

9 Two samples of copper oxide were made by different methods. The oxides were then converted to copper. The following results were obtained:

	Sample 1	Sample 2
Mass of copper oxide	16.0 g	32.0 g
Mass of copper obtained	12.8 g	25.6 g

a Calculate the percentage composition of copper in each sample.
b Were the two oxides of the same or different composition?

10 In a reaction to make manganese from manganese oxide, the following results were obtained:
174 g of manganese oxide produced 110 g of manganese. (RAM's: Mn = 55, O = 16)
a What mass of oxygen is there in 174 g of manganese oxide?
b How many moles of oxygen atoms is this?
c How many moles of manganese atoms are there in 110 g of manganese?
d What is the simplest formula of manganese oxide?

11 The following results were obtained in an experiment to find the formula of magnesium oxide:
Mass of crucible = 12.5 g
Mass of crucible + magnesium = 14.9 g
Mass of crucible + magnesium oxide = 16.5 g
(The RAM's are: Mg = 24, O = 16)
a What mass of magnesium was used in the experiment?
b How many moles of magnesium atoms is this?
c What mass of oxygen combined with the magnesium?
d How many moles of oxygen atoms is this?
e Use your answers to parts **b** and **d** to find the formula of magnesium oxide.

12 27 g of aluminium burn in a stream of chlorine to form 133.5 g of aluminium chloride.
(RAM's: Al = 27, Cl = 35.5)
a What mass of chlorine is present in 133.5 g of aluminium chloride?
b How many moles of chlorine atoms is this?
c How many moles of aluminium atoms are present in 27 g of aluminium?
d Use your answers for parts **b** and **c** to find the simplest formula of aluminium chloride.
e 133.5 g of aluminium chloride are dissolved to form 1 dm^3 of an aqueous solution. What is the concentration in mol per dm^3 of this aluminium chloride solution? (1 dm^3 is the same as 1 litre.)
f 1 dm^3 of an aqueous solution is made using 13.35 g of aluminium chloride. What is its concentration in mol per dm^3?

13 Ammonia (NH_3), carbon dioxide (CO_2), hydrogen (H_2), and oxygen (O_2) are all gases at room temperature.
a Which of the gases are compounds?
b Which are diatomic?
c For each gas, calculate the mass of one mole of molecules.
d How many moles of ammonia molecules are there in 34 g of ammonia?
e How many moles of hydrogen atoms are there in 34 g of ammonia?
f What mass of carbon dioxide contains the same number of oxygen atoms as 16 g of oxygen?

14 An oxide of copper can be converted to copper by heating it in a stream of hydrogen in the apparatus shown below:

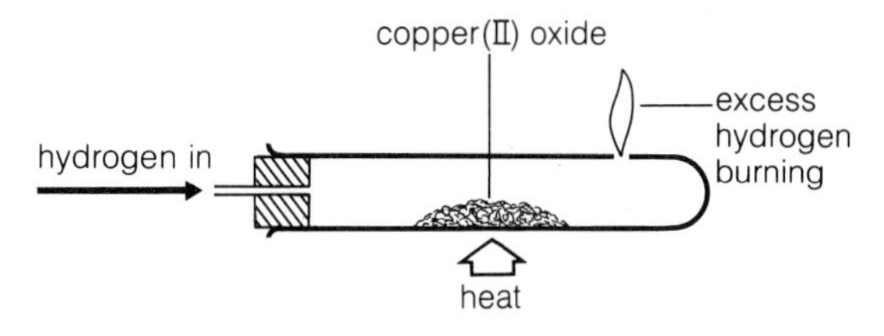

The hydrogen supply was turned on, and hydrogen was allowed to pass through the test tube for some time before the excess gas was lit. The test tube was heated until all the copper oxide was converted to copper. The apparatus was allowed to cool, with hydrogen still passing through, before it was dismantled.
a Copy and complete the word equation for the reaction:
copper oxide + hydrogen → copper + ...
b Why was the hydrogen allowed to pass through the apparatus for some time, before the excess was lit? Why was the excess hydrogen burned?

The experiment was repeated several times, by different groups in the class. Each group used a different mass of oxide. The results are shown below.

Group	Mass of copper oxide/g	Mass of copper produced/g	Mass of oxygen lost/g
1	0.62	0.55	0.07
2	0.90	0.80	0.10
3	1.12	1.00	0.12
4	1.69	1.50	
5	1.80	1.60	

c Work out the missing figures for the table.
d On graph paper, plot the mass of copper against the mass of oxygen. (Show the mass of copper along the x-axis and the mass of oxygen along the y-axis.) Then draw the best straight line through the origin and the set of points.
e From the graph, find the mass of oxygen which would combine with 1.28 g of copper.
f Calculate the mass of oxygen which would combine with 128 g of copper.
(Remember, $1.28 \times 100 = 128$)
g How many moles of copper atoms are there in 128 g of copper? (Cu = 64)
h How many moles of oxygen atoms combine with 128 g of copper? (O = 16)
i What is the simplest formula of this oxide of copper?
j What is another name for the *simplest formula* found in this way?

5.1 Physical and chemical change

A substance can be changed by heating it, adding water to it, mixing another substance with it, and so on. The change that takes place will be either a **chemical** change or a **physical** one.

Chemical change

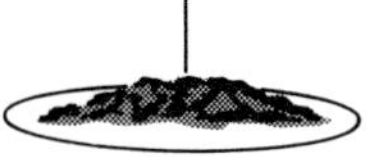

1 Some yellow sulphur and black iron filings are mixed together.

2 The mixture is easily separated again, by using a magnet to attract the iron . . .

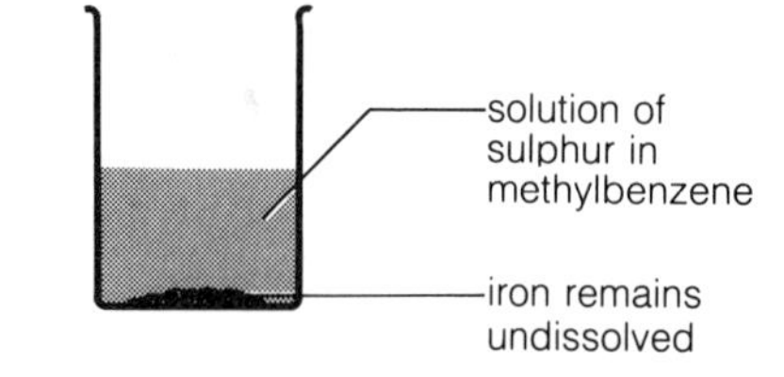

3 . . . or by dissolving the sulphur in methylbenzene (a solvent).

4 But when the mixture is *heated*, it glows brightly. The yellow specks of sulphur disappear. A black solid forms.

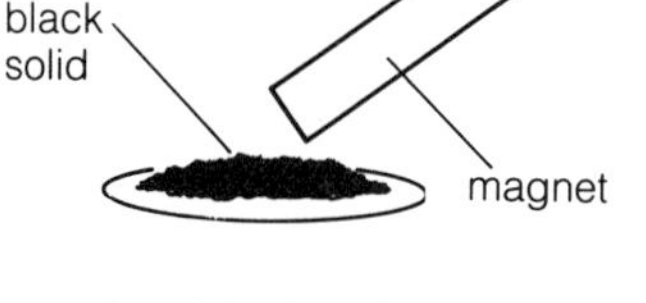

5 This black solid is not at all like the mixture. It is not affected by a magnet . . .

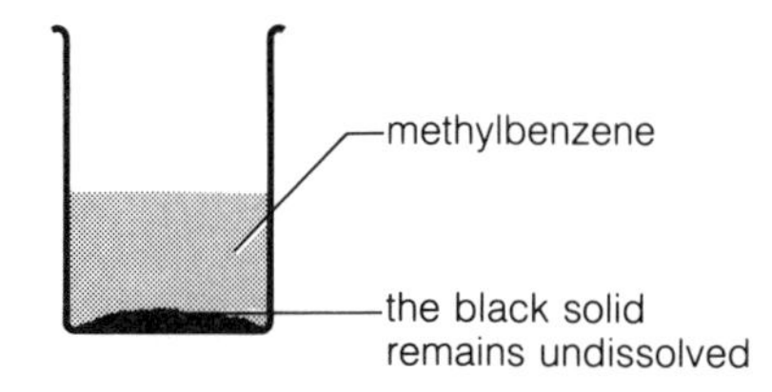

6 . . . and none of it dissolves in methylbenzene.

The black solid is obviously a new chemical substance.
When it produces a new chemical substance, a change is called a chemical change.
So in step 4, a chemical change has taken place. The iron and sulphur have **reacted** together, to form the compound iron sulphide.

The difference between the mixture and the compound In the mixture above, iron and sulphur particles are mixed closely together, but they are not bonded to each other—the iron particles still behave like iron, and the sulphur particles like sulphur. During the reaction, however, iron and sulphur atoms form ions which bond to each other. The magnet and solvent now have no effect.

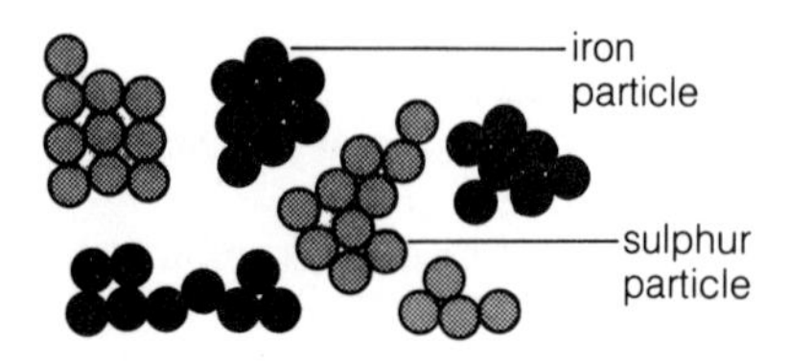

Mixture—iron and sulphur particles mixed together. Each particle contains many atoms.

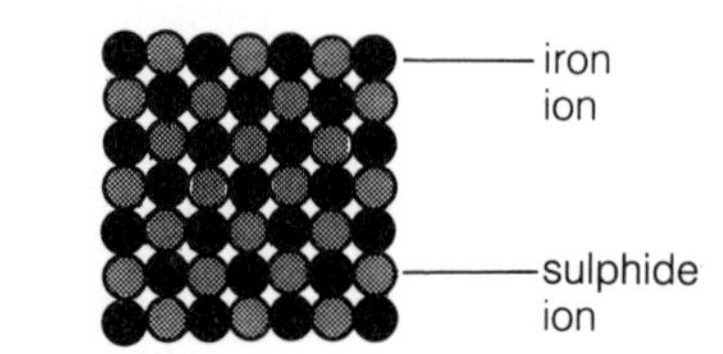

Compound—iron and sulphide ions bonded together.

The compound iron sulphide occurs in the earth as **iron pyrites**.

The signs of a chemical change

A chemical change is usually called a **chemical reaction**. You can tell when a chemical reaction has taken place, by these signs:

1 **One or more new chemical substances are formed.**
The new substances usually look quite different from the starting substances. For example:

iron (black filings)	+	sulphur (yellow powder)	⟶	iron sulphide (black solid)

2 **Energy is taken in or given out, during the reaction.**
In step 4, a little heat from a bunsen is needed to start off the reaction between iron and sulphur. But the reaction gives out heat once it begins.
A reaction that gives out heat energy is **exothermic**.
A reaction that takes in heat energy is **endothermic**.
So the reaction between iron and sulphur is exothermic. The reactions that take place when you fry an egg are endothermic.
Some reactions give out energy in the form of light or sound. For example magnesium burns in air with a bright white light and a hiss.

3 **The change is usually difficult to reverse.**
You would need to carry out several other reactions, to get back the iron and sulphur from iron sulphide.

Fireworks contain magnesium and other substances. When they burn the reactions give out energy in the form of heat, light, and sound.

The reactions that take place when you fry an egg are endothermic.

Physical change

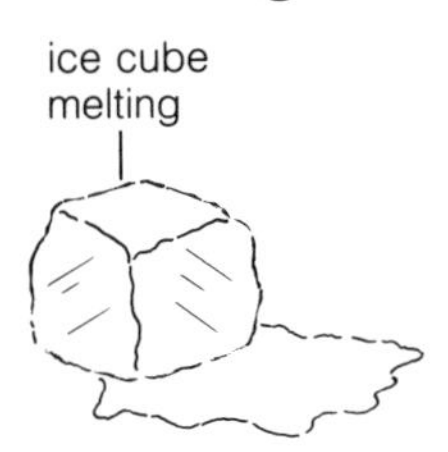

Ice turns to water at 0 °C. It is easy to change the water back to ice again, by cooling it.

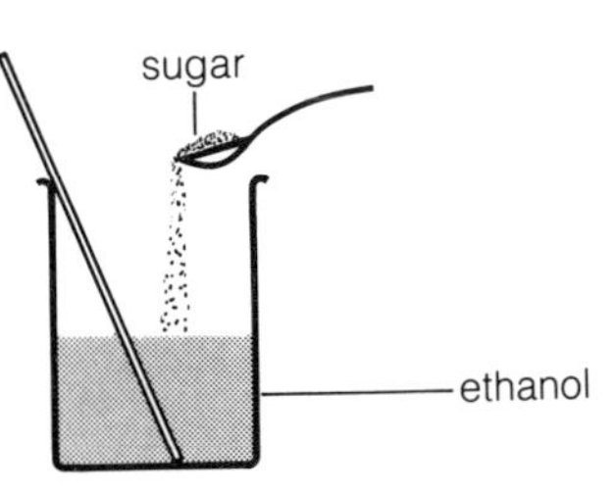

Sugar dissolves in ethanol. You can separate the two again by distilling the solution.

No new chemical substances are formed in these changes.
For example, although ice and water *look* different, they are both made of water molecules, and have the formula H_2O.
When no new chemical substance is formed, a change is called a physical change.
So the changes above are both physical changes. Physical changes are usually easy to reverse.

Questions

1 Explain the difference between a *mixture* of iron and sulphur and the *compound* iron sulphide.
2 What are the signs of a chemical change?
3 What is: **a** an exothermic reaction?
b an endothermic reaction?
4 Is the change chemical or physical? Give reasons.
a a glass bottle breaking
b butter and sugar being made into toffee
c wool being knitted into a sweater
d coal burning in air

5.2 Equations for chemical reactions

The reaction between carbon and oxygen When carbon is heated in oxygen, they react together, and carbon dioxide is formed. The carbon and oxygen are called **reactants**, because they react together. Carbon dioxide is the **product** of the reaction.
You could show the reaction by a diagram, like this:

+

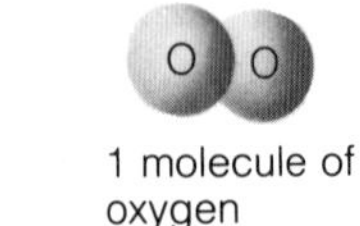

1 atom of carbon — 1 molecule of oxygen — 1 molecule of carbon dioxide

or in a shorter way, by using symbols, like this:

$$C + O_2 \longrightarrow CO_2$$

This short way to describe the reaction is called a **chemical equation.**

In a coal fire, the main reaction is $C + O_2 \rightarrow CO_2$

The reaction between hydrogen and oxygen When hydrogen and oxygen react together, the product is water. The diagram is:

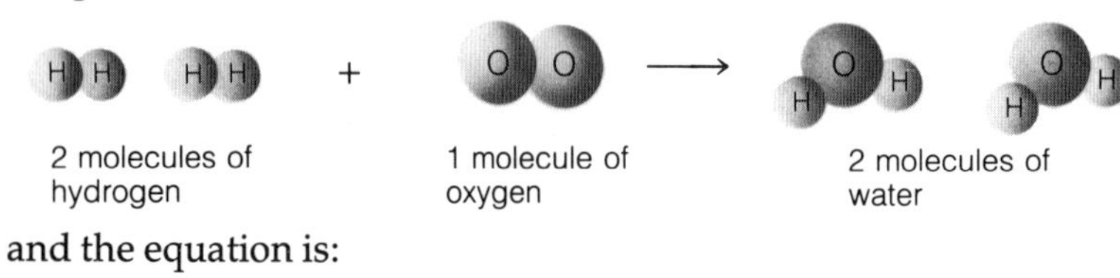

and the equation is:

$$2H_2 + O_2 \longrightarrow 2H_2O$$

Can you see why there is a 2 in front of H_2 and H_2O, in the equation? Now look at the number of atoms on each side of the equation:

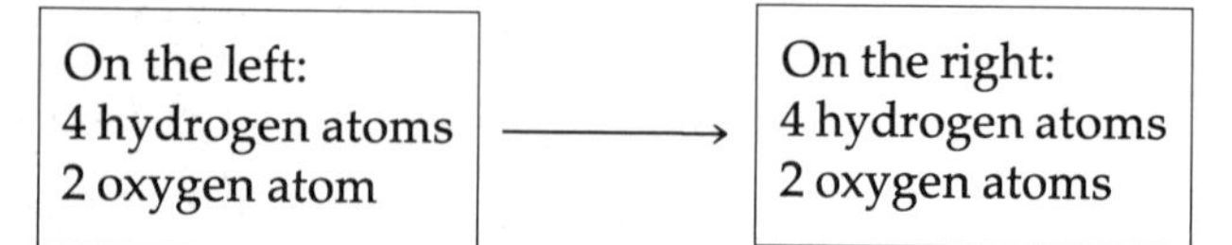

The numbers of hydrogen and oxygen atoms are the same on both sides of the equation. This is because atoms do not *disappear* during a reaction—they are just *rearranged*, as shown in the diagram.
When the numbers of different atoms are the same on both sides, an equation is said to be **balanced**. An equation which is not balanced is not correct. Check the equation for the reaction between carbon and oxygen above. Is it balanced?

Adding more information to equations Reactants and products may be solids, liquids, gases or solutions. You can show their states by adding **state symbols** to the equations. The state symbols are:

(*s*) for solid (*l*) for liquid
(*g*) for gas (*aq*) for aqueous solution (solution in water)

For the two reactions above, the equations with state symbols are:

$$C(s) + O_2(g) \longrightarrow CO_2(g)$$
$$2H_2(g) + O_2(g) \longrightarrow 2H_2O(l)$$

The reaction between hydrogen and oxygen gives out so much energy that it is used to power rockets. The hydrogen and oxygen are carried as liquids in the fuel tanks.

How to write the equation for a reaction

These are the steps to follow, when writing an equation:

1 Write the equation in words.
2 Now write the equation using symbols. Make sure all the formulae are correct.
3 Check that the equation is balanced, for each type of atom in turn. *Make sure you do not change any formulae.*
4 Add the state symbols.

Example 1 Calcium burns in chlorine to form calcium chloride, a solid. Write an equation for the reaction, using the steps above.

1 Calcium + chlorine $\longrightarrow$ calcium chloride
2 $Ca + Cl_2 \longrightarrow CaCl_2$
3 Ca: 1 atom on the left and 1 atom on the right.
Cl: 2 atoms on the left and 2 atoms on the right.
The equation is balanced.
4 $Ca\,(s) + Cl_2\,(g) \longrightarrow CaCl_2\,(s)$

Example 2 In industry, hydrogen chloride is formed by burning hydrogen in chlorine. Write an equation for the reaction.

1 Hydrogen + chlorine $\longrightarrow$ hydrogen chloride
2 $H_2 + Cl_2 \longrightarrow HCl$
3 H: 2 atoms on the left and 1 atom on the right.
Cl: 2 atoms on the left and 1 atom on the right.
The equation is *not* balanced. It needs another molecule of hydrogen chloride on the right. So a 2 is put *in front of* the HCl.
$H_2 + Cl_2 \longrightarrow 2HCl$
The equation is now balanced. Do you agree?
4 $H_2\,(g) + Cl_2\,(g) \longrightarrow 2HCl\,(g)$

Example 3 Magnesium burns in oxygen to form magnesium oxide, a white solid. Write an equation for the reaction.

1 Magnesium + oxygen $\longrightarrow$ magnesium oxide
2 $Mg + O_2 \longrightarrow MgO$
3 Mg: 1 atom on the left and 1 atom on the right.
O: 2 atoms on the left and 1 atom on the right.
The equation is *not* balanced. Try this:
$Mg + O_2 \longrightarrow 2MgO$ (Note, the 2 goes *in front of* the MgO.)
Another magnesium atom is now needed on the left:
$2Mg + O_2 \longrightarrow 2MgO$
The equation is balanced.
4 $2Mg\,(s) + O_2\,(g) \longrightarrow 2MgO\,(s)$

Magnesium burning in oxygen.

Questions

1 What do + and $\longrightarrow$ mean, in an equation?
2 Balance the equation, and add the state symbols:
a $Na + Cl_2 \longrightarrow 2NaCl$
b $H_2 + I_2 \longrightarrow HI$
c $Na + H_2O \longrightarrow NaOH + H_2$
3 Balance the following equations.
a $2NH_3\,(g) \longrightarrow N_2\,(g) + H_2\,(g)$
b $C(s) + CO_2(g) \longrightarrow CO(g)$
4 Aluminium burns in chlorine to form aluminium chloride, $AlCl_3(s)$. Write an equation for the reaction.

5.3 Calculations from equations

What an equation tells you

When carbon burns in oxygen, the equation for the reaction is:

$C\,(s) + O_2\,(g) \longrightarrow CO_2\,(g)$

This equation tells you that:

1 carbon atom	reacts with	1 molecule of oxygen	to give	1 molecule of carbon dioxide

Now suppose there was 1 *mole* of carbon atoms. These would react with 1 *mole* of oxygen molecules:

1 mole of carbon atoms	reacts with	1 mole of oxygen molecules	to give	1 mole of carbon dioxide molecules

Moles can be changed to grams, using RAM's and formula masses. The RAM's are: C = 12, O = 16. So the formula mass of CO_2 is (12 + 16 + 16) = 44, and we can write:

12 g of carbon	reacts with	32 g of oxygen	to give	44 g of carbon dioxide

You can find out the same kind of information from any equation:
The equation for a reaction tells you how many moles and how many grams of each substance takes part in the reaction.

But that's not all . . .

The reaction above involves gases. Thanks to Avogadro and other scientists, we know this fact about gases:
At room temperature and pressure, one mole of any gas has a volume of 24 dm^3 (24 litres or 24000 cm^3).
24 dm^3 is the **molar gas volume** at room temperature and pressure **(rtp)**. So from the equation above you can tell that:

12 g of carbon	reacts with	24 dm^3 of oxygen	to give	24 dm^3 of carbon dioxide

the gas volumes being measured at rtp.

Does the mass change, during a reaction?

Now look what happens to the total mass, during the above reaction:

Mass of carbon and oxygen at the start: 12 g + 32 g = **44 g**
Mass of carbon dioxide at the end: **44 g**

The total mass has not changed, during the reaction. This is because

the atoms taking part have not changed either.
During a chemical reaction atoms do not disappear and new atoms do not form. The atoms just get rearranged.

These models show how the atoms are rearranged, during the reaction between hydrogen and water. The equation is $2H_2 + O_2 \rightarrow 2H_2O$.

Calculations from equations

Example 1 Hydrogen burns in oxygen to form water. The equation for the reaction is: $2H_2(g) + O_2(g) \longrightarrow 2H_2O(l)$
How much oxygen is needed to burn 1 gram of hydrogen?

1 The RAM's are: H = 1, O = 16. So $H_2 = 2$ and $O_2 = 32$.
2 $2H_2(g) + O_2(g) \longrightarrow 2H_2O(l)$
2 moles of hydrogen molecules need 1 mole of oxygen molecules
4 g of hydrogen needs 32 g of oxygen (moles changed to grams)
1 g of hydrogen needs 8 g of oxygen
3 The reaction needs **8 g** of oxygen.

Example 2 The equation for the reaction between iron and sulphur is: $Fe(s) + S(s) \longrightarrow FeS(s)$. When 7 g of iron is heated with excess sulphur, how much iron(II) sulphide is formed?
(*Excess* sulphur means *more than enough* sulphur for the reaction.)

1 The RAM's are: Fe = 56, S = 32. So FeS = 56 + 32 = 88.
2 $Fe(s) + S(s) \longrightarrow FeS(s)$
1 mole of iron atoms gives 1 mole of iron sulphide units so
56 g of iron gives 88 g of iron sulphide
1 g of iron gives $\frac{88}{56}$ g of iron sulphide
7 g of iron gives $7 \times \frac{88}{56}$ g of iron sulphide
3 $7 \times \frac{88}{56} = 11$ so **11 g** of iron(II) sulphide are produced.

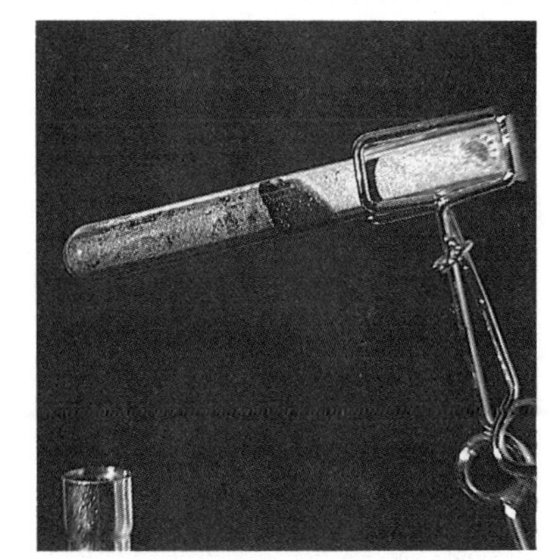
The reaction between iron and sulphur. It is an exothermic reaction but heat is needed to get it started.

Example 3 What volume of hydrogen will react with 24 dm^3 of oxygen to form water? (Gas volumes measured at rtp.)

1 $2H_2(g) + O_2(g) \longrightarrow 2H_2O(l)$
2 2 moles of hydrogen gas react with 1 mole of oxygen gas.
1 mole of any gas has a volume of 24 dm^3 at rtp, so
2×24 dm^3 of hydrogen will react with 24 dm^3 of oxygen.
3 **48 dm^3** of hydrogen are needed.

Questions

1 Explain why the total mass does not change, during a reaction.

2 The reaction between magnesium and oxygen is:
$2Mg(s) + O_2(g) \longrightarrow 2MgO(s)$
a Write a word equation for the reaction.
b How many moles of magnesium atoms react with 1 mole of oxygen molecules?
c The RAM's are: Mg = 24, O = 16.
How many grams of oxygen react with:
i 48 g of magnesium? **ii** 12 g of magnesium?

3 Copper(II) carbonate breaks down on heating:
$CuCO_3(s) \xrightarrow{heat} CuO(s) + CO_2(g)$
a Write a word equation for the reaction.
b Find the mass of 1 mole of each substance in the reaction. (Cu = 64, C = 12, O = 16)
c **i** If 31 g of copper(II) carbonate are used, how many grams of carbon dioxide will form?
ii How many moles of carbon dioxide is this?
iii What volume will it occupy at rtp?

5.4 Different types of chemical reaction (I)

Combination or synthesis

Often, two or more substances combine together, to form a single substance. This type of reaction is called a **combination** or a **synthesis**, and it has *only one product*. One example is the reaction between iron and sulphur, that you met on page 64:

iron + sulphur ⟶ iron sulphide

$Fe(s) + S(s) \longrightarrow FeS(s)$

Decomposition

In some reactions, a single substance breaks down into two or more simpler substances. This is called **decomposition**. A decomposition reaction has *only one reactant*. An example is the decomposition of calcium carbonate or **limestone**:

calcium carbonate (limestone) $\xrightarrow{heat}$ calcium oxide (quicklime) + carbon dioxide

$CaCO_3(s) \xrightarrow{heat} CaO(s) + CO_2(g)$

The label on the arrow shows that the limestone must be heated, to make it decompose. Decomposition caused by heat is called **thermal decomposition**.

In cement works, limestone is decomposed to quicklime and used to make cement. The decomposition is carried out in rotating kilns. There are five of them at the front of this photograph.

Some decomposition reactions are caused by **light**. For example silver chloride is a white solid. It breaks down in light to give tiny black crystals of silver:

silver chloride $\xrightarrow{light}$ silver + chlorine

$2AgCl(s) \xrightarrow{light} 2Ag(s) \quad Cl_2(g)$

Silver bromide and silver iodide decompose in the same way. These reactions are used in photography. Photographic film and paper have a coating of silver chloride or bromide or iodide in gelatine. The silver compound decomposes where light strikes it, giving a dark image. The rest of the compound is washed away during processing.

Precipitation

Sometimes when two solutions are mixed together, they react to give an insoluble product. The product appears as a suspension or **precipitate**, and the reaction is called a **precipitation**.

For example when aqueous solutions of silver chloride and silver nitrate are mixed, a white precipitate of silver chloride forms:

$AgNO_3(aq) + NaCl(aq) \longrightarrow AgCl(s) + NaNO_3(aq)$

Sodium nitrate is soluble in water, so it remains in solution. The silver chloride precipitates because it is insoluble. Soon the precipitate turns black. Can you explain why?

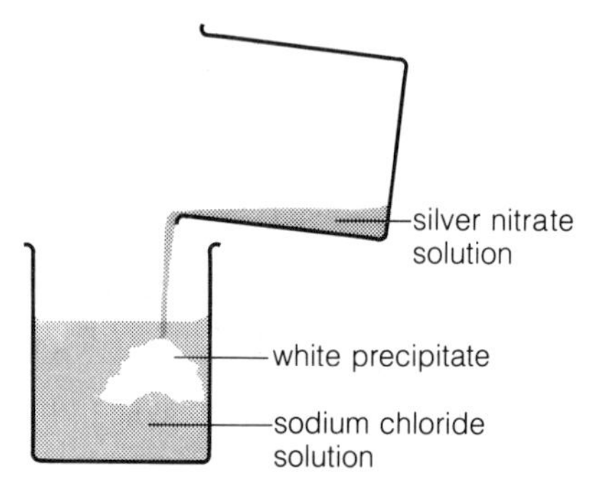

Combustion of petrol is what makes cars move. A mixture of petrol and air is burned in the engine.

One way to stop combustion is to use cold water to lower the temperature.

Combustion

Combustion is often just called **burning**. It usually means the reaction of a substance with oxygen in the air, giving out heat and light. For example:

magnesium + oxygen $\longrightarrow$ magnesium oxide
$2Mg(s) + O_2(g) \longrightarrow 2MgO(s)$

This reaction gives out heat and white light.
Combustion reactions are an essential part of our lives: the burning of gas, coal, petrol and oil are all combustion reactions. The heat they give out is used to cook food, warm houses and drive engines.

Help, fire! Combustion is not always welcome. People are not usually happy when their homes catch fire. Burning needs three things: fuel, oxygen (air), and heat. Remove any one of them and the burning will stop. So, to put out a fire:

1 **Cut off the fuel.** Turn off the gas or electricity. Cover puddles of oil or petrol with sand or soil.
2 **Get rid of the heat.** Cool things down with water.
3 **Cut off the air supply.** Cover the burning things with foam, carbon dioxide, or a fire blanket.

Water and foam should *not* be used on fires in electrical appliances, because they can conduct electricity and give people shocks. And water should not be used for oil or petrol fires, because oil and petrol will float on water and spread the fire even further.

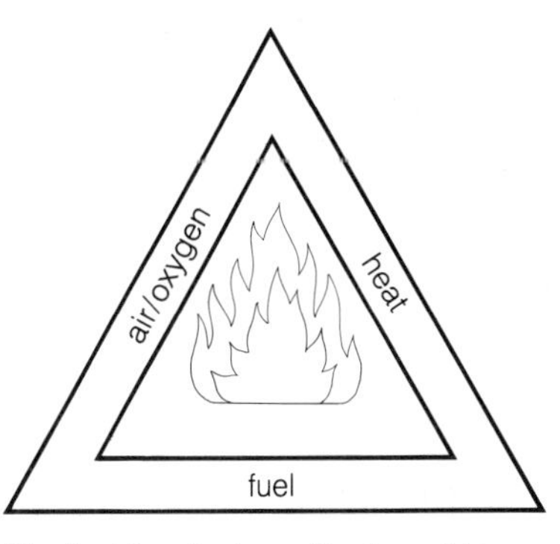

The fire triangle shows the three things needed for burning. Removing any of them will put out a fire.

Questions

1 Give one example of a combination reaction.
2 Explain how decomposition by light is used in photography.
3 Write down the three steps you could take to put out a fire.
4 What is a precipitation reaction?
5 Say which type of reaction each is:
  a $CaCO_3(s) \longrightarrow CaO(s) + CO_2(g)$
  b $NaOH(aq) + FeSO_4(aq) \longrightarrow Fe(OH)_2(s) + Na_2SO_4(aq)$
  c $C_2H_4(g) + Br_2(l) \longrightarrow C_2H_4Br_2(l)$

5.5 Different types of chemical reaction (II)

Oxidation and reduction

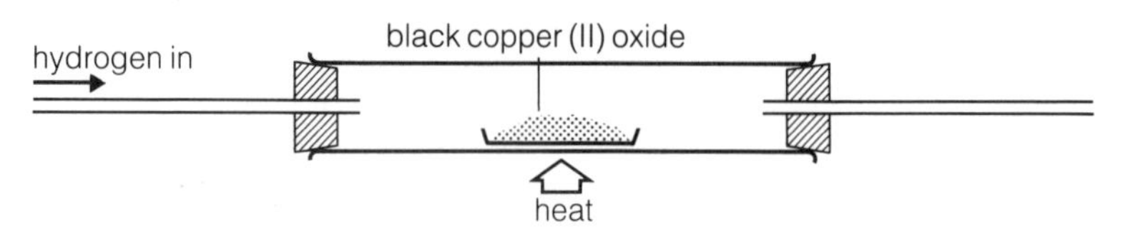

When hydrogen is passed over black copper(II) oxide in the apparatus above, the black powder turns pink. This reaction is taking place:

copper(II) oxide + hydrogen ⟶ copper + water

$CuO(s) + H_2(g) \longrightarrow Cu(s) + H_2O(g)$

The copper(II) oxide is losing oxygen. It is being **reduced**.
The hydrogen is gaining oxygen. It is being **oxidised**.
If a substance loses oxygen during a reaction, it is reduced.
If a substance gains oxygen during a reaction, it is oxidised.

Reduction and oxidation always take place together in a reaction. So the reaction is called a **redox reaction**.

In the reaction above, copper(II) oxide gets reduced because hydrogen takes its oxygen away. So hydrogen is the **reducing agent**. Or you could say that hydrogen gets oxidised because copper(II) oxide gives it oxygen. So copper(II) oxide is the **oxidising agent**.

Making iron from its ore. This is a redox reaction. Iron ore is mainly iron(III) oxide. It is reduced to iron in the blast furnace. Molten iron runs out the bottom of the furnace.

Another definition for oxidation and reduction

When magnesium burns in oxygen, magnesium oxide is formed:

$2Mg(s) + O_2(g) \longrightarrow 2MgO(s)$

This is an example of combustion. But it is also a redox reaction. It is easy to see that the magnesium has been oxidised. It is not so easy to see that the oxygen has been reduced. We need to look more closely at the reaction:

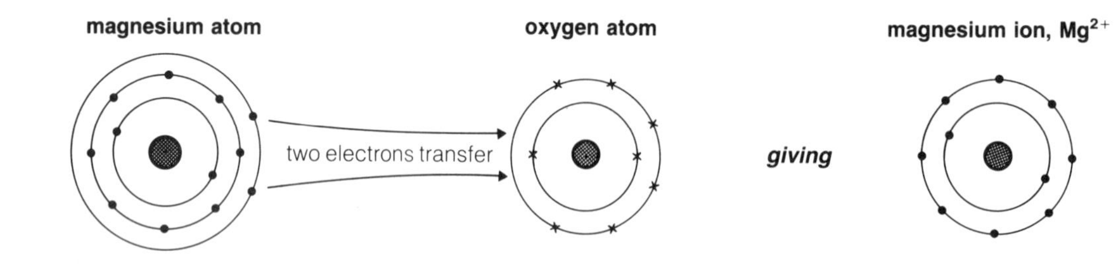

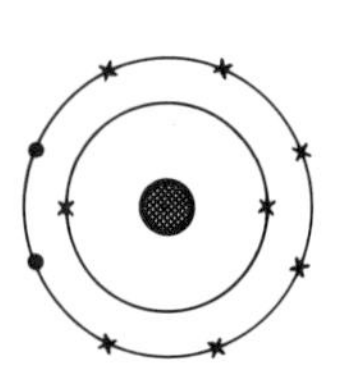

During the reaction, a magnesium atom loses two electrons and an oxygen atom gains them. This gives us another definition for oxidation and reduction:
A substance is oxidised if it loses electrons during a reaction.
A substance is reduced if it gains electrons during a reaction.
So magnesium has been oxidised and oxygen has been reduced.

Reversible reactions

Blue copper(II) sulphate has the formula $CuSO_4.5H_2O$. This shows that its crystals contain water molecules as well as copper and sulphate ions. The crystals are said to be **hydrated**. The water is called **water of crystallisation.**
Now look what happens when blue copper(II) sulphate is heated:

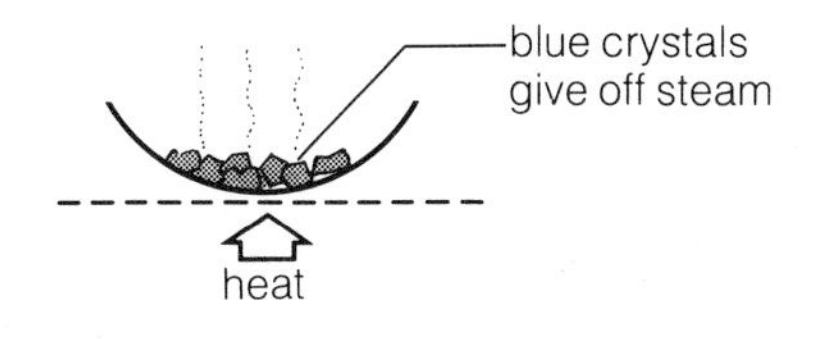

When the blue crystals are heated, the water of crystallisation is driven off as steam, leaving a white powder behind.

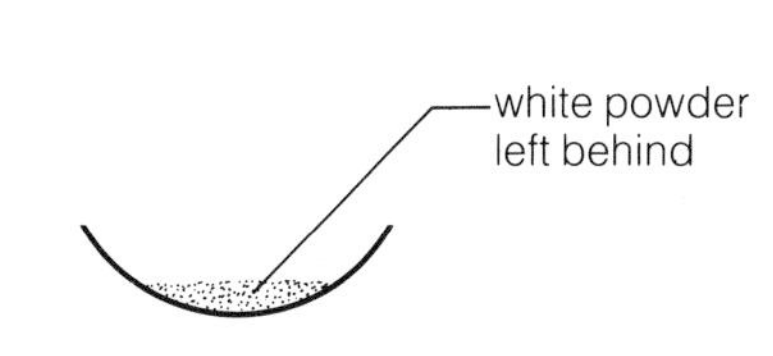

The white powder is **anhydrous copper(II) sulphate**. Its formula is $CuSO_4$. This reaction has taken place:
$CuSO_4.5H_2O \longrightarrow CuSO_4 + 5H_2O$

dropper with water

powder gets hot and turns blue

The reaction is easy to reverse. When water is added to the white powder, the powder gets very hot and turns blue:
$CuSO_4 + 5H_2O \longrightarrow CuSO_4.5H_2O$

The reaction between copper(II) sulphate and water is described as a **reversible** reaction.
A reversible reaction is one that can go in either direction, depending on the conditions.
To show a reaction is reversible, the symbol $\rightleftharpoons$ is used in the equation:

$CuSO_4.5H_2O \rightleftharpoons CuSO_4 + 5H_2O$
blue white

Because of the colour change, this reaction is used to test for water.

Anhydrous copper(II) sulphate is used in the laboratory to test for water. Water makes the white powder turn blue.

Reversible reactions in industry

Many important reactions in industry are reversible. An example is the reaction between nitrogen and hydrogen to make ammonia:

$N_2(g) + 3H_2(g) \rightleftharpoons 2NH_3(g)$

Because it is reversible, the reaction poses problems. Some ammonia molecules will be breaking down while others form.
Chemical engineers therefore choose conditions that will give a good yield of ammonia as quickly and economically as possible. For example the ammonia is removed from the reaction vessel before it can break down again. You can read more about this on page 158.

Questions

1 $2Mg(s) + SO_2(g) \longrightarrow 2MgO(s) + S(s)$
In this reaction, which substance is:
a oxidised? **b** the oxidising agent?
c reduced? **d** the reducing agent?

2 Where does the word *redox* come from?

3 What is a reversible reaction? Give an example.

4 Explain how copper(II) sulphate is used to test whether a liquid contains water.

5 Explain why the reaction conditions for making ammonia must be chosen carefully.

Questions on Chapter 5

1 Decide whether each change below is a physical change or a chemical change. Give reasons for your answers.

a ice melting
b iron rusting
c petrol burning
d candle wax melting
e a candle burning
f wet hair drying
g milk souring
h perfume evaporating
i a lump of roll sulphur being crushed
j copper being obtained from copper(II) oxide
k clothes being ironed
l custard being made
m a cigarette being smoked

2 Write a chemical equation for each of the following. Example.

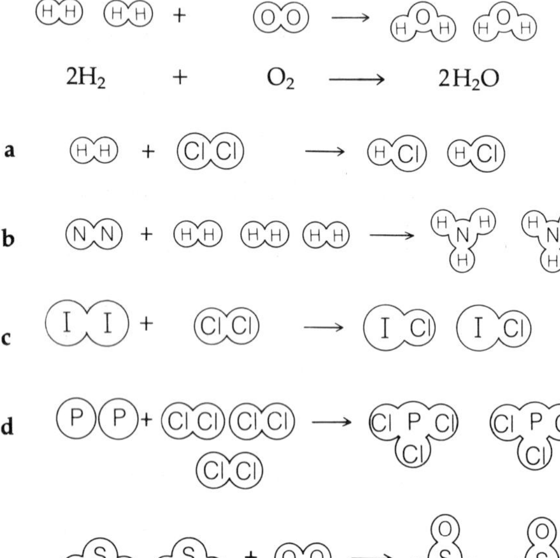

3 Write equations for the following reactions:

a 1 mole of copper atoms combine with 1 mole of sulphur atoms to form 1 mole of copper(II) sulphide, CuS.

b 3 moles of lead atoms combine with 2 moles of oxygen molecules to form 1 mole of lead oxide, Pb_3O_4.

c 1 mole of ethanol molecules, C_2H_5OH, burns in 3 moles of oxygen molecules to form 2 moles of carbon dioxide molecules and 3 moles of water molecules.

d 1 mole of iron(III) oxide, Fe_2O_3, is reduced by 3 moles of hydrogen molecules to form 2 moles of iron atoms and 3 moles of water molecules.

4 Balance the following equations:

a Synthesis of hydrogen bromide
$H_2(g) + Br_2(g) \longrightarrow HBr(g)$

b Combustion of carbon monoxide
$CO(g) + O_2(g) \longrightarrow CO_2(g)$

c Combustion of methane (natural gas)
$CH_4(g) + O_2(g) \longrightarrow CO_2(g) + H_2O(g)$

d Reduction of carbon dioxide
$Mg(s) + CO_2(g) \longrightarrow MgO(s) + C(s)$

e Decomposition of lead nitrate
$Pb(NO_3)_2(s) \xrightarrow{heat} PbO(s) + NO_2(g) + O_2(g)$

f Displacement of silver
$Zn(s) + AgNO_3(aq) \longrightarrow Zn(NO_3)_2(aq) + Ag(s)$

g Precipitation of barium sulphate
$BaCl_2(aq) + Na_2SO_4(aq) \longrightarrow BaSO_4(s) + NaCl(aq)$

5 Water at 25 °C was used to dissolve two compounds. Immediately after the compounds had dissolved, the temperature of each solution was measured:

Compound	**Temperature/°C**	
	Water	**Solution**
NH_4Cl	25	15
$CaCl_2$	25	45

a Name the two compounds.
b What was the temperature change for each?
c Which compound dissolves exothermically?
d Which compound dissolves endothermically?

6 Mercury(II) oxide breaks down into mercury and oxygen when heated. The equation for the reaction is:
$2HgO(s) \longrightarrow 2Hg(l) + O_2(g)$

a Calculate the mass of 1 mole of mercury(II) oxide. (O = 16, Hg = 201)

b Find how much mercury and oxygen are produced when 21.7 g of mercury(II) oxide is heated.

7 Iron(II) sulphide is formed when iron and sulphur react together:
$Fe(s) + S(s) \longrightarrow FeS(s)$

a How many grams of sulphur will react with 56 g of iron? (The RAM's on page 55 will help you.)

b If 7 g of iron and 10 g of sulphur are used, which substance is in excess?

c If 7 g of iron and 10 g of sulphur are used, name the substances present when the reaction is complete, and find the mass of each.

d Is the reaction endothermic or exothermic? Explain.

For questions 8 to 11 you will need to remember that: **the volume of one mole of any gas measured at room temperature and pressure is 24 dm^3 or 24 000 cm^3.**

8 What is the volume at rtp in dm^3 and cm^3, of:
   a 2 moles of hydrogen, H_2?
   b 0.5 moles of carbon dioxide, CO_2?
   c 0.01 moles of nitrogen, N_2?
   d 0.3 moles of oxygen, O_2?

9 The compound sodium hydrogen carbonate, $NaHCO_3$, decomposes as follows when heated:

$2NaHCO_3(s) \longrightarrow Na_2CO_3(s) + H_2O(l) + CO_2(g)$

   a Write a word equation for the reaction.
   b i How many moles of sodium hydrogen carbonate are there on the left side of the equation?
      ii What is the mass of this amount? (Na = 23, H = 1, C = 12, O = 16)
   c i How many moles of carbon dioxide are there, on the right side of the equation?
      ii What is the volume of this amount at rtp?
   d What volume at rtp of carbon dioxide would be obtained if:
      i 84 g of sodium hydrogen carbonate were completely decomposed?
      ii 8.4 g of sodium hydrogen carbonate were completely decomposed?

10 When calcium carbonate is heated strongly, the following chemical change occurs:

$CaCO_3(s) \longrightarrow CaO(s) + CO_2(g)$

(Ca = 40, C = 12, O = 16)
   a Write a word equation for the chemical change.
   b How many moles of $CaCO_3$ are there in 50 g of calcium carbonate?
   c i What mass of calcium oxide is obtained from the thermal decomposition of 50 g of calcium carbonate?
      ii What mass of carbon dioxide would be given off at the same time?
      iii What volume would the gas occupy at rtp?

11 The synthesis of ammonia is shown by the equation:

$N_2(g) + 3H_2(g) \rightleftharpoons 2NH_3(g)$

   a Write a word equation for the reaction.
   b Why is the reaction called a **synthesis**?
   c What does the sign $\rightleftharpoons$ mean?
   d What is the volume of 1 mole of N_2 (at rtp)?
   e i How many moles of hydrogen react with one mole of nitrogen?
      ii What volume of hydrogen is this (at rtp)?
   f What volume of ammonia (at rtp) would be formed if all the N_2 and H_2 reacted together?
   g i What volumes of N_2 and H_2 need to react together to form 100 cm^3 of ammonia?
      ii In fact it is not possible to form 100 cm^3 of ammonia by reacting these volumes together. Explain why.

12 The following equation represents a reaction in which iron is obtained from iron(III) oxide:

$Fe_2O_3(s) + 3CO(g) \xrightarrow{heat} 2Fe(s) + 3CO_2(g)$

   a Write a word equation for the reaction.
   b The reaction is a redox reaction.
      i Which substance is reduced?
      ii Which substance is oxidised?
      iii Name the reducing agent.
      iv Name the oxidising agent.
   c What is the formula mass of iron(III) oxide? (Fe = 56, O = 16)
   d How many moles of Fe_2O_3 are present in 320 kg of iron(III) oxide? (1 kg = 1000 g)
   e How many moles of Fe are obtained from 1 mole of Fe_2O_3?
   f From **d** and **e**, find the number of moles of iron obtained from 320 kg of iron(III) oxide.
   g Find the mass of iron obtained from 320 kg of iron(III) oxide.

13 When solutions of potassium sulphate and barium chloride are mixed, a white *precipitate* forms. The equation for the reaction is:

$K_2SO_4(aq) + BaCl_2(aq) \longrightarrow BaSO_4(s) + 2KCl(aq)$

   a What is a precipitate?
   b Which compound above is the precipitate?
   c How would you separate the precipitate from the solution?
   d i What would remain if the precipitate was removed?
      ii How would you obtain this substance as a dry solid?
   e A precipitate also forms if sodium sulphate is used instead of potassium sulphate, above. Write an equation for the reaction.

14 23.3 g of barium sulphate was obtained from the precipitation reaction in question 13.
   a What is the formula mass of barium sulphate? (Ba = 137, S = 32, O = 16)
   b How many moles of $BaSO_4$ is 23.3 g of barium sulphate?
   c How many moles of K_2SO_4 and $BaCl_2$ must have reacted, to form 23.3 g of barium sulphate?
   d The concentrations of the two reacting solutions were 0.1 M (that is, 0.1 mole of solute in 1 litre of solution). What volume of each solution was needed to form 23.3 g of barium sulphate?

15 When 6.5 g of zinc (Zn) were added to a solution of copper(II) sulphate ($CuSO_4$), 6.4 g of copper were obtained. (The RAM's are: Zn = 65, Cu = 64.)
   a What type of chemical reaction is this?
   b How many moles of zinc atoms were used?
   c How many moles of copper atoms were obtained?
   d Write a word equation for the reaction.
   e Use the information from **b** and **c** to write a balanced equation for the reaction.

6.1 Conductors and non-conductors

Batteries and electric current

The photograph below shows a battery, a bulb and a rod of graphite (carbon) joined or **connected** to each other by copper wires. The arrangement is called an **electric circuit**. The bult is lit; this shows that electricity must be flowing in the circuit.
Electricity is a stream of moving electrons.
Do you remember what an electron is? It is a tiny particle with a negative charge and almost no mass.

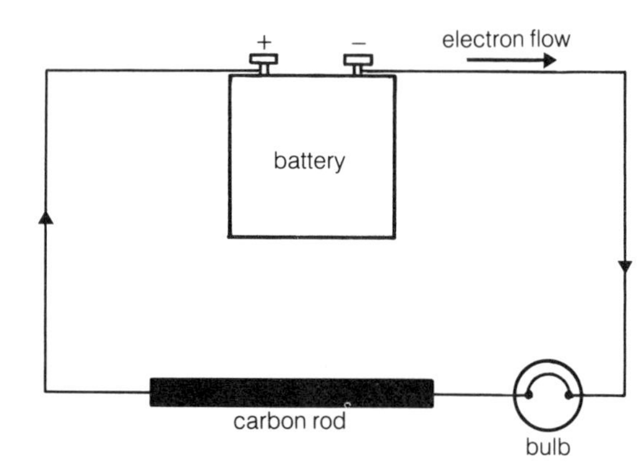

The diagram above shows how the electrons move through the circuit. The battery acts like an electron pump. Electrons leave it through one terminal, called the **negative terminal**. They are pumped through the wires, the bulb and the rod, and enter the battery again through the **positive terminal**. When the electrons stream through the fine wire in the bulb, they cause the wire to heat up so much that it gets white hot and gives out light.

Conductors

In the circuit above, the light will go out if:
1 you disconnect a wire, so that everything no longer joins up *or*
2 you connect something into the circuit that prevents electricity from flowing through.

The copper wires and graphite rod obviously do allow electricity to flow through them—they **conduct** electricity. Copper and graphite are therefore called **conductors.**
A conductor of electricity is a substance that allows electricity to flow through it.
A substance that does not conduct electricity is called a **non-conductor** or **insulator.**

Copper is the conductor inside this electric drill. But plastic (an insulator) is used to connect the bit to the motor, and for the outer case. Why is that?

Testing substances to see if they conduct

The circuit above can be used to test any substance to see if it conducts electricity. The substance is simply connected into the circuit, like the graphite rod above. Some examples are given on the next page.

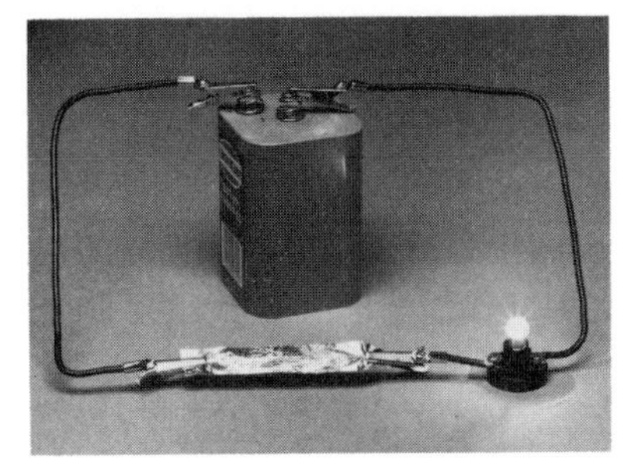

Testing tin. A strip of tin is connected into the circuit. The bulb lights, so tin must be a conductor.

Testing ethanol. The liquid is connected into the circuit by dipping graphite rods into it. The bulb does not light—ethanol is a non-conductor.

Testing molten lead bromide. A bunsen is used to melt it. It conducts, and at the same time gives off a choking brown vapour.

The results Many tests have been carried out, with these results:

1 **The only solids that conduct are the metals and graphite.**
These conduct because of their free electrons (pages 50 and 51). The free electrons get pumped out one end of the solid, by the battery. Electrons then flow in the other end, and through the spaces left behind.
For the same reason, *molten* metals conduct. (It is not possible to test molten graphite, because graphite sublimes.)

metal ion moving electron

2 **Molecular or covalent substances are non-conductors.**
This is because they contain no free electrons, or other charged particles, that can flow out through them.
Ethanol above is one example of a molecular substance. Others are petrol, paraffin, sulphur, sugar and plastic. These never conduct, whether solid or molten.

3 **Ionic substances do not conduct when solid. However, they conduct when melted or dissolved in water, and they decompose at the same time.**
An ionic substance contains no free electrons. However, it does contain ions, which are also charged particles. The ions become free to move when the substance is melted or dissolved, and it is they that conduct the electricity.
Lead bromide above is an example. It is a non-conductor when solid. But it begins to conduct the moment it is melted, and a brown vapour bubbles off at the same time. The vapour is bromine, and it forms because electricity causes the lead bromide to **decompose.**
Decomposition caused by electricity is called electrolysis, and the liquid that decomposes is called an electrolyte.
Molten lead bromide is therefore an electrolyte. Ethanol is a **non-electrolyte** because it does not conduct at all.

Like other metals, aluminium conducts electricity. It is used for electricity cables because it is so light.

Questions

1 What is a **conductor** of electricity?
2 Draw a circuit to show how you would test whether mercury conducts.
3 Explain why metals are able to conduct electricity.
4 Naphthalene is a molecular substance. Do you think it conducts when molten? Explain why.
5 What is: **a** an electrolyte? **b** a non-electrolyte? Give *three* examples of each.

6.2 A closer look at electrolysis

The electrolysis of lead bromide

On the last page, you saw that molten lead bromide decomposes when it conducts electricity. Decomposition caused by electricity is called **electrolysis**, and molten lead bromide is an **electrolyte.**

The apparatus This is shown on the right. The graphite rods carry the current into and out of the molten lead bromide. Conducting rods like these are called **electrodes**. The electrode joined to the negative terminal of the battery is called the **cathode**. It is also negative, because the electrons from the battery flow to it. The other electrode is positive, and is called the **anode**. Notice the switch. When it is open, no electricity can flow.

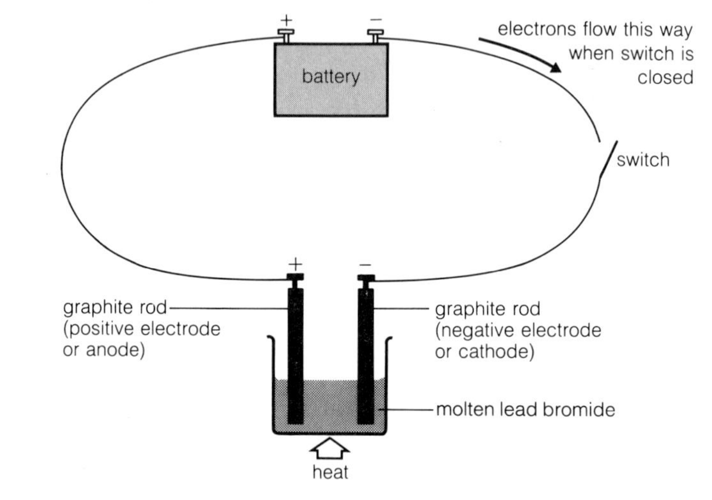

The electrolysis Once the switch is closed, **bromine vapour** starts to bubble out of the molten lead bromide, around the anode. After some time a bead of **molten lead** forms below the cathode. The electrical energy from the battery has caused **a chemical change:**

lead bromide ⟶ lead + bromine
$PbBr_2(l) \longrightarrow Pb(l) + Br_2(g)$

Why the molten lead bromide decomposes When lead bromide melts, its lead ions (Pb^{2+}) and bromide ions (Br^-) become free to move about. When the switch is closed, the electrodes become charged, and the ions are immediately attracted to them:

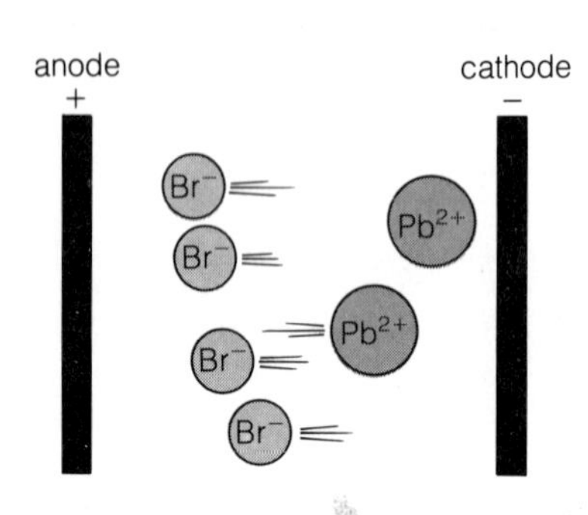

Opposite charges attract, so the lead ions are attracted to the cathode, and the bromide ions to the anode.

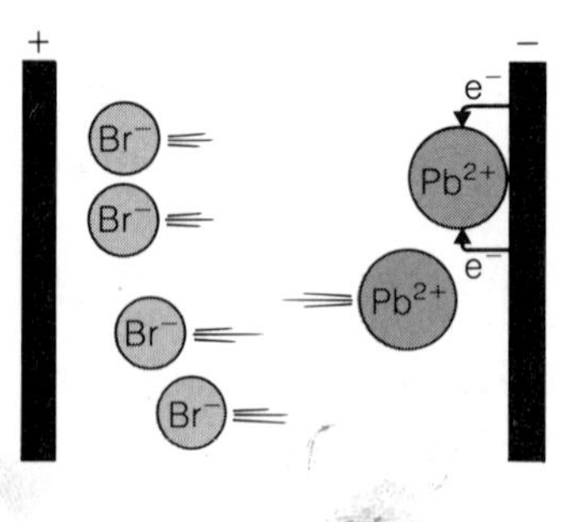

At the cathode, the lead ions each receive 2 electrons and become lead atoms:
$Pb^{2+} + 2e^- \longrightarrow Pb$
The lead atoms collect together on the cathode, and in time fall to the bottom of the beaker.

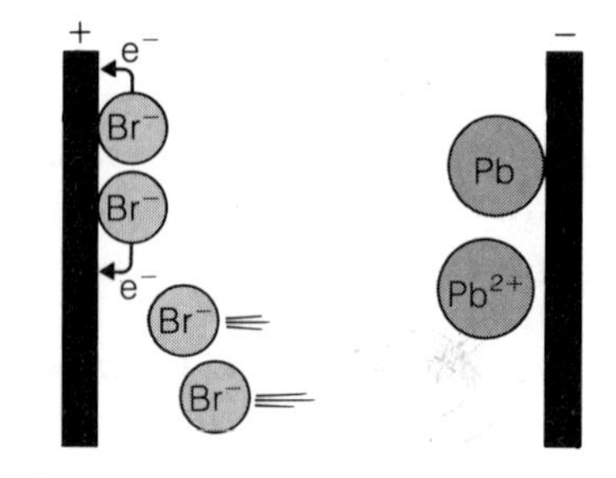

At the anode, the bromide ions each give up 1 electron to become bromine atoms. These pair together as molecules:
$2\,Br^- \longrightarrow Br_2 + 2e^-$
The bromine bubbles off as a gas.

Why the molten lead bromide conducts During the electrolysis, each lead ion takes two electrons from the cathode, as shown on the right. At the same time two bromide ions each give an electron to the anode. The effect is the same as if two electrons *flowed through the liquid* from the cathode to the anode. In other words, the lead bromide is acting as a conductor of electricity.

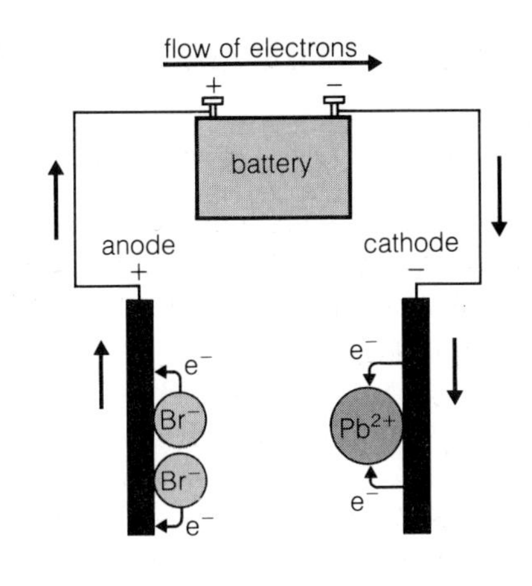

The electrolysis of other compounds

All ionic compounds can be electrolysed, when they are molten. (Another word for molten is **fused**.) These are some points to remember about the process:

1. The electrolyte always **decomposes**. So electrical energy is causing **a chemical change**.
2. The electrodes are usually made of graphite or platinum. These substances are unreactive or **inert**. That means they will not react with the electrolyte or the products of the electrolysis.
3. Positive ions always go to the cathode. Because of this they are called **cations**.
4. Negative ions always go to the anode, so are called **anions**.

This table shows the results when some other ionic compounds are electrolysed:

Electrolyte	The decomposition	At the cathode	At the anode
Sodium chloride NaCl	sodium chloride ⟶ sodium + chlorine $2NaCl(l) \longrightarrow 2Na(l) + Cl_2(g)$	$2Na^+ + 2e^- \longrightarrow 2Na$	$2Cl^- \longrightarrow Cl_2 + 2e^-$
Potassium iodide KI	potassium iodide ⟶ potassium + iodine $2KI(l) \longrightarrow 2K(l) + I_2(g)$	$2K^+ + 2e^- \longrightarrow 2K$	$2I^- \longrightarrow I_2 + 2e^-$
Copper(II) bromide $CuBr_2$	copper(II) bromide ⟶ copper + bromine $CuBr_2(l) \longrightarrow Cu(l) + Br_2(g)$	$Cu^{2+} + 2e^- \longrightarrow Cu$	$2Br^- \longrightarrow Br_2 + 2e^-$

The difference between electrolytes and conductors

Electrolytes and conductors both conduct electricity. The difference between them is this:

Conductors are elements, and remain unchanged when they conduct. Electrolytes are molten compounds, or solutions of compounds, and they decompose when they conduct.

Questions

1. Explain what each of these terms means:
   electrolysis anode cathode
2. For the electrolysis of molten lead bromide, draw diagrams to show:
   a how the ions move when the switch is closed
   b what happens at the anode
   c what happens at the cathode
3. What is: a cation? an anion?
4. Molten copper and molten copper(II) bromide both conduct. What changes would you expect to see when they conduct?
5. Write equations for the overall reaction, and the reaction at each electrode, when fused magnesium chloride ($MgCl_2$) is electrolysed.

6.3 The electrolysis of solutions

Sodium chloride solution

When sodium chloride dissolves in water, its ions become free to move. So the solution can be electrolysed.
In the laboratory, the electrolysis is carried out with graphite or platinum electrodes, and a concentrated solution of the salt. You might expect sodium and chlorine to be obtained at the electrodes. But this does not happen—hydrogen and chlorine are obtained instead.
When a concentrated solution of sodium chloride is electrolysed, using graphite or platinum electrodes, hydrogen and chlorine are obtained at the electrodes.
The apparatus is shown on the right. It allows the gases to be collected as they bubble off. Note the symbol for the battery.

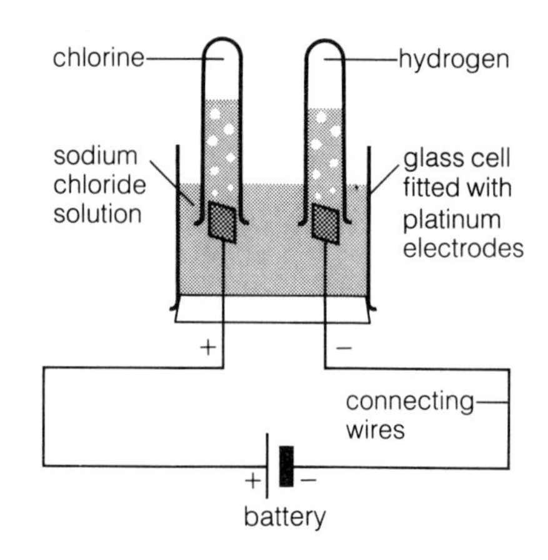

Where the hydrogen comes from The hydrogen can only come from the *water* in the solution.
Pure water is molecular, so it should not conduct at all. However, it does conduct *very slightly*, because it always contains a tiny fraction of water molecules that are split into ions:

some water molecules $\longrightarrow$ hydrogen ions + hydroxide ions

$$H_2O(l) \longrightarrow H^+(aq) + OH^-(aq)$$

So a solution of sodium chloride contains four types of ions:

Na^+ and Cl^- from sodium chloride, H^+ and OH^- from water.

During the electrolysis, this is what happens:

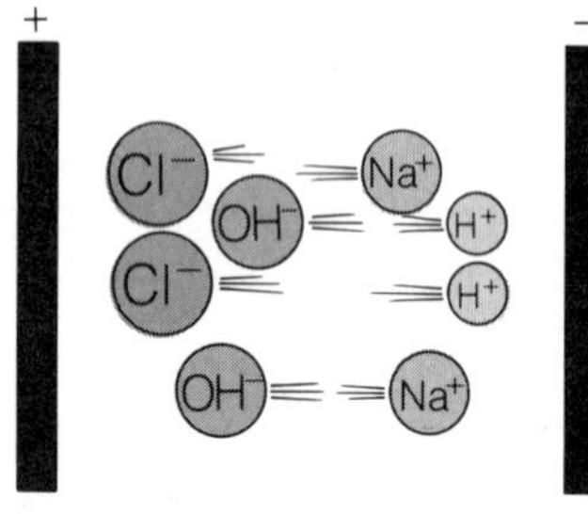

The sodium and hydrogen ions are attracted to the cathode. The chloride and hydroxide ions are attracted to the anode.

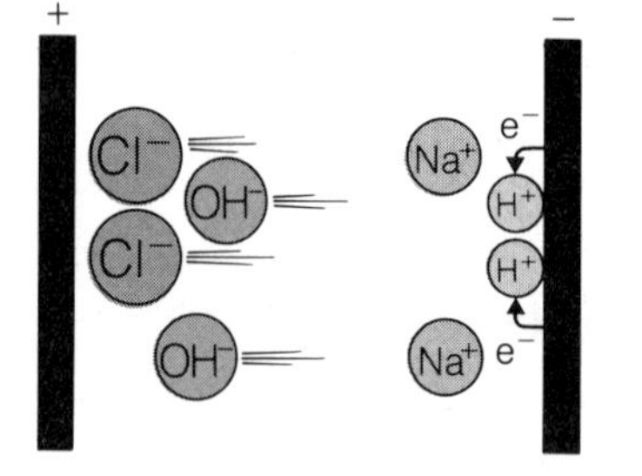

At the cathode, hydrogen ions accept electrons more readily than sodium ions do. They become atoms, then form molecules:

$$2H^+ + 2e^- \longrightarrow H_2$$

The hydrogen bubbles off. The sodium ions remain in solution.

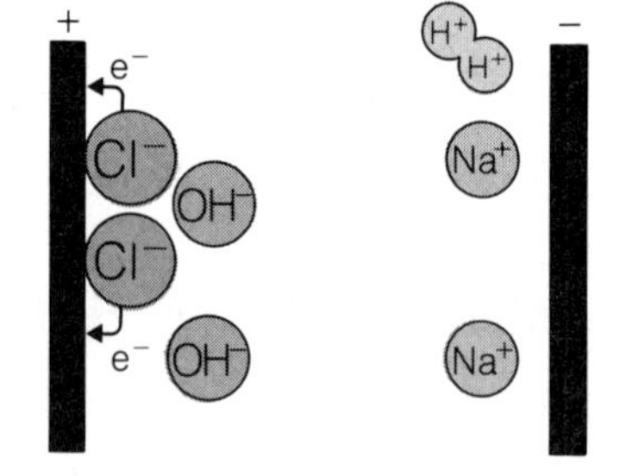

At the anode, chloride ions give up electrons more readily than hydroxide ions do. Chlorine gas bubbles off:

$$2Cl^- \longrightarrow Cl_2 + 2e^-$$

The hydroxide ions remain in solution.

At the cathode, the electrons are accepted by hydrogen ions rather than by sodium ions, because sodium is **more reactive** than hydrogen, with a stronger tendency to exist as ions.

Dilute sulphuric acid

When dilute sulphuric acid is electrolysed using platinum electrodes, hydrogen and oxygen are obtained at the electrodes. The apparatus is usually carried out in the **Hofmann voltameter**, shown on the right. It allows the two gases to be collected.
Sulphuric acid has the formula H_2SO_4. In water it forms ions:

$H_2SO_4(aq) \longrightarrow 2H^+(aq) + SO_4^{2-}(aq)$

So, in the dilute solution of sulphuric acid, these ions are present:

H^+ and SO_4^{2-} from sulphuric acid, H^+ and OH^- from water.

The positive ions go to the cathode, the negative ions to the anode.

At the cathode Hydrogen gas bubbles off, because of this reaction:

$4H^+ + 4e^- \longrightarrow 2H_2$

At the anode Hydroxide ions give up their electrons more readily than sulphate ions do, so oxygen and water are formed.

$4OH^- \longrightarrow 2H_2O + O_2 + 4e^-$

The oxygen bubbles off. Sulphate ions remain in solution.
The overall result is that *water* decomposes, rather than acid:

water $\longrightarrow$ hydrogen + oxygen
$2H_2O(l) \longrightarrow 2H_2(g) + O_2(g)$

So this electrolysis is often called **the electrolysis of acidified water.**

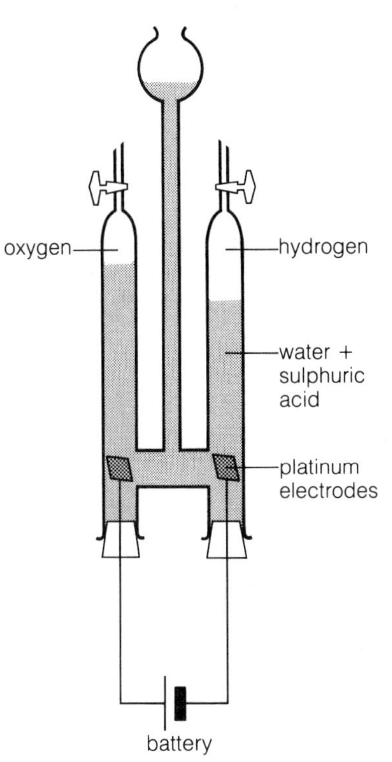

The Hofmann voltameter. Note that the volume of hydrogen collected is twice the volume of oxygen.

Copper(II) sulphate solution

When copper(II) sulphate solution is electrolysed using platinum electrodes, copper and oxygen are obtained at the electrodes.
A solution of copper(II) sulphate contains these ions:

Cu^{2+} and SO_4^{2-} from copper(II) sulphate, H^+ and OH^- from water.

The positive ions go to the cathode, the negative ions to the anode.

At the cathode Copper is formed rather than hydrogen, because copper is **less reactive** than hydrogen. Its ions accept electrons more readily than hydrogen ions do:

$2Cu^{2+} + 4e^- \longrightarrow 2Cu$

The copper atoms cling to the platinum electrode. It becomes coated with copper, as shown on the right. The hydrogen ions stay in solution.

At the anode Since hydroxide ions give up electrons more readily than sulphate ions do, oxygen and water are formed:

$4OH^- \longrightarrow 2H_2O + O_2 + 4e^-$

The sulphate ions remain in solution.

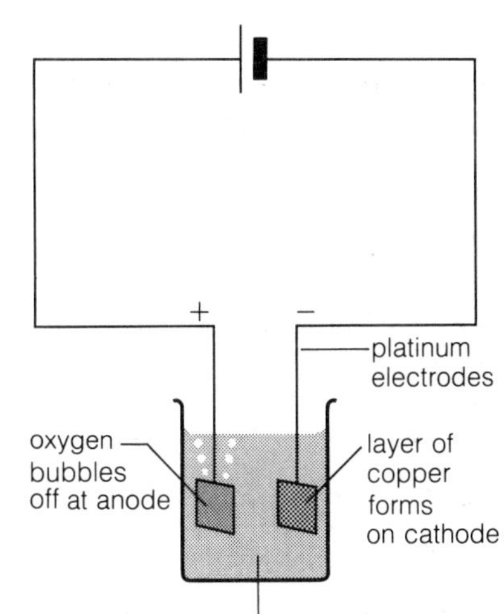

Questions

1. Explain why water conducts electricity *very* slightly.
2. Describe what happens during the electrolysis of concentrated sodium chloride solution.
3. Write down: **a** the anions present **b** the cations present **c** the reaction at each electrode, when dilute sulphuric acid is electrolysed.

6.4 Some uses of electrolysis

1 Making sodium hydroxide

On page 80 you saw that when a solution of sodium chloride is electrolysed, hydrogen and chlorine bubble off at the electrodes. Sodium ions and hydroxide ions are left in solution. This method is used in industry for making sodium hydroxide. The starting material is **brine**, a concentrated solution of sodium chloride in water, pumped up from salt mines.

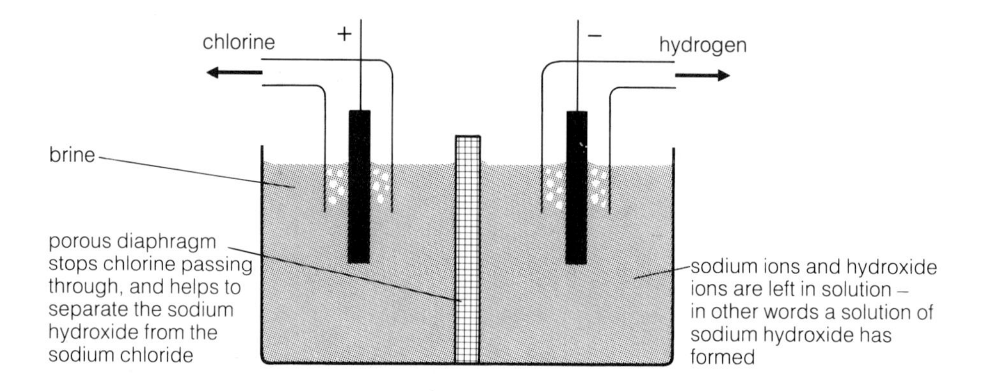

Sodium hydroxide has many uses. It is used for making soap and detergents, paper, rayon, and other fibres, and in dyeing textiles. It is used in purifying bauxite from which aluminium is extracted. It is also the starting point for making many other chemicals.

The chlorine and hydrogen from the electrolysis are important too. For example chlorine is used as a bleach and as a disinfectant for water supplies and swimming pools. Hydrogen is used as fuel for pumping the brine and for heating the sodium hydroxide solution to concentrate it.

2 Purifying metals

On page 81 you saw that when copper(II) sulphate solution is electrolysed, using platinum electrodes, copper is formed at the cathode and oxygen at the anode.
The result is a bit different when *copper* electrodes are used.

At the cathode Copper ions become copper atoms, as before:

$Cu^{2+} + 2e^- \longrightarrow Cu$

The copper atoms cling to the copper cathode.

At the anode The copper anode dissolves, forming copper ions in solution, as shown on the right:

$Cu \longrightarrow Cu^{2+} + 2e^-$

The result is that the cathode grows thicker, while the anode wears away. This idea is used to **purify** copper, as shown on the next page. (Copper must be very pure for use in electrical wires.)

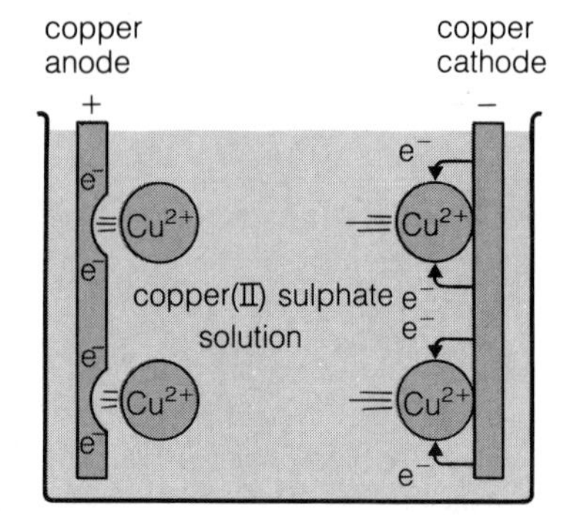

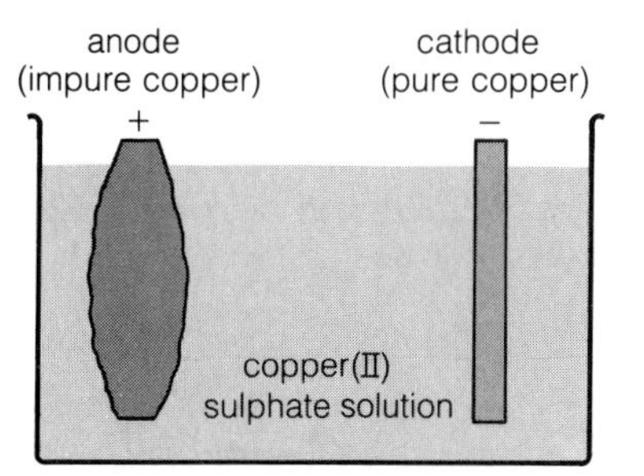

The impure copper is used as the anode of an electrolysis cell. The cathode is pure copper, and the electrolyte is copper(II) sulphate solution.

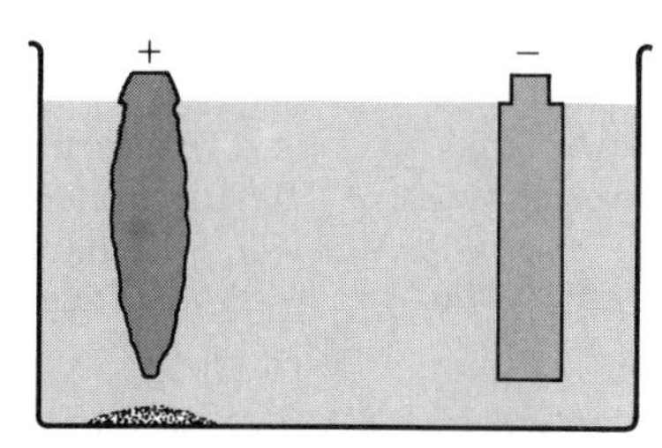

During electrolysis, the anode dissolves, and the impurities drop to the floor of the cell. A layer of pure copper builds up on the cathode.

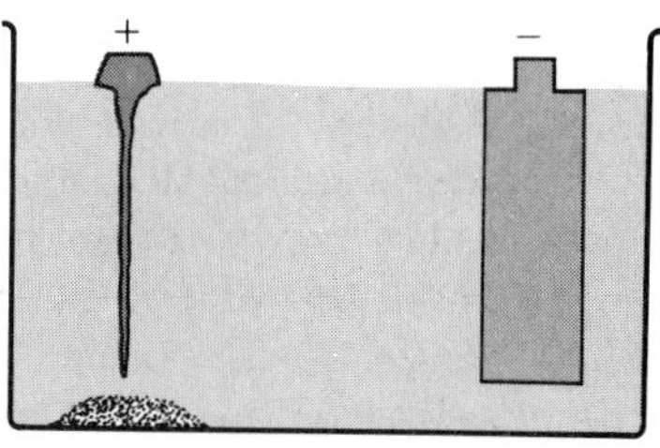

Eventually the cathode is removed. The impurities at the bottom of the cell are then checked—they sometimes contain gold and silver.

Metals such as nickel and lead can also be purified in this way.

3 Electroplating

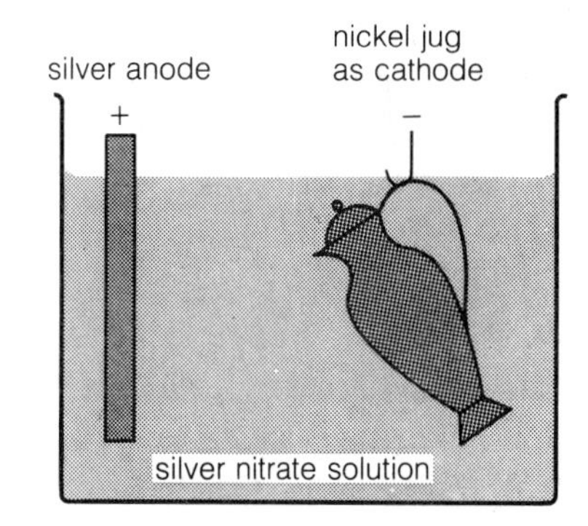

On page 81, you saw how a platinum electrode became coated with copper during electrolysis. Electrolysis is often used to coat one metal with another – usually to make the first metal look more attractive, or to prevent it from rusting. The process is called **electroplating**.

The diagram on the right shows how a nickel jug could be electroplated with silver. The jug is made the cathode of an electrolysis cell. The anode is made of silver. The electrolyte is a solution of a silver compound, for example silver nitrate.

At the anode The silver dissolves, forming silver ions in solution:

$Ag \longrightarrow Ag^+ + e^-$

At the cathode Silver ions receive electrons, and form a layer of silver on the jug:

$Ag^+ + e^- \longrightarrow Ag$

When the layer of silver is thick enough, the jug is removed.

In general, to electroplate an object with a metal X:
the object is used as the cathode of a cell
metal X is used as the anode of the cell
a solution of a compound of X is used as the electrolyte.
Electroplating is used a great deal in industry. For example, car bumpers are electroplated with chromium, to make them shiny and protect them from rust. Steel is electroplated with tin, to make 'tins' for food.

A nickel jug electroplated with silver (silverplated).

Questions

1 a What is brine?
  b When brine is electrolysed, which gas is given off: i at the cathode? ii at the anode?
  c Where does the sodium hydroxide come from?
  d Why does the cell contain a diaphragm?
  e Why is sodium hydroxide an important chemical?

2 Copper can be purified by electrolysis.
  a What electrolyte is used?
  b Explain what happens at each electrode.

3 a What is electroplating?
  b Draw a labelled diagram to show how you would electroplate an iron nail with copper.

Questions on Chapter 6

1 a What does the term *electrolysis* mean?

b These words are all connected with electrolysis: *anode, cathode, electrolyte, anion, cation.*

Copy the diagram below, and label it using these words:

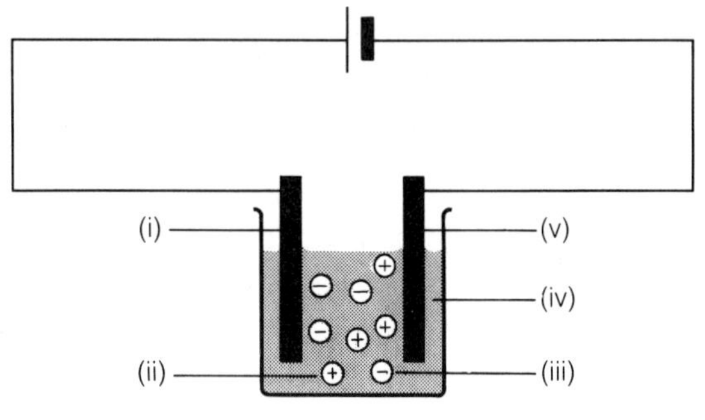

2 In which of these would the bulb light?

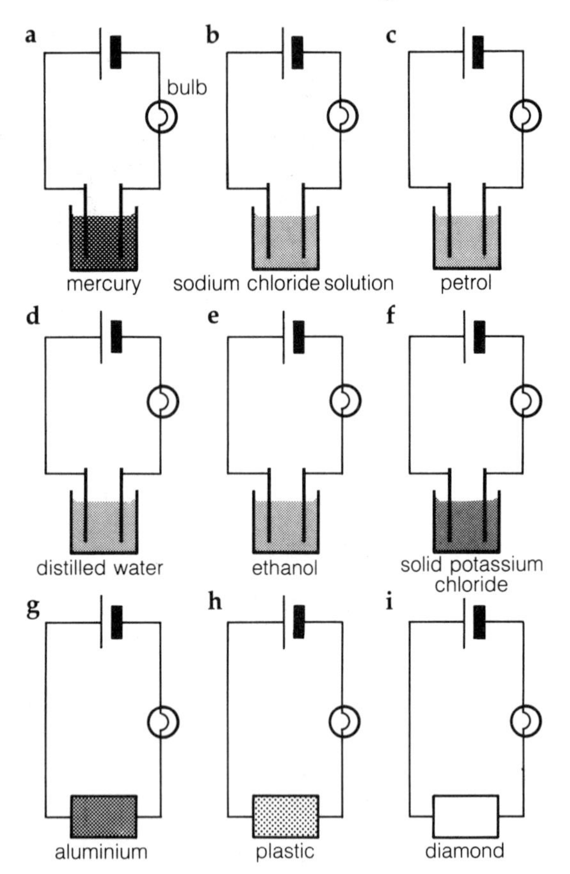

3 a Which of the substances in question 2 are:

i conductors? **ii** non-conductors?

iii electrolytes? **iv** non-electrolytes?

b What is the difference between a conductor and an electrolyte?

c For which substances above would you expect to see changes taking place at the electrodes?

4 The electrolysis of lead bromide can be investigated using the following apparatus.

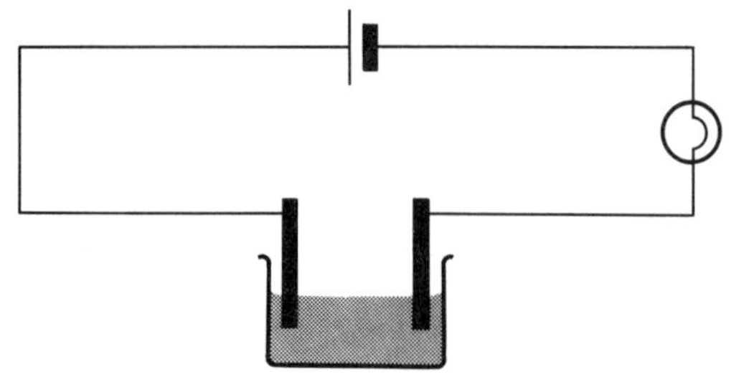

a What must be done to the lead bromide before the bulb will light?

b What would be *seen* at the positive electrode during the experiment?

c Name the substance in **b**.

d What is formed at the negative electrode?

e Write an equation for the reaction at each electrode.

5 This question is about the electrolysis of *molten* lithium chloride. Lithium chloride is ionic, and contains lithium ions (Li^+) and chloride ions (Cl^-).

a Which ion is the anion?

b Which ion is the cation?

c Copy the following diagram and use arrows to show which way:

i the ions flow when the switch is closed

ii the electrons flow in the wires

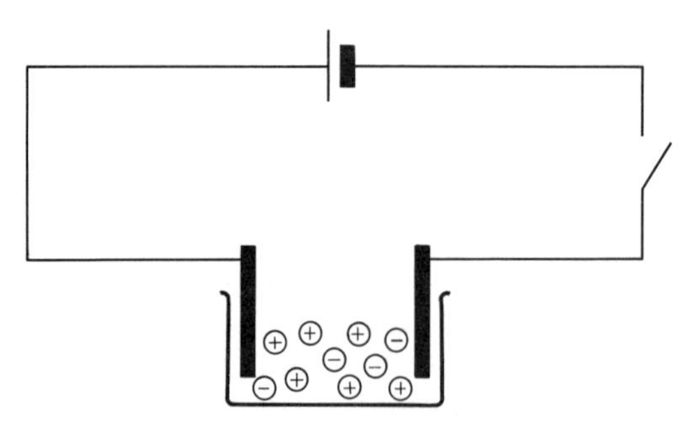

d Write equations for the reaction at each electrode, and the overall reaction.

6 This question is about the electrolysis of an aqueous solution of lithium chloride.

a Write down the names and symbols of all the ions present in the solution.

b Lithium is a reactive metal, like sodium. What will be formed at the cathode?

c What will be formed at the anode?

d Write an equation for the reaction at each electrode.

e Name two other electrolytes that will give the same electrolysis products as this one.

7 Write an equation for:
a the overall decomposition
b the reaction at each electrode
when molten sodium chloride is electrolysed.

8 a List the anions and cations present in:
i sodium chloride solution
ii copper(II) chloride solution
b Write down the reaction you would expect at:
i the anode ii the cathode
when each solution in **a** is electrolysed, using platinum electrodes.
c Explain why the anode reactions in **b** are both the same.
d Explain why copper is obtained at the cathode, but not sodium.

9 Six substances A to F were dissolved in water, and connected in turn into the circuit below. The symbol Ⓐ represents an ammeter, which is an instrument for measuring the current.

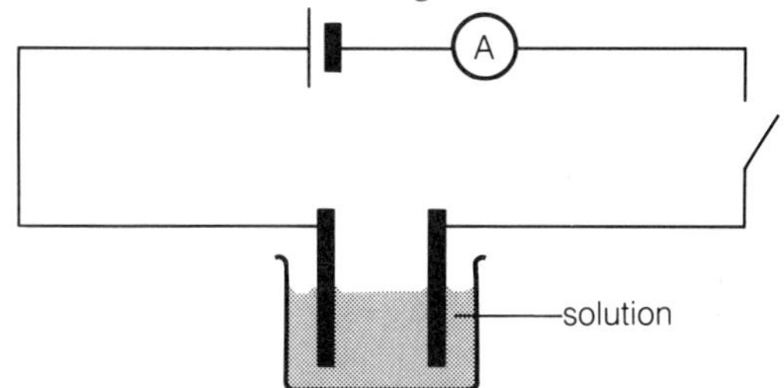

The results are shown in this table:

Substance	Current/ amps	Cathode (+)	Anode (−)
A	0.8	copper	chlorine
B	1.0	hydrogen	chlorine
C	0.0	———	———
D	0.8	copper	oxygen
E	1.2	hydrogen	oxygen
F	0.7	silver	oxygen

a Which solution conducts best?
b Which solution is a non-electrolyte?
c Which solution could be:
i silver nitrate?
ii copper(II) sulphate?
iii copper(II) chloride?
iv sodium hydroxide?
v sugar?
vi potassium chloride?

10 Hydrogen chloride is a molecular substance. However, it dissolves in water to form *hydrochloric acid*, which exists as ions:
$HCl(g) \longrightarrow H^+(aq) + Cl^-(aq)$
List the ions present in a solution of hydrochloric acid. What result would you expect, when the solution is electrolysed with platinum electrodes?

11 Two beakers of copper(II) sulphate solution are set up in a circuit, as shown below. A and B are platinum electrodes, C and D are copper electrodes.

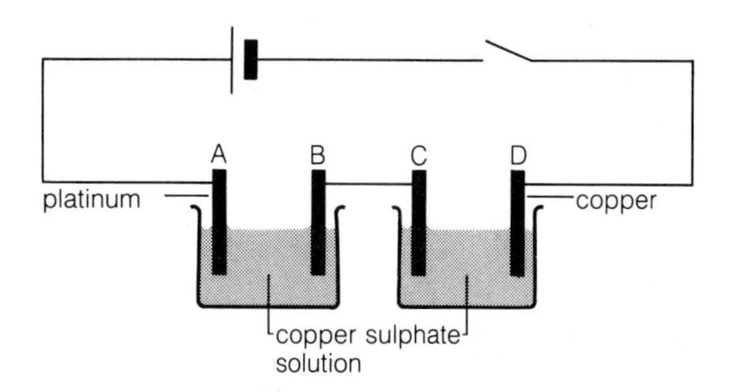

a Which electrodes are the *anodes*?
b What would happen at each of the electrodes when the switch is closed?
c Use your answer to **b** to explain what electroplating is.
d Draw a diagram to show how you would plate a metal object with nickel.
e Suppose electrode C is made of impure copper. Explain how the copper becomes purified, during the process above.

12 The diagram shows an apparatus which can be used for the electrolysis of acidified water, producing hydrogen and oxygen.

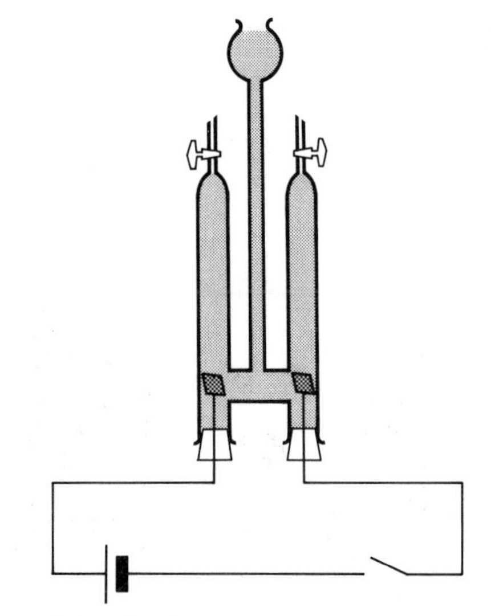

a Copy the diagram.
b Label the cathode and the anode.
c Draw in the level of water in each tube after the apparatus has been switched on for some time.
d Label the two gases (hydrogen and oxygen).
e Which acid is usually used to acidify water?
f Write a chemical equation to represent the decomposition of water to hydrogen and oxygen.
g What is the relative molecular mass of water?
h What mass of hydrogen can be obtained from 54 g of water, by electrolysis?

7.1 Rates of reaction

Fast and slow

Some reactions are **fast** and some are **slow**. Look at these examples:

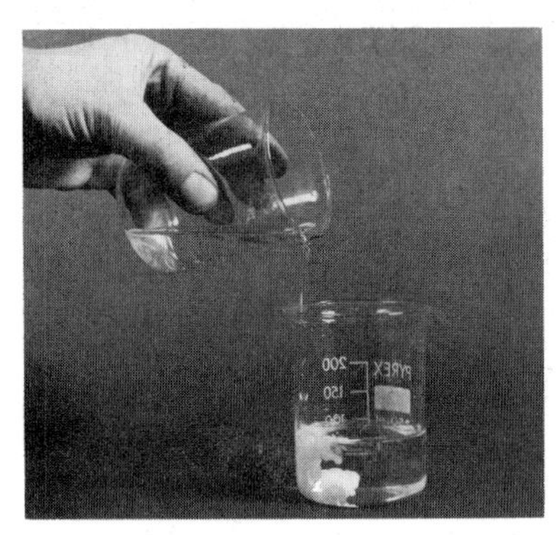

Silver chloride precipitating, when solutions of silver nitrate and sodium chloride are mixed. This is a very fast reaction.

Concrete setting. This reaction is quite slow. It will take a couple of days for the concrete to harden.

Rust forming. This is usually a very slow reaction. The pipe flange above has been rusting for fifteen years.

It is not always enough to know just that a reaction is fast or slow. For example, in a factory that makes products from chemicals, the chemical engineers need to know *exactly* how fast each reaction is going, and how long it takes to complete. In other words, they need to know the **rate** of each reaction.

What is rate?

Rate is a measure of how fast or slow something is. Here are some everyday examples:

This plane has just flown 2000 kilometres in 1 hour. It flew at a **rate** of 2000 kilometres per hour.

This petrol pump can pump petrol at a **rate** of 50 litres per minute.

This machine can print newspapers at a **rate** of 10 copies per second.

From these examples you can see that:
Rate is a measure of the change that happens in a single unit of time.
Any suitable unit of time can be used—a second, a minute, an hour, even a day.

Rate of a chemical reaction

When zinc is added to dilute sulphuric acid, they react together. The zinc disappears slowly, and a gas bubbles off at the same time.

After a time, the bubbles of gas form less quickly. The reaction is slowing down.

Finally, no more bubbles appear. The reaction is over, because all the acid has been used up. Some zinc remains behind.

In this example, the gas that forms is hydrogen. The equation for the reaction is:

zinc + sulphuric acid ⟶ zinc sulphate + hydrogen

$Zn(s) + H_2SO_4(aq) \longrightarrow ZnSO_4(aq) + H_2(g)$

Both zinc and sulphuric acid get used up in the reaction. At the same time, zinc sulphate and hydrogen form.

You could measure the rate of the reaction, by measuring either:

(i) the amount of zinc used up per minute *or*

(ii) the amount of sulphuric used up per minute *or*

(iii) the amount of zinc sulphate produced per minute *or*

(iv) the amount of hydrogen produced per minute.

For this reaction, it is easiest to measure the amount of hydrogen produced per minute—the hydrogen can be collected as it bubbles off, and its volume can then be measured. On the next page you will find out how this is done.

In general, to find the rate of a reaction, you should measure:
the amount of a reactant used up per unit of time *or*
the amount of a product produced per unit of time.

Questions

1 Here are some reactions that take place in the home. Put them in order of decreasing rate (the fastest one first).
  a Gloss paint drying
  b Fruit going rotten
  c Cooking gas burning
  d A cake baking
  e A metal bath rusting

2 Which of these rates of travel is slowest?
  5 kilometres per second
  20 kilometres per minute
  60 kilometres per hour

3 Suppose you had to measure the rate at which zinc is used up, in the reaction above. Which of these units would be suitable?
  a litres per minute
  b grams per minute
  c centimetres per minute.
  Explain your choice.

4 Iron reacts with sulphuric acid like this:
  $Fe(s) + H_2SO_4(aq) \longrightarrow FeSO_4(aq) + H_2(g)$
  a Write a word equation for this reaction.
  b Write down four different ways in which the rate of the reaction could be measured.

7.2 Measuring the rate of a reaction

Some reactions are so fast that their rates would be very difficult to measure—like this detonation of an old mine.

On the last page you saw that the rate of a reaction is found by measuring the amount of a **reactant** used up per unit of time or the amount of a **product** produced per unit of time.
Take for example the reaction between magnesium and excess dilute hydrochloric acid. Its equation is:

magnesium + hydrochloric acid ⟶ magnesium chloride + hydrogen

$$Mg(s) + 2HCl(aq) \longrightarrow MgCl_2(aq) + H_2(g)$$

In this reaction, hydrogen is the easiest substance to measure. This is because it is the only gas in the reaction. It bubbles off and can be collected in a **gas syringe**, where its volume is measured.

The method This apparatus is suitable:

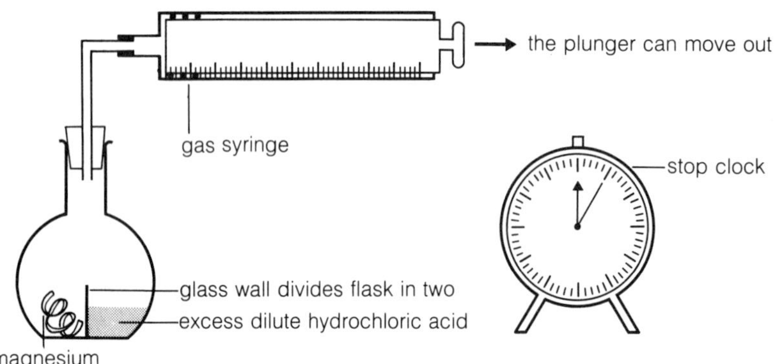

The magnesium is cleaned with sandpaper and put into one part of the flask. Dilute hydrochloric acid is put into the other part. The flask is tipped up to let the two reactants mix, and the clock is started at the same time. Hydrogen begins to bubble off. It rises up the flask, and pushes its way into the gas syringe. The plunger is forced to move out:

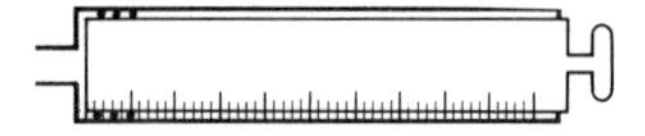

At the start the plunger is fully in. No gas has yet been collected.

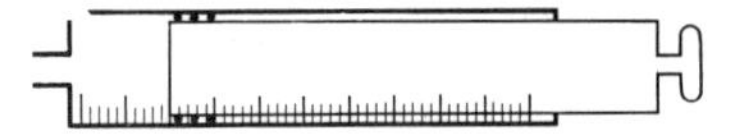

Now the plunger has moved out to the 20 cm^3 mark. 20 cm^3 of gas has been collected.

The volume of gas in the syringe is noted at intervals, for example at the end of each half-minute. How would you know when the reaction is complete?

The results Here are some typical results:

Time/minutes	0	$\frac{1}{2}$	1	$1\frac{1}{2}$	2	$2\frac{1}{2}$	3	$3\frac{1}{2}$	4	$4\frac{1}{2}$	5	$5\frac{1}{2}$	6	$6\frac{1}{2}$
Volume of hydrogen/cm^3	0	8	14	20	25	29	33	36	38	39	40	40	40	40

These results can be plotted on a graph, as shown on the next page.

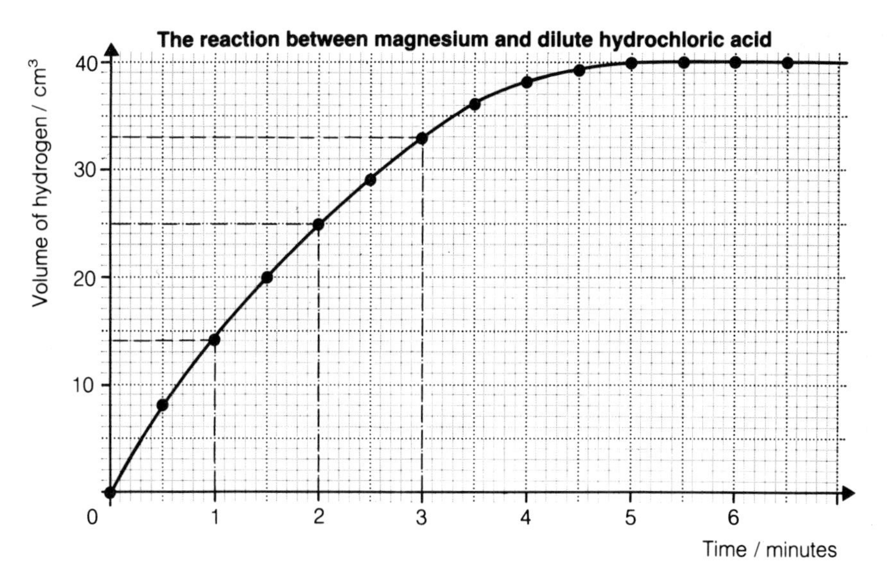

Notice these things about the results:

1 In the first minute, 14 cm^3 of hydrogen is produced.
So the rate for the first minute is 14 cm^3 of hydrogen per minute.
In the second minute, only 11 cm^3 is produced. (25 – 14 = 11)
So the rate for the second minute is 11 cm^3 of hydrogen per minute.
The rate for the third minute is 8 cm^3 of hydrogen per minute.
So you can see that the rate decreases as time goes on.
The rate changes all through the reaction. It is greatest at the start, but gets less as the reaction proceeds.

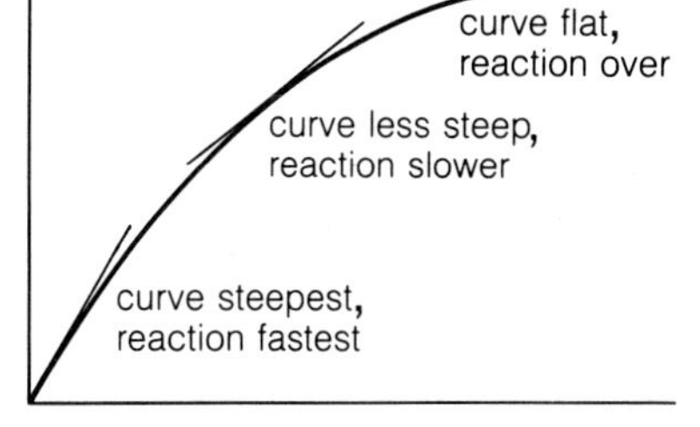

2 The reaction is fastest in the first minute, and the curve is steepest then. It gets less steep as the reaction gets slower.
The faster the reaction, the steeper the curve.

3 After 5 minutes, no more hydrogen is produced, so the volume no longer changes. The reaction is over, and the curve goes flat.
When the reaction is over, the curve goes flat.

4 Altogether, 40 cm^3 of hydrogen are produced in 5 minutes.

$$\text{The } \textit{average} \text{ rate for the reaction} = \frac{\text{total volume of hydrogen}}{\text{total time for the reaction}}$$

$$= \frac{40\,\text{cm}^3}{5\text{ minutes}}$$

$$= \mathbf{8\,cm^3 \text{ of hydrogen per minute.}}$$

Note that this method can be used to measure the rate of *any* reaction in which one product is a gas – like the reaction shown on page 87.

Questions

1 For this experiment, can you explain why:
  a a divided flask is used?
  b the magnesium ribbon is first cleaned?
  c the clock is started the moment the reactants are mixed?

2 From the graph above, how can you tell when the reaction is over?

3 This question is about the graph above.
  a How much hydrogen is produced in:
    i 2.5 minutes? **ii** 4.5 minutes?
  b How many minutes does it take to produce:
    i 10 cm^3 **ii** 20 cm^3 of hydrogen?
  c What is the rate of the reaction during:
    i the fourth minute? **ii** the fifth minute?

7.3 Changing the rate of a reaction (I)

The effect of concentration

A reaction can be made to go faster or slower by changing the **concentration** of a reactant.
Suppose the experiment with magnesium and excess hydrochloric acid is repeated twice (A and B below). Everything is kept the same each time, *except* the concentration of the acid:

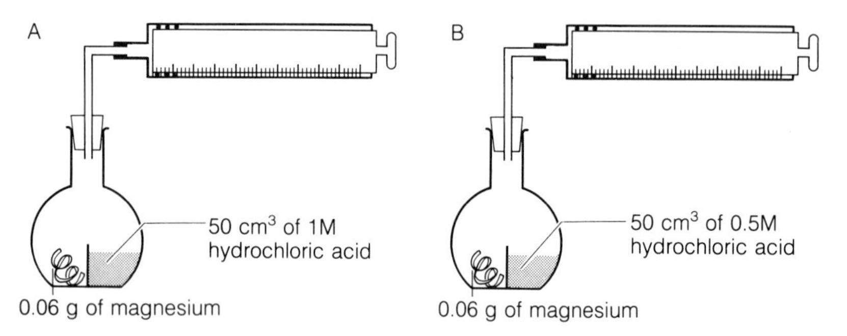

The acid in A is *twice as concentrated* as the acid in B.
Here are both sets of results shown on the same graph.

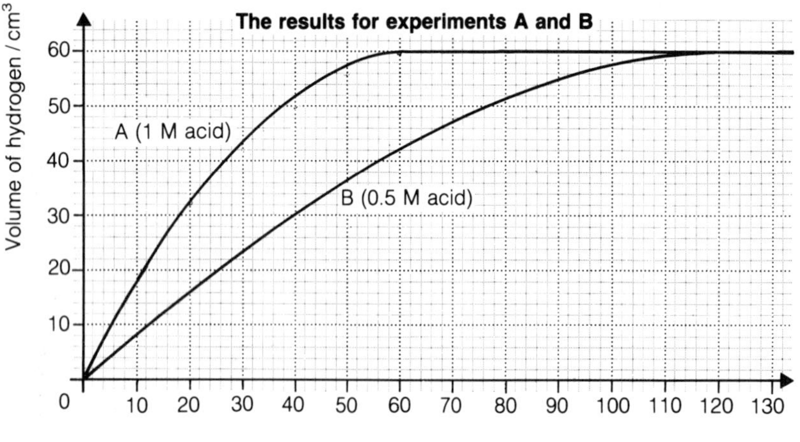

Notice these things about the results:

1. Curve A is steeper than curve B. From this you can tell straight away that the reaction was faster in A than in B.
2. In A, the reaction lasts 60 seconds. In B it lasts for 120 seconds.
3. Both reactions produced 60 cm^3 of hydrogen. Do you agree?
   In A it was produced in 60 seconds, so the average rate was 1 cm^3 of hydrogen per second. In B it was produced in 120 seconds, so the average rate was 0.5 cm^3 of hydrogen per second.
   The average rate in A was twice the average rate in B.

These results show that:
A reaction goes faster when the concentration of a reactant is increased.
For the reaction above, the rate doubles when the concentration of acid is doubled.

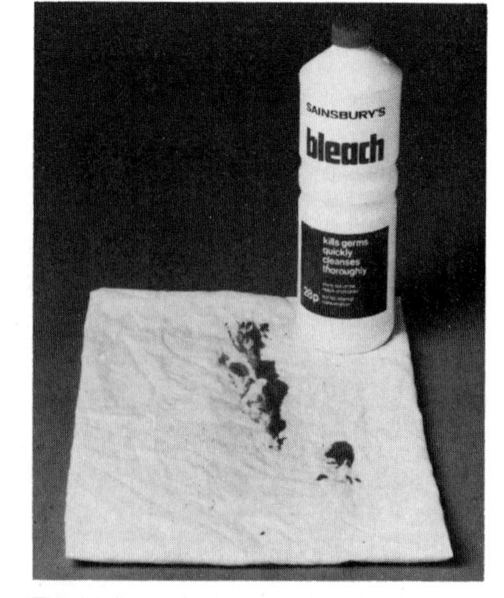

This stain could be removed with a solution of bleach. The more concentrated the solution, the faster the stain will disappear.

The effect of temperature

A reaction can also be made to go faster or slower by changing the **temperature** of the reactants.
This time, a different reaction is used: when dilute hydrochloric acid is mixed with sodium thiosulphate solution, a fine yellow precipitate of sulphur forms. The rate can be followed like this:

1 A cross is marked on a piece of paper.
2 A beaker containing some sodium thiosulphate solution is put on top of the paper. The cross should be easy to see through the solution, from above.
3 Hydrochloric acid is added quickly, and a clock started at the same time. The cross grows fainter as the precipitate forms.
4 The clock is stopped the moment the cross can no longer be seen from above.

View from above the beaker:

The cross grows fainter with time

The experiment is repeated several times. The quantity of each reactant is kept exactly the same each time. Only the temperature of the reactants is changed. This table shows the results:

Temperature/°C	20	30	40	50	60
Time for cross to disappear/seconds	200	125	50	33	24
	The higher the temperature the faster the cross disappears				

The cross disappears when enough sulphur forms to blot it out. Notice that this takes 200 seconds at 20 °C, but only 50 seconds at 40 °C. So the reaction is *four times faster* at 40 °C than at 20 °C.
A reaction goes faster when the temperature is raised. When the temperature increases by 10 °C, the rate approximately doubles.
This fact is used a great deal in everyday life. For example, food is kept in the fridge to slow down decomposition reactions and keep it fresh for longer. Can you think of any other examples?

The low temperature in the fridge slows down decomposition reactions.

Questions

1 Look at the graph on the opposite page.
  a After 2 minutes, how much hydrogen was produced in:
    i experiment A? **ii** experiment B?
  b From the shape of the curves, how can you tell which reaction was faster?
2 Explain why experiments A and B both produce the same amount of hydrogen.
3 Copy and complete: A reaction goes .... when the concentration of a .... is increased. It also goes .... when the .... is raised.
4 Why does the cross disappear, in the experiment with sodium thiosulphate and hydrochloric acid?
5 What will happen to the rate of a reaction when the temperature is *lowered*? Use this to explain why milk is stored in a fridge.

7.4 Changing the rate of a reaction (II)

The effect of surface area

In many reactions, one of the reactants is a solid. The reaction between hydrochloric acid and calcium carbonate (marble chips) is one example. Carbon dioxide gas is produced:

$CaCO_3(s) + 2HCl(aq) \longrightarrow CaCl_2(aq) + H_2O(l) + CO_2(g)$

The rate can be measured using the apparatus on the right.

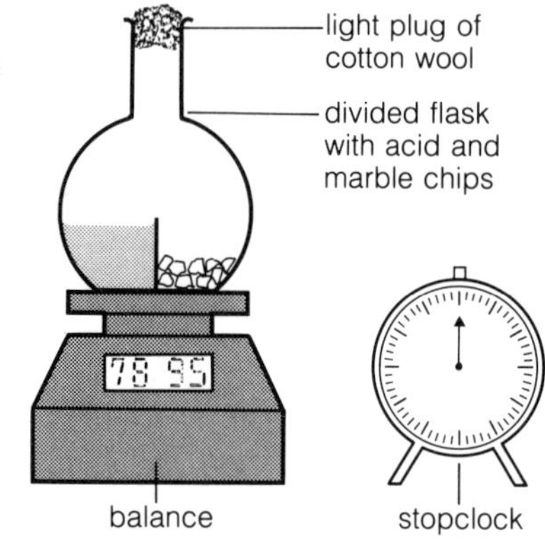

The method Marble chips and acid are placed in the flask, which is then plugged with cotton wool. This prevents any liquid from splashing out during the reaction. Next the flask is weighed. Then it is tipped up, to let the reactants mix, and a clock is started at the same time. The mass is noted at regular intervals, until the reaction is complete.

Since carbon dioxide can escape through the cotton wool, the flask gets lighter as the reaction proceeds. So by weighing the flask, you can follow the rate of the reaction.

The experiment is repeated twice. Everything is kept exactly the same each time, except the **surface area** of the marble chips:

For experiment 1, large chips of marble are used. The surface area is the total area of the surface of these chips.

For experiment 2, the same *mass* of marble is used. But this time it is in small chips, so its surface area is greater.

The results The results of the two experiments are plotted below:

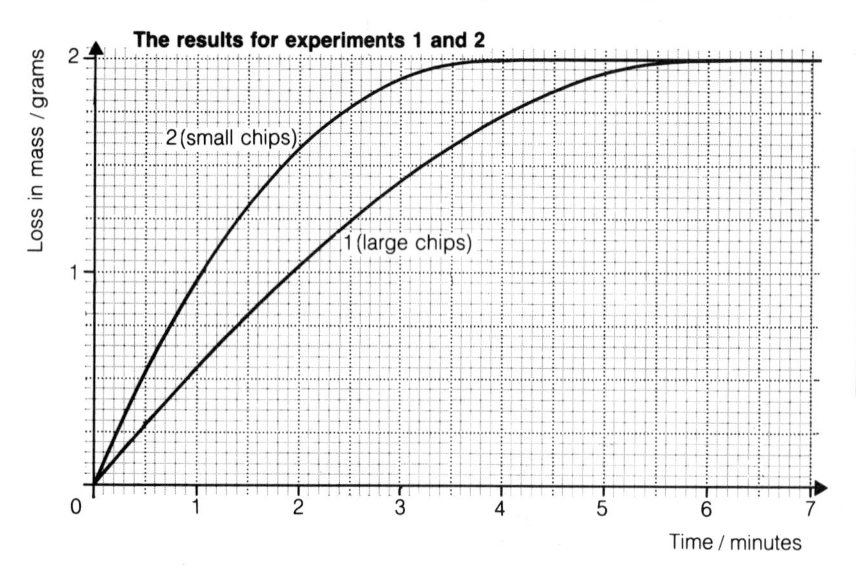

How to draw the graph:

First you have to find the *loss in mass* at different times:
loss in mass at a given time = mass at start − mass at that time
Then you plot the values for loss in mass against the times.

You should notice these things about the results:

1 Curve 2 is steeper than curve 1. This shows immediately that the reaction is faster for the small chips.
2 In both experiments, the final loss in mass is 2.0 grams. In other words, 2.0 grams of carbon dioxide are produced each time.
3 For the small chips, the reaction is complete in 4 minutes. For the large chips, it lasts for 6 minutes.

These results show that:

The rate of a reaction increases when the surface area of a solid reactant is increased.

The effect of a catalyst

Hydrogen peroxide is a clear colourless liquid with the formula H_2O_2. It can decompose to water and oxygen:

hydrogen peroxide ⟶ water + oxygen

$$2H_2O_2(aq) \longrightarrow 2H_2O(l) + O_2(g)$$

The rate of the reaction can be followed by collecting the oxygen:

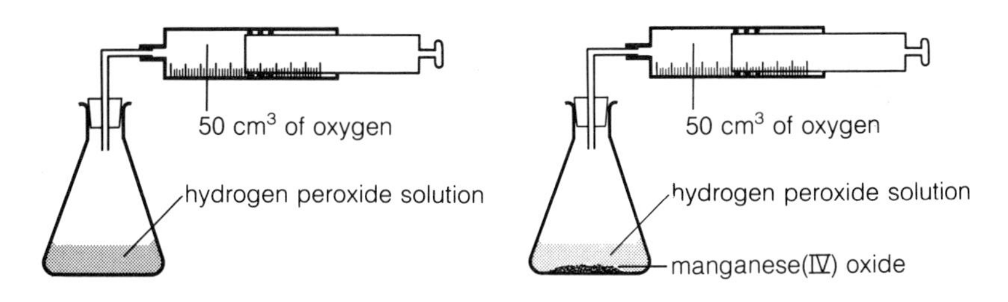

The reaction is in fact very slow. It could take 500 days to collect 50 cm^3 of oxygen.

However, if 1 gram of black manganese(IV) oxide is added, the reaction goes much faster. 50 cm^3 of oxygen is produced in a few minutes.

After the reaction, the black powder is removed by filtering. It is then dried, weighed and tested. It is still manganese(IV) oxide, and still weighs 1 gram.

The manganese(IV) oxide speeds up the reaction without being used up itself. It is called a **catalyst** for the reaction.

A catalyst is a substance that changes the rate of a chemical reaction but remains chemically unchanged itself.

Catalysts have been discovered for many reactions. They are usually **transition metals** or **compounds of transition metals**. For example, iron speeds up the reaction between nitrogen and hydrogen, to make ammonia (page 158). The human body produces its own catalysts called **enzymes**. For example a gland called the pancreas, below your stomach, makes pancreatic juice which contains enzymes that speed up digestion.

Questions

1 This question is about the graph on the opposite page. For each experiment find:
 a the loss in mass in the first minute
 b the mass of carbon dioxide produced during the first minute
 c the average rate of production of the gas

2 What is a catalyst? Give two examples of reactions and their catalysts.

3 Look again at the decomposition of hydrogen peroxide, above. How would you show that:
 a the reaction goes even faster if *more than* 1 gram of catalyst is used? **b** the catalyst is not used up?

7.5 Explaining rates

A closer look at a reaction

On page 88, you saw that magnesium and dilute hydrochloric acid react together:

magnesium + hydrochloric acid ⟶ magnesium chloride + hydrogen

$$Mg(s) + 2HCl(aq) \longrightarrow MgCl_2(aq) + H_2(g)$$

In order for the magnesium and acid particles to react together:

(i) they must collide with each other

(ii) the collision must have enough energy

This is shown by the drawings below.

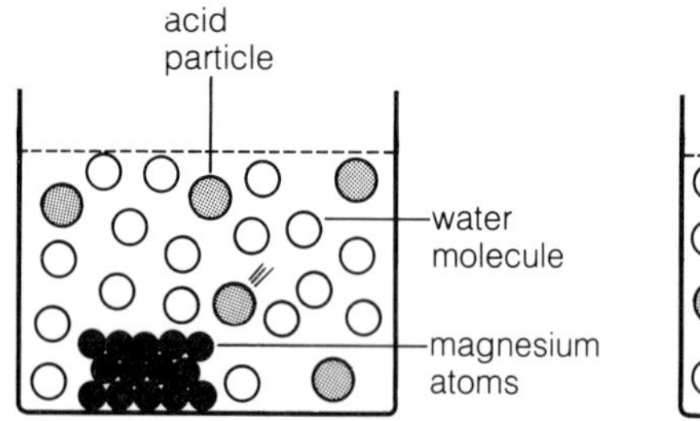

The particles in the liquid move around continually. Here an acid particle is about to collide with a magnesium atom.

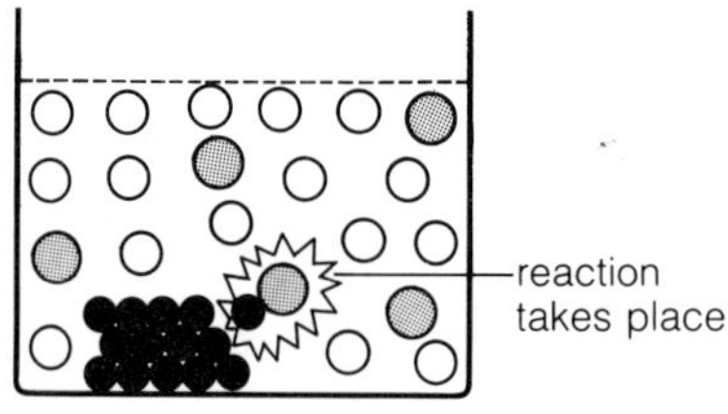

If the collision has enough energy, reaction takes place. Magnesium chloride and hydrogen are formed.

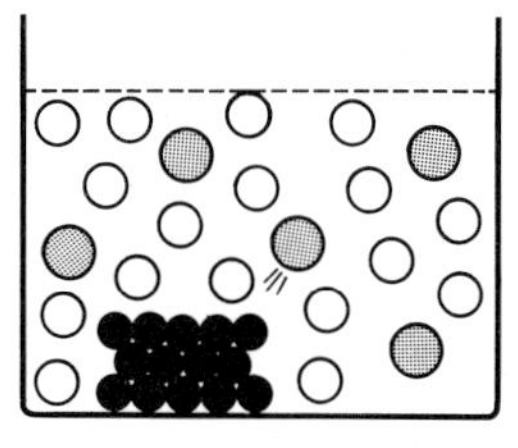

If the collision does not have enough energy, no reaction occurs. The acid particle bounces away again.

If there are lots of successful collisions in a given minute, then a lot of hydrogen is produced in that minute. In other words, the reaction goes quickly—its rate is high. If there are not many, its rate is low. **The rate of a reaction depends on how many successful collisions there are in a given unit of time.**

Changing the rate of a reaction

Why rate increases with concentration If the concentration of the acid is increased, the reaction goes faster. It is easy to see why:

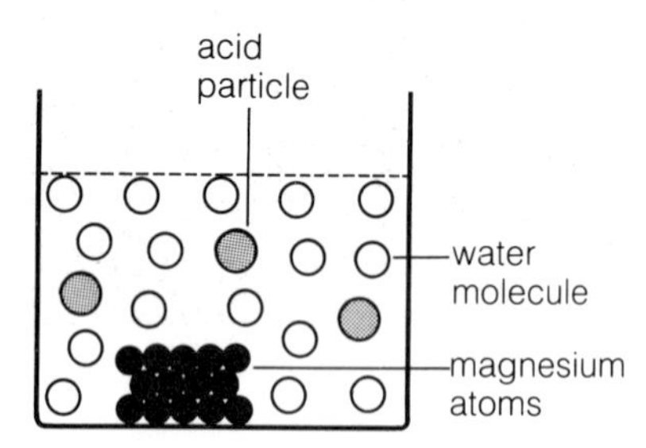

In dilute acid, there are not so many acid particles. This means there is not much chance of an acid particle hitting a magnesium atom.

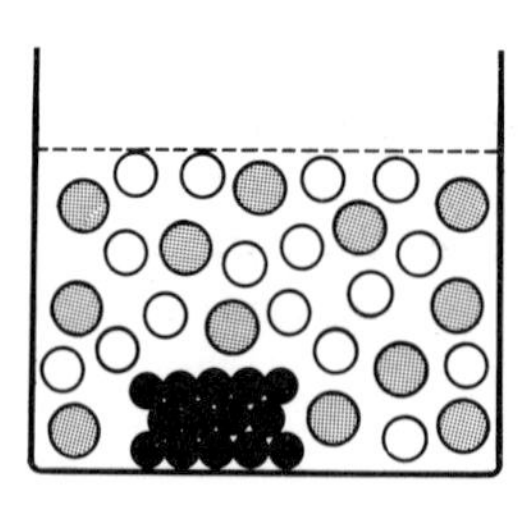

Here the acid is more concentrated—there are more acid particles in it. There is now more chance of a successful collision occurring.

The more successful collisions there are, the faster the reaction.

This idea also explains why the reaction between magnesium and hydrochloric acid slows down as time goes on:

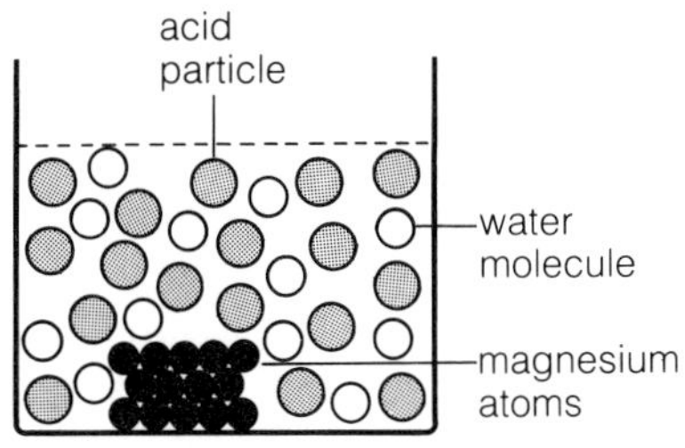

At the start, there are plenty of magnesium atoms and acid particles. But they get used up during successful collisions.

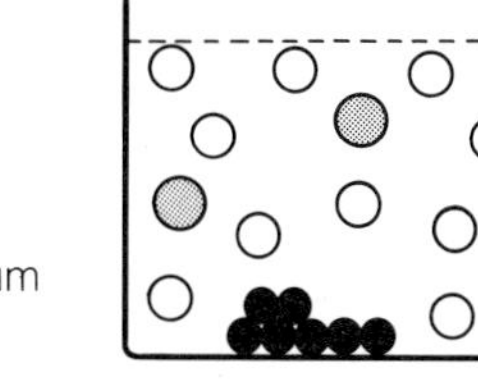

After a time, there are fewer magnesium atoms, and the acid is less concentrated. So the reaction slows down.

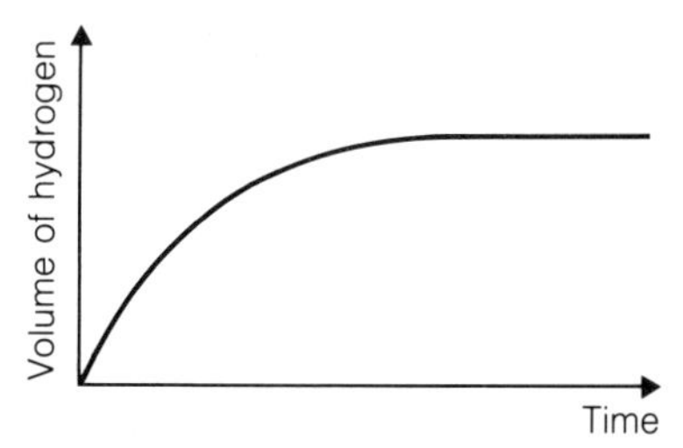

This means that the slope of the reaction curve decreases with time, as shown above.

Why rate increases with temperature At low temperatures, particles of reacting substances do not have much energy. However, when the substances are heated, the particles take in energy. This causes them to move faster and collide more often. The collisions have more energy, so more of them are successful. Therefore the rate of the reaction increases.

Why rate increases with surface area The reaction between the magnesium and acid is much faster when the metal is powdered:

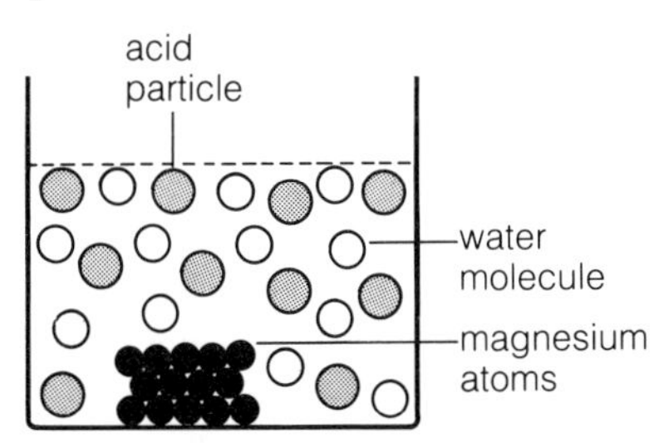

Acid particles can collide only with magnesium atoms on the *outside* of the metal.

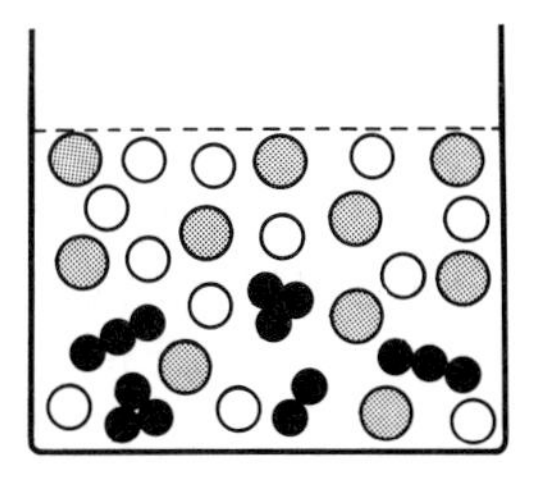

When the metal is powdered, many more atoms are exposed. So there is a greater chance of successful collisions.

Why a catalyst increases the rate Some reactions can be speeded up by adding a catalyst. *In the presence of a catalyst, a collision needs less energy in order to be successful.* The result is that more collisions become successful, so the reaction goes faster. Catalysts are very important in industry, because they speed up reactions even at low temperatures. This means that less fuel is needed, so money is saved.

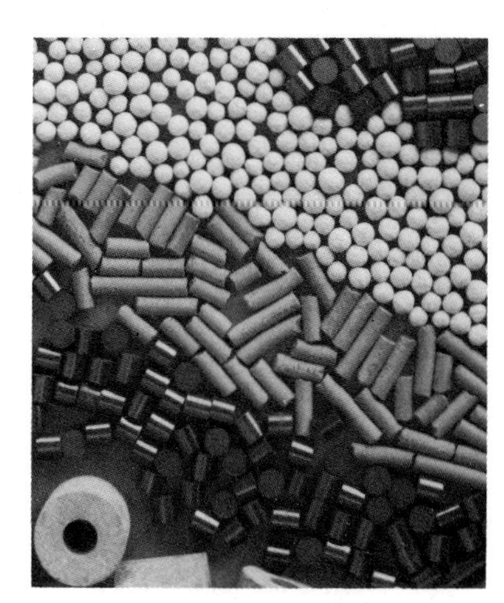

Some catalysts used in industry.

Questions

1. Copy and complete: Two particles can only react together if they .... and if the .... has enough ....
2. What is:
   - **a** a successful collision?
   - **b** an unsuccessful collision?
3. In your own words, explain why the reaction between magnesium and acid goes faster when:
   - **a** the temperature is raised
   - **b** the magnesium is powdered
4. Explain why a catalyst can speed up a reaction, even at low temperatures.

Questions on Chapter 7

1 The rate of the reaction between magnesium and dilute hydrochloric acid could be measured using this apparatus:

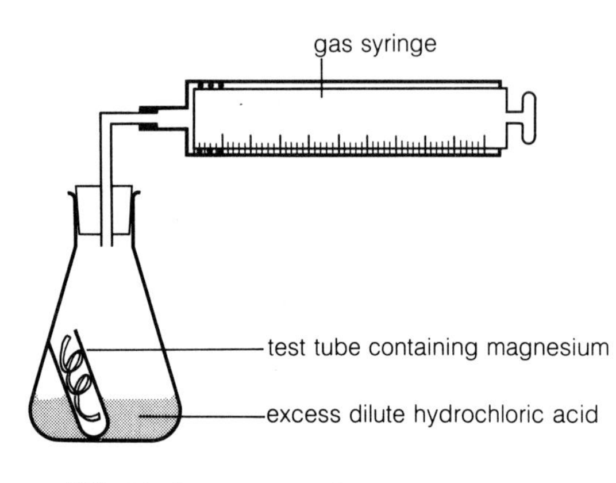

a What is the purpose of:
i the test tube?
ii the gas syringe?
b How would you get the reaction to start?

2 Some magnesium and an *excess* of dilute hydrochloric acid were reacted together. The volume of hydrogen produced was recorded every minute, as shown in the table:

Time/min	0	1	2	3	4	5	6	7
Volume of hydrogen/cm^3	0	14	23	31	38	40	40	40

a What does an *excess* of acid mean?
b Plot a graph of the results, labelling the axes as on page 89.
c How much hydrogen was produced in:
i the first minute?
ii the second minute?
iii the third minute?
iv the fourth minute?
v the fifth minute?
d What is the *rate of the reaction* (cm^3 of hydrogen per minute) during each minute?
e What is the total volume of hydrogen produced in the reaction?
f How many minutes pass before the reaction finishes?
g What is the *average rate* of the reaction?
h A similar reaction had a rate of 15 cm^3 of hydrogen in the first minute. Is this a slower or faster reaction than the one above?
i How could you make the above reaction go slower, while still using the same quantities of metal and acid?

3 For this question you will need the graph you drew for question 2.
The experiment with magnesium and an excess of dilute hydrochloric acid was repeated. This time a different concentration of hydrochloric acid was used. The results were:

Time/min	0	1	2	3	4	5	6
Volume of hydrogen/cm^3	0	22	34	39	40	40	40

a Plot these results on the graph you drew for question 2.
b Which reaction was faster? How can you tell?
c In which experiment was the acid more concentrated? Give a reason for your answer.
d The same volume of hydrogen was produced in each experiment. What does that tell you about the mass of magnesium used?

4 Name three factors that effect the rate of a reaction, and describes the effect of changing each factor.

5 Suggest a reason for each of the following observations:
a Magnesium powder reacts faster than magnesium ribbon, with dilute sulphuric acid.
b Hydrogen peroxide decomposes much faster in the presence of the enzyme *catalase*.
c The reaction between manganese carbonate and dilute hydrochloric acid speeds up when some concentrated hydrochloric acid is added.
d Zinc powder burns much more vigorously in oxygen than zinc foil does.
e The reaction between sodium thiosulphate and hydrochloric acid takes a very long time if carried out in an ice bath.
f Zinc and dilute sulphuric acid react much more quickly when a few drops of copper(II) sulphate solution are added.
g Drenching with water prevents too much damage from spilt acid.
h A car's exhaust pipe will rust faster if the car is used a lot.
i In fireworks, powdered magnesium is used rather than magnesium ribbon.
j In this country, dead animals decay quite quickly. But in Siberia, bodies of mammoths that died 30 000 years ago have been found fully preserved in ice.
k The more sweet things you eat, the faster your teeth decay.
l Food cooks much faster in a pressure cooker than in an ordinary saucepan.

6 When sodium thiosulphate reacts with hydrochloric acid, a precipitate forms. In an investigation, the time taken for the solution to become opaque was recorded. (*Opaque* means that you cannot see through it.)
Four experiments (A to D) were carried out. Only the concentration of the sodium thiosulphate solution was changed each time.
The following results were obtained:

Experiment	A	B	C	D
Time taken/seconds	42	71	124	63

a Draw a diagram of suitable apparatus for this experiment.
b Name the precipitate that forms.
c What would be *observed* during the experiment?
d In which experiment was the reaction: fastest? slowest?
e In which experiment was the sodium thiosulphate solution most concentrated? How can you tell?
f Suggest two other ways of speeding up this reaction.

7 Copper(II) oxide catalyses the decomposition of hydrogen peroxide. 0.5 g of the oxide was added to a flask containing 100 cm^3 of hydrogen peroxide solution. A gas was released. It was collected and its volume noted every 10 seconds. This table shows the results:

Time/sec	0	10	20	30	40	50	60	70	80	90
Volume/cm^3	0	18	30	40	48	53	57	58	58	58

a What is a catalyst?
b Draw a diagram of suitable apparatus for this experiment.
c Name the gas that is formed.
d Write a balanced equation for the decomposition of hydrogen peroxide.
e Plot a graph of the volume of gas (vertical axis) against time (horizontal axis).
f Describe how the rate changes during the reaction.
g What happens to the concentration of hydrogen peroxide as the reaction proceeds?
h What chemicals are present in the flask after 90 seconds?
i What mass of copper(II) oxide would be left in the flask at the end of the reaction?
j Sketch on your graph the curve that might be obtained for 1.0 g of copper(II) oxide.
k Name one other chemical that catalyses this decomposition.

8 A small quantity of hydrochloric acid was added to a large quantity of marble chips in an evaporating dish, which was placed on the pan of a balance. The mass of the dish and its contents was recorded every half minute. The results are shown in the graph below.

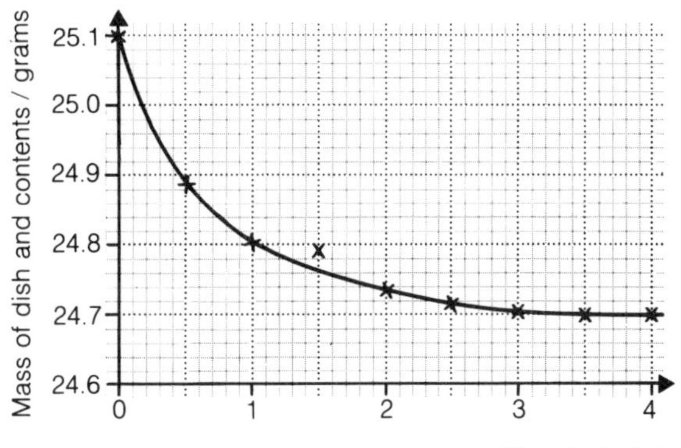

a Explain why the curve slopes *down*.
b Which result would seem to be incorrect? Why?
c What was the mass of the evaporating dish and its contents (i) at the start and (ii) at the end of the experiment?
d What mass of carbon dioxide was produced?
e How long did the reaction last?
f During which of these times in the experiment was the rate of reaction most rapid:
$0-\frac{1}{2}$ min? $1-1\frac{1}{2}$ min? $3-3\frac{1}{2}$ min?
g Make a copy of the graph. On your copy, sketch the curve you would expect if the acid had first been chilled.
h Now sketch the curve you would expect if the chips had first been crushed to powder.
i What apparatus could be used instead of an evaporating dish, to prevent any loss of acid by splashing?

9 Potassium chlorate decomposes when heated, like this:
$2KClO_3(s) \longrightarrow 2KCl(s) + 3O_2(g)$
a Write a word equation for the reaction.
b What gas is given off when potassium chlorate decomposes?
c Manganese(IV) oxide acts as a catalyst for this reaction. What would you expect if two test tubes, one of potassium chlorate and the other a mixture of potassium chlorate and manganese(IV) oxide, were heated?
d Potassium chloride is soluble in water, and manganese(IV) oxide is insoluble. How could you show that the manganese(IV) oxide is not used up during the reaction?
e How many moles of oxygen gas are obtained from two moles of potassium chlorate?
f Will there be *more* or *less* oxygen produced, when the catalyst is used? Explain your answer.

8.1 Air

What's in air?

Air is a mixture of gases:

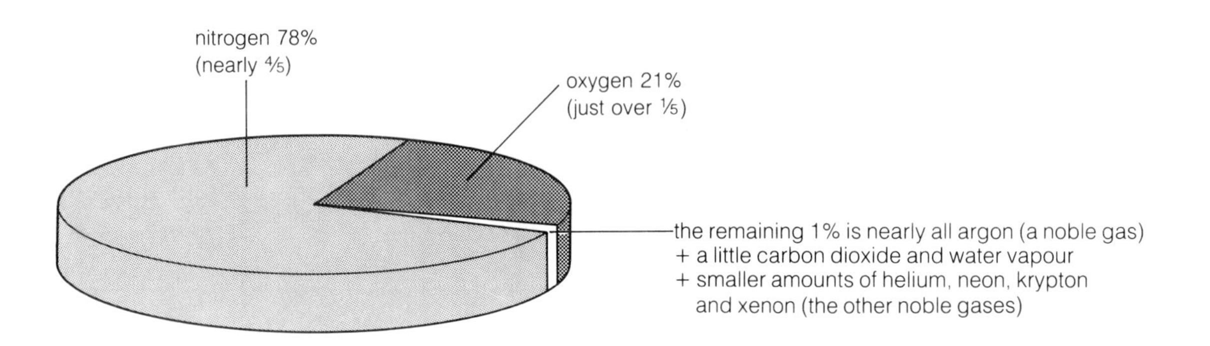

The composition of air is not always exactly the same. For example, the percentage of water vapour in the air is much greater on a damp day than on a dry day. Besides, the air over a busy city also contains gases such as **carbon monoxide** and **sulphur dioxide**, which come from car exhausts and factory chimneys (page 102).

Some unpleasant things being added to air.

The importance of air

Air is essential to life. The reasons are given briefly below, but you will find out more about them later.

For building living things All plants and animals (including humans) are built from compounds containing carbon, nitrogen, hydrogen and oxygen. Air provides the starting materials for these compounds—carbon dioxide, nitrogen and water vapour. Plants act as the 'factory' for making them:

1 The **nitrogen** from the air is used by man to make compounds called nitrates. These are added to the soil as fertilisers. Some nitrogen is also changed naturally to nitrates in the soil by bacteria, and by lightning.
2 The **water vapour** condenses and falls as rain, which is soaked up by the soil.
3 Plants take in the **carbon dioxide** through holes in their leaves. They take in the **nitrates** and **water** from the soil through their roots. They use these to build up the compounds they need for their roots, stems and leaves.
4 Then animals obtain the compounds they need by eating plants. And some animals get eaten in turn by others.

For energy All living things need **oxygen** for a process called **respiration.** Respiration produces the energy they need:

oxygen + glucose ⟶ carbon dioxide + water + energy

In your body, the glucose comes from food; the energy is needed for growing, moving and so on. How does the oxygen get into your body?

Fish use the oxygen dissolved in water. They take it in through their gills.

Measuring the percentage of oxygen in the air

Oxygen is the **reactive gas** of the air. It reacts easily with a great many other elements and compounds. For example, when copper is heated in air, only the oxygen combines with it:

copper + oxygen ⟶ copper(II) oxide
$2Cu(s) + O_2(g) \longrightarrow 2CuO(s)$
pink **black**

This reaction is used in the following experiment to measure the percentage of oxygen in the air.

The apparatus The apparatus is a tube of hard glass, connected to two gas syringes, A and B. The tube is packed with small pieces of pink copper wire. One syringe contains 100 cm^3 of air. The other is empty:

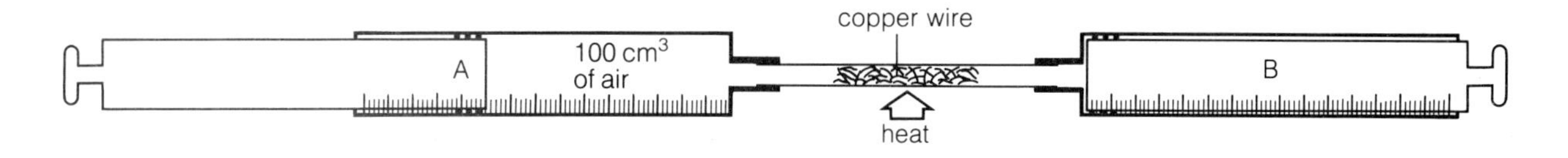

These are the steps in the experiment:

1. The tube is heated by a bunsen. When the plunger of syringe A is pushed in, the air gets forced through the heated tube, into B. As the air passes over the hot copper wire, the oxygen in it reacts with the wire, which starts to turn black. Next the plunger of B is pushed in, so the air is forced back to A. This is repeated several times.
2. After about three minutes, heating is stopped. The apparatus is allowed to cool down. Then all the gas is pushed into one of the syringes and its volume measured. (The volume is now less than 100 cm^3.)
3. Steps 1 and 2 are repeated until the volume of gas no longer decreases. This is a sign that all the oxygen in the syringes has been used up. The final volume is noted.

The results Here are some typical results:

Volume of gas at the start = 100 cm^3
Volume of gas at the end = 79 cm^3
Volume of gas used up (oxygen) = 21 cm^3

So 100 cm^3 of air contains 21 cm^3 of oxygen.

The **fraction** that is oxygen $= \frac{21}{100}$

The **percentage** that is oxygen $= \frac{21}{100} \times 100 = $ **21%**

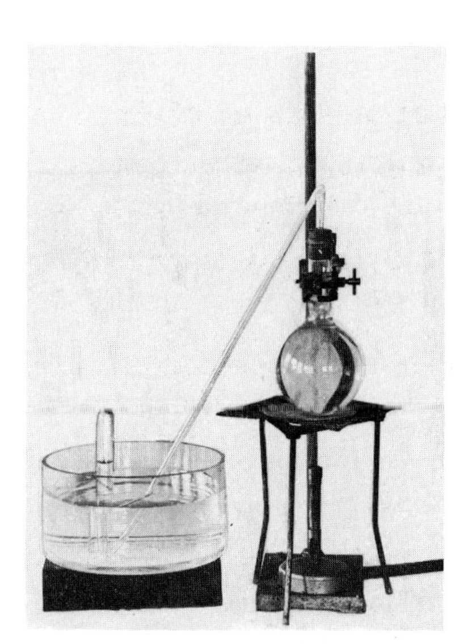

Oxygen is more soluble than nitrogen in water. If you tested the air being boiled out of this water, you would find it contained 33% oxygen. Normal air contains only 21% oxygen.

Questions

1. List the gases that make up the air.
2. What is respiration? Write an equation to describe the process.
3. At 0°C, 0.07 grams of oxygen can dissolve in every 100 grams of water. Explain why this is important for fish in the winter.
4. In the experiment above:
   a why is copper wire used instead of a lump of copper? (Think about rates.)
   b why does the copper wire turn black?
   c why is the gas allowed to cool, before its volume is measured?

8.2 Making use of air

Separating gases from the air

You have seen that air is essential for living things—without it, plants and animals could not survive.
Humans have other uses for it too. For example, the oxygen from air is needed to make steel, and the nitrogen can be turned into fertilisers. But first these gases must be **separated** from the air. The separation is carried out like this:

first the air is cooled down until it becomes liquid;
then the liquid air is slowly warmed up. All the substances in it have different boiling points, so they boil off one by one, and are collected separately. This is an example of **fractional distillation**.

Boiling points of gases distilled from liquid air:

xenon	–108°C
krypton	–153°C
oxygen	–183°C
argon	–186°C
nitrogen	–196°C

Here is a diagram of an air separation plant:

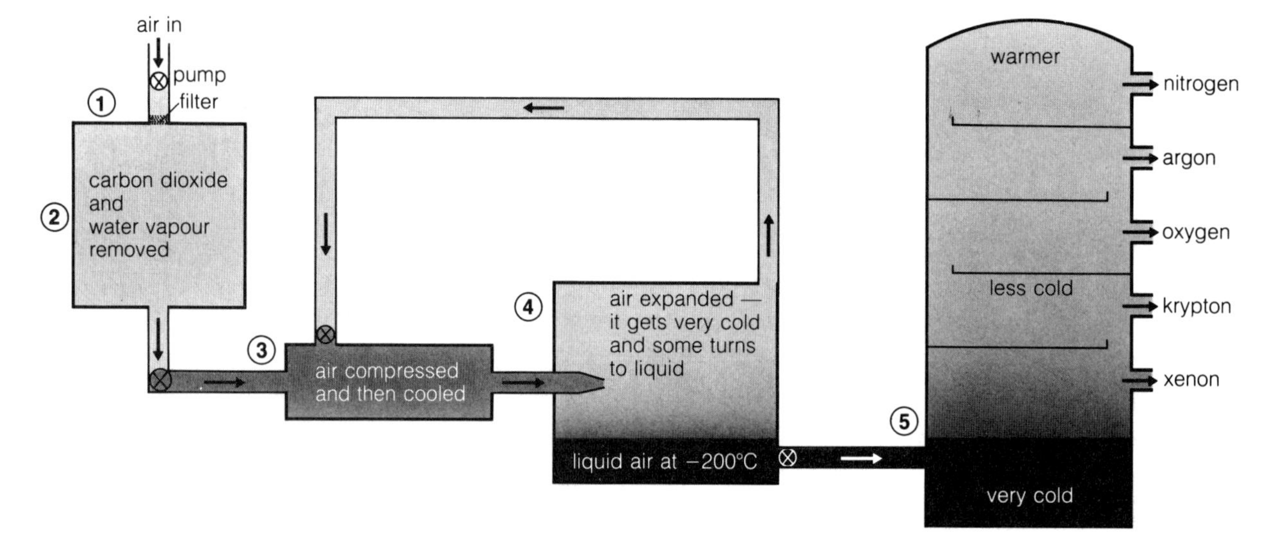

The steps in the process are:

1 Air from outside is pumped into the plant. The filter gets rid of any dust.
2 Next the carbon dioxide and water vapour are removed. Otherwise they would freeze later, when the air is cooled, and block the pipes.
3 Next the air is forced into a small space, or **compressed**. That makes it hot (just like air gets hot when you pump it into a bicycle tyre). It is then cooled down again.
4 The cold compressed air is passed through a jet, into a larger space. It expands, and that makes it very cold.

Steps 3 and 4 are repeated several times, and each time the air gets colder. By the time it reaches –200 °C, all its gases have become liquid, except for neon and helium. These two are removed.

5 The liquid air is pumped into the fractionating column. There it is slowly warmed up. The gases boil off one by one, and are collected in tanks or cylinders. Nitrogen boils off first. Can you explain why?

Liquid nitrogen.

Uses of oxygen

Breathing There is no air in space, so astronauts have to carry cylinders of oxygen with them. So do deep-sea divers. It is also needed in places where there is not enough air for proper breathing, for example in high altitude aircraft, and by firemen fighting fierce fires. In hospitals it is piped to patients with breathing problems.

Making steel In steel works, oxygen is blown through molten steel to purify it. Impurities float to the top of the steel and burn away. (There is more about this on page 147.)

Cutting and welding metals When oxygen is mixed with another gas called **acetylene**, or **ethyne**, the mixture burns strongly. The flame gets so hot that it is able to cut through metals by melting them. The flame can also be used to join or **weld** metals to each other. The metals to be joined are melted, then brought together and allowed to solidify.

This cot has its own oxygen supply for sick infants.

Uses of nitrogen

Making ammonia Nitrogen is needed for making ammonia (NH_3). Some of the ammonia is then used to make **fertilisers**.

Freezing things Liquid nitrogen is very cold (below $-196\,°C$). This makes it useful for quick-freezing food, and for keeping it frozen during transportation. It is also used in engineering. For example it is used to freeze the liquid in damaged pipes, before repairing them, and to freeze soggy earth, before digging it up.

Inert atmosphere Nitrogen gas is unreactive, so it is pumped into oil storage tanks to prevent fires. It is also used in food packaging, instead of air, to keep the food fresh.

Frozen pizzas coming out of a liquid nitrogen freezing tunnel.

Uses of the noble gases

The noble gases are also unreactive. This is why they are used for:

Lighting Argon is used to fill ordinary light bulbs. Neon is used in advertising signs, because it glows red when electricity is passed through it. Krypton and xenon are used in powerful lamps, such as light-house lamps.

Inert atmosphere Some welding uses electricity to melt the metals, rather than a hot flame. This type of welding is often carried out in an argon atmosphere. The argon protects the metals from air, which might react with them.

Balloons Helium is very light, so it is used to fill balloons. (Hydrogen is lighter but it is dangerous to use. Do you know why?)

Liquid nitrogen is used to freeze liquid in damaged pipes, before repairs.

Questions

1 In the fractional distillation of liquid air:
 a why is the air compressed and expanded?
 b why is argon obtained *before* oxygen?

2 State three uses of oxygen.

3 What is an oxy-acetylene flame?

4 Give two reasons why nitrogen is so useful.

8.3 Air pollution

Pollution over a city. The main pollutants are carbon monoxide, sulphur dioxide, oxides of nitrogen, hydrocarbons, smoke and dust.

The pie chart on page 98 shows the gases that make up the air: nitrogen, oxygen, carbon dioxide, water vapour and the noble gases. This mixture gives **clean air**. Unfortunately, much of the air we breathe is not clean—it contains substances that are unpleasant or even dangerous. These are called **pollutants** and they cause **air pollution**.

Where pollution comes from

Most air pollution is the result of burning things—fuels or just rubbish. **Petrol** is a good example. It is made of **hydrocarbons**, which are compounds of hydrogen and carbon. If it burns in plenty of air, the main products are carbon dioxide and water vapour. But in a car engine there is not enough air. The petrol burns, but not completely, and harmful products stream out in the exhaust:

1 The main harmful product is **carbon monoxide**. If this gas gets into your bloodstream, it stops red blood cells from carrying oxygen from the lungs to the rest of the body. Too much carbon monoxide causes death.
2 **Hydrocarbons** make up the next largest harmful part of the exhaust. They are unburnt gases from the petrol. They smell horrible, and can cause liver damage and even cancer.
3 **Sulphur dioxide** forms because petrol contains a little sulphur. It attacks the lungs and breathing tubes, causing bronchitis and other diseases. It also dissolves in rain, making it acidic. The acid rain damages trees, plants, buildings, and metalwork.
4 **Nitrogen oxides** form because the air in the engine gets so hot that its nitrogen and oxygen react together. Like sulphur dioxide, nitrogen oxides attack the breathing system, and cause acid rain.
5 The **smoke** is mainly tiny particles of carbon and lead. (The lead comes from **tetraethyl lead**, a compound which is added to petrol to make it burn better.) Smoke makes everything black and grimy. Besides, it attacks lungs, and the lead in it can cause brain damage, especially to children.

Tokyo suffers badly from pollution. Here a shopper stops for a breath of oxygen.

Cars are not the only problem Over half our air pollution is due to cars and lorries. Nearly all the rest comes from those factories, power stations and homes that burn **coal** and **oil**. Like petrol, these fuels contain hydrocarbons, and a little sulphur. When they burn in a limited supply of air they produce carbon monoxide, hydrocarbons, sulphur dioxide and smoke. If the temperature is high, nitrogen oxides form too.
Many factories also produce other pollutants—solvent fumes, iron dust, cement dust and so on.

Stonework worn away by acid rain.

The fight against air pollution

Air pollution costs money. For example, British farmers lose crops worth around £25 million every year, because of acid rain. These are some of the ways to cut down air pollution:

From cars Cars can be fitted with exhaust systems containing special catalysts, that make the harmful gases react together to produce safe gases. In the United States, all new cars must have these, by law.
The amount of lead in petrol can also be cut down. The government has placed a limit on the amount of lead allowed per litre of petrol. But many people think lead is so dangerous that it should be banned from petrol altogether, and in some countries lead-free petrol is already on sale.

From homes Many areas were turned into **smokeless zones** after a Government Clean Air Act in 1956. People in these areas cannot burn ordinary coal. They must use special smokeless fuels on their fires.

From factories and power stations The methods include:
Tall chimneys. These carry smoke, dust and waste gases high up into the air, where the wind blows them away. But this is not a very good solution. Can you explain why?
Scrubbers. These are tanks where the waste gets sprayed by jets of water, before it reaches the chimneys. The water washes away smoke and dust, and dissolves some of the harmful gases.
Electrostatic precipitators. In some factories, the waste is led into a chamber called an electrostatic precipitator, where an electric current is passed through it. The smoke and dust particles get charged, and cling to electrodes. In time, they fall to the bottom of the precipitator and are removed.

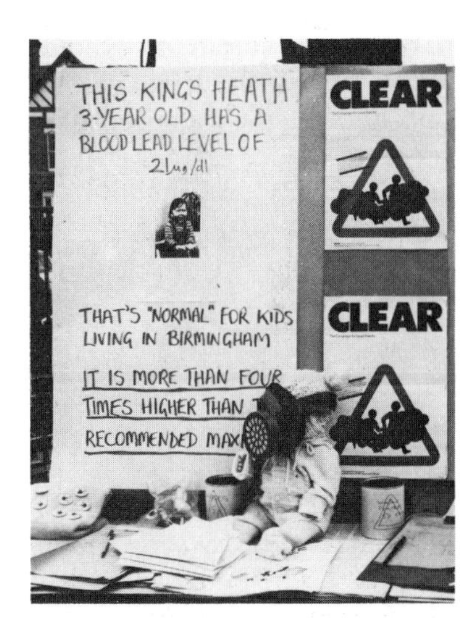

Part of a demonstration against the use of lead in petrol. Lead is especially harmful to children.

Questions

1. List the six main substances that cause air pollution.
2. Copy and complete: When hydrocarbons .... in a limited supply of air, carbon .... is formed instead of carbon ....
3. What harm is done by sulphur dioxide?
4. Why do nitrogen oxides form in a car engine?
5. Why are power stations a source of pollution?
6. Explain the purpose of:
   a. catalytic exhausts on cars
   b. tall chimneys on factories
   c. scrubbers in factories
7. How do electrostatic precipitators work?

8.4 Water and the water cycle

What is water?

Water is a compound of hydrogen and oxygen. Its formula is $\mathbf{H_2O}$. You could make it in the laboratory by burning a jet of hydrogen in air, as shown on the right. The reaction is fast and may be dangerous:

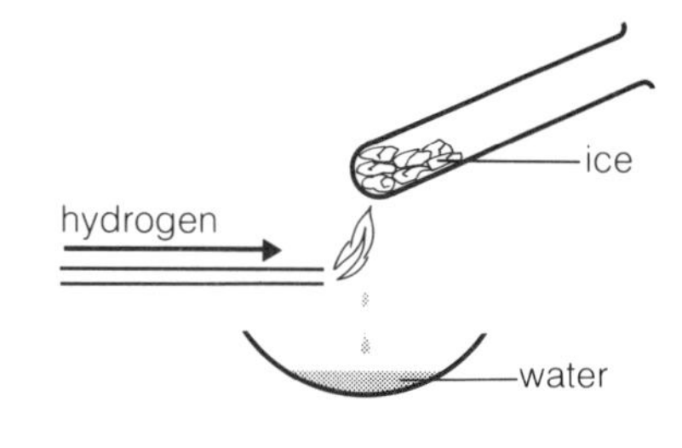

hydrogen + oxygen ⟶ water

$2H_2(g) + O_2(g) \longrightarrow 2H_2O(g)$

The water forms as a gas. It condenses to liquid on an ice-cold tube.

Tests for water

If a liquid is water, it will:

1 turn blue when you add white anhydrous copper(II) sulphate
2 turn blue cobalt(II) chloride paper pink.

If a liquid is *pure* water it will boil at 100°C and freeze at 0°C, at normal pressure.

The water cycle

There is no need to make water in the laboratory, or in factories, because there is so much of it around already. Nearly three-quarters of the earth's surface is covered by water. It rains from the sky and pours to the sea as streams and rivers. Why does the supply not run out? Because the water is continually recycled in the **water cycle**:

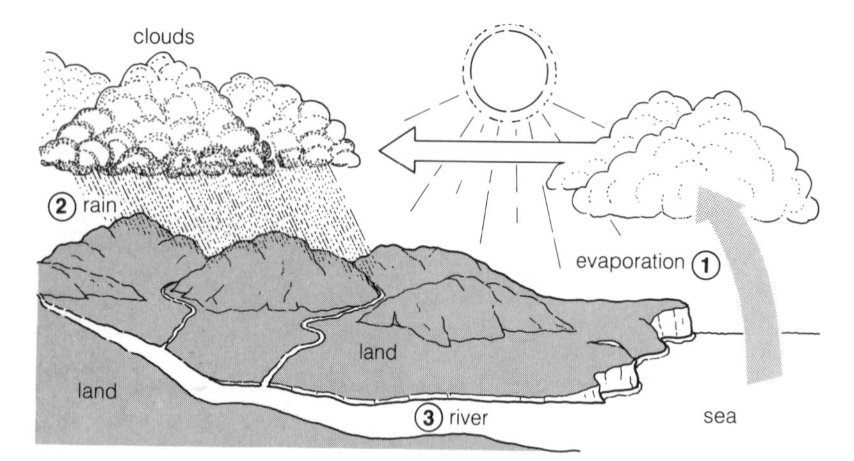

1 Heat from the sun causes water to evaporate from seas and oceans. The vapour rises, cools, and condenses to form tiny water droplets. The droplets form **clouds**.
2 The clouds get carried along by currents of air. They cool, and the droplets join to form larger drops. These fall as rain. (If the air is very cold, they fall as hail, sleet or snow.)
3 Some of the rainwater flows along the ground as streams. Some soaks through the ground and then reappears as springs. The streams and springs join up to form rivers. These flow back to the sea, and the water cycle is complete.

The water cycle in action.

Hidden water

As well as the water on the earth's surface, there is also a lot of hidden water around: living things are mostly water.
All living things are built up of tiny **cells**, and each cell holds some water. The body of a fifteen-year-old contains about 35 dm^3 of water altogether. Most of this is in the cells; the rest is in blood and the other body fluids.

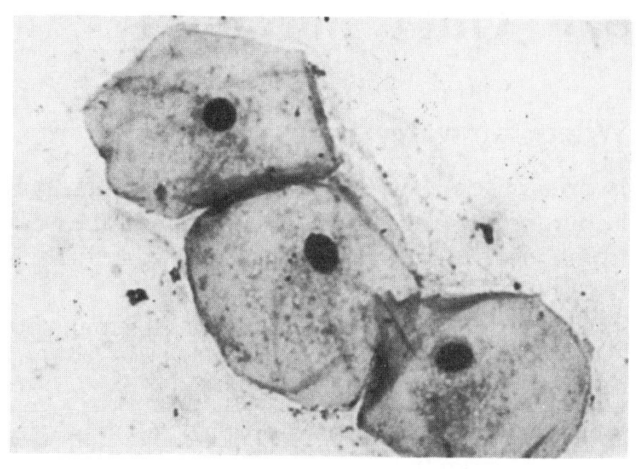

These are cells from the human body, enlarged 1000 times. Each cell contains some water.

All living things need water

Water is essential for living things.
Each day, the human body loses about 2½ dm^3 of water. Most of this leaves the body in urine and sweat; the rest is breathed out as water vapour. This loss *must* be replaced by water from food and drink. If not, the blood thickens and the flesh grows stiff. The sufferer dies after 4 or 5 days.
Other animals depend on water in the same way. So do plants.
A plant needs water for **photosynthesis**, which is a process that takes place in green leaves, in daylight:

$$\text{water} + \text{carbon dioxide} \xrightarrow[\text{chlorophyll}]{\text{light}} \text{glucose} + \text{oxygen}$$

The water is soaked up from the soil through the plant's roots, and the carbon dioxide is obtained from the air. **Chlorophyll** is the green substance in leaves. It acts as a catalyst for the reaction.
The plant then uses the glucose it produces, with nitrates and other compounds from the soil, to grow new roots, stems and leaves. The oxygen goes into the air.

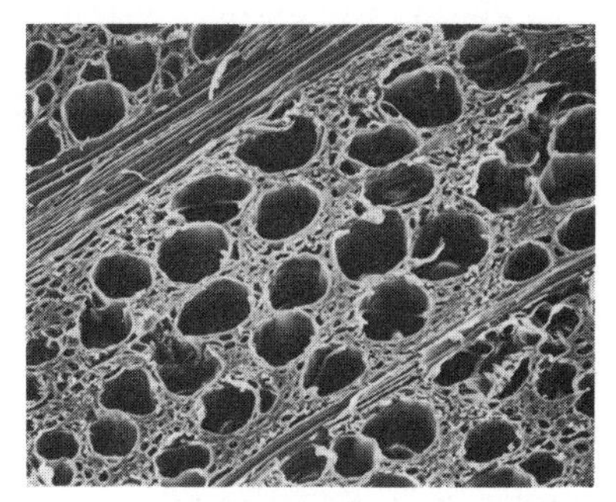

The water rises up a tree trunk through tubes like these (enlarged 1200 times).

Other uses of water

A huge amount of water is used every day in Britain—and not just for drinking. At home it is used for cooking, washing, and flushing the lavatory. Outside the home it is used for nearly everything you can think of: for making cement, beer, steel and electricity, for printing newspapers, fighting fires and filling swimming pools. Taking everything into account we use around 300 dm^3 a day, per person—that's about 3 bathfuls.

On a Summer day, a mature oak tree can use up to 250 litres of water.

Questions

1. Write the equation for the reaction that takes place when hydrogen burns in oxygen.
2. Look at the diagram of the apparatus for the reaction. What is the purpose of the ice? How would you test that the liquid was water?
3. Explain why the formula for water is H_2O.
4. Draw a diagram to show the water cycle.
5. What percentage of a human is water? Where in the body is most of this water?
6. What is photosynthesis? Write an equation to describe it.
7. What is chlorophyll? Where is it found?
8. In photosynthesis, sugars are produced. What are these sugars used for?

8.5 Our water supply

Part of a modern waterworks.

Where tap water comes from

In Britain, some tap water comes from **rivers**, some from **underground wells**, and some from **mountain reservoirs**. Water from these sources is never completely pure, especially river water. It may contain:

(i) **bacteria**—tiny living organisms, so small you would need a microscope to see them. Most bacteria are harmless, but some can cause disease.

(ii) **dissolved substances**—nitrates and sulphates from the soil, gases from the air, and sometimes calcium and magnesium compounds from rocks.

(iii) **solid substances**—particles of mud, sand, grit, twigs, dead plants, and perhaps tins and rags that people have dumped.

Before the water is safe to drink, the bacteria and solid substances must be removed. This is done at the **waterworks**.

The waterworks

This diagram shows what happens at waterworks:

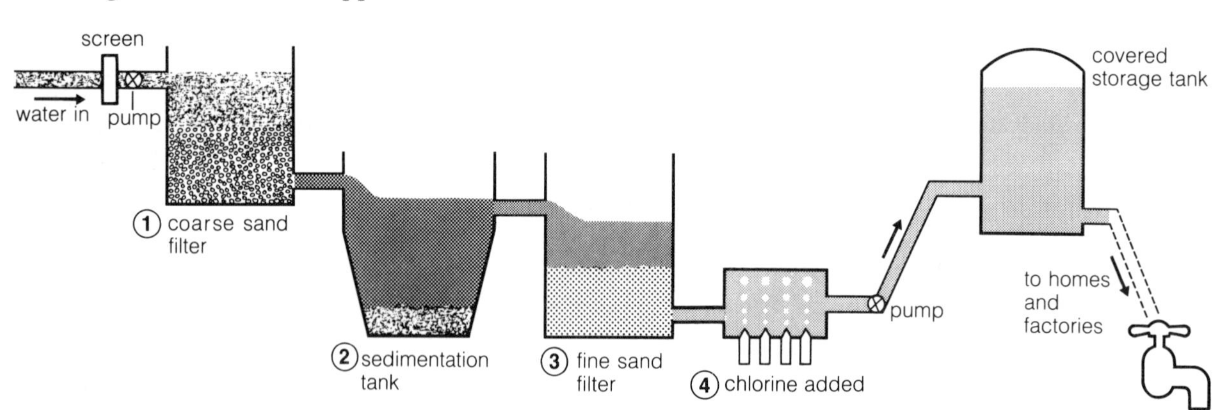

The water is pumped in through a screen, which gets rid of the larger bits of rubbish. Then it goes through these stages:

1. It is filtered through a bed of coarse sand, which traps the larger particles of solid.
2. Next it flows into a **sedimentation tank**. Here chemicals are added to it, to make the smaller particles stick together. These particles then settle to the bottom of the tank.
3. Water flows from the top of the sedimentation tank, into a filter of fine sand. This traps any remaining particles.
4. Finally a little chlorine gas is added. It dissolves, and kills any remaining bacteria. This is called **disinfecting** or **sterilising** the water.

In some places, a fluoride compound is also added to the water, to help prevent tooth decay.
The water is now fit to drink. It is pumped into high storage tanks and from there piped to homes and factories.

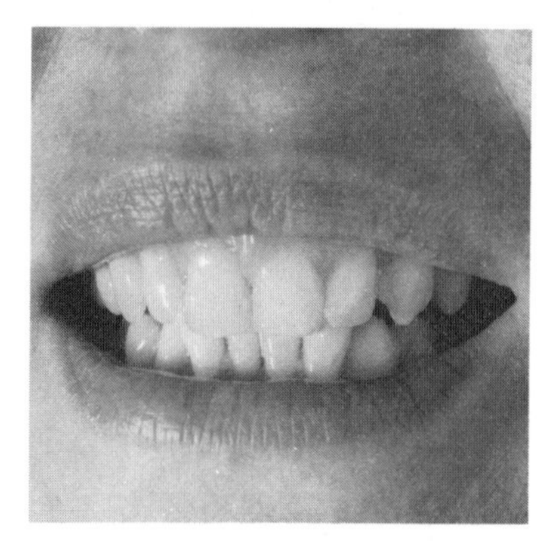
Fluoride added to the water supply helps to ensure healthy teeth.

Clean but not pure The water that flows from taps is *clean* but not quite *pure*. It still contains dissolved substances which were not removed at the waterworks. These substances do not harm health. But some of them make the water **hard**, as you will see on page 108.

Waste water and sewage plants

All sorts of things get mixed with tap water: shampoo, toothpaste, detergents, grease, body waste, food, grit, sand, and waste from factories. This mixture goes down the drain, and is called **sewage**. But it does not disappear for ever. Instead, it flows underground to a **sewage plant**, where the water in it is cleaned up and fed back to the river. Below is a diagram of the plant:

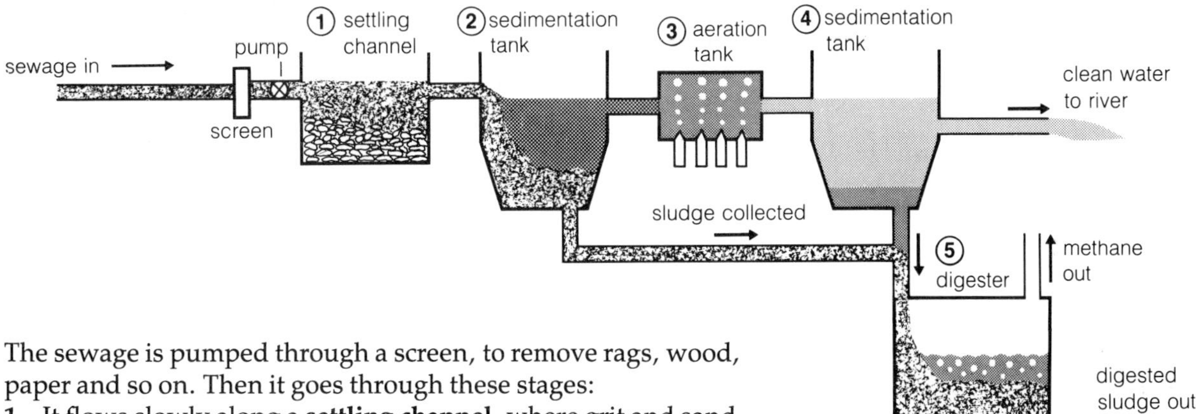

The sewage is pumped through a screen, to remove rags, wood, paper and so on. Then it goes through these stages:

1. It flows slowly along a **settling channel**, where grit and sand settle out.
2. Next it passes into a **sedimentation tank**. Here smaller pieces of waste sink slowly to the bottom. This waste is called **sludge**. It is grey and evil-smelling, and contains many harmful things.
3. The water now looks cleaner. It flows into an **aeration tank**, which contains special bacteria growing on sludge. These bacteria feed on harmful things in the water, and make them harmless. For this they need a lot of oxygen, so air is continually pumped through the sludge, from the bottom of the tank.
   Instead of aeration tanks, some plants use **percolating filters**, where the bacteria live on stones, and the water trickles over them.
4. Next comes another **sedimentation tank**, where any remaining sludge settles out. The water is now safe to put into the river.
5. All the sludge is collected into tanks called **digesters**. Here it is mixed with bacteria which destroy the harmful substances, and at the same time produce **methane gas**. Methane is a good fuel, so it is often used to make electricity for the sewage plant.
   The digested sludge is burned to ash, or dumped at sea, or sold to farmers as fertiliser.

A sludge ship carries sludge out to sea.

Questions

1. List 10 impurities that you might find in river water.
2. What happens in the sedimentation tanks at water works?
3. Tap water is clean but not pure. Explain.
4. What is: **a** sewage? **b** sludge?
5. At a sewage plant, describe what happens in: **a** aeration tanks **b** digesters
6. What happens to the digested sludge?

8.6 Soft and hard water

Different types of tap water

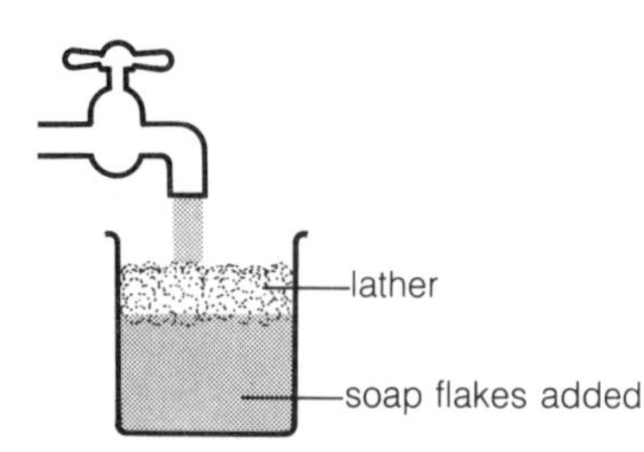

In some places, the tap water lathers easily with soap. It is called **soft water**.

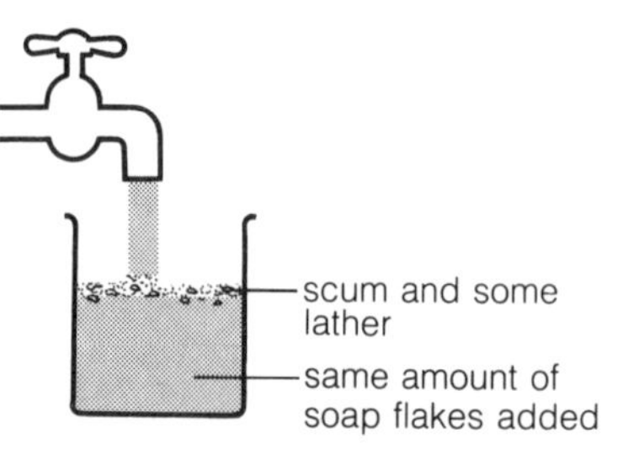

In other places, the same amount of soap gives a scum, and hardly any lather. This is a sign of **hard water**.

Washday blues. Soap caused a scum in hard water which was difficult to wash out of clothes. Modern detergents do not contain soap, so do not make scum.

Hardness in water is caused by dissolved calcium and magnesium compounds, which do not get removed at the waterworks—mainly:

calcium hydrogen carbonate
calcium sulphate
magnesium hydrogen carbonate
magnesium sulphate

The scum forms because these compounds react with soap, giving an insoluble product that floats on the water. For example:

calcium sulphate + sodium stearate (soap) ⟶ calcium stearate (scum) + sodium sulphate

A proper lather cannot form until the soap has reacted with all the dissolved calcium and magnesium compounds in the water. That means hardness in water wastes soap.

Where hardness comes from

Calcium hydrogen carbonate is the most common cause of hard water. It forms when rain falls on rocks of **limestone** and **chalk**. Limestone and chalk are mostly **calcium carbonate**, which is not soluble in water. However, rain is not pure water. As it falls through the air, it dissolves carbon dioxide, forming a weak acidic solution. This solution then attacks the calcium carbonate, forming calcium hydrogen carbonate which *is* soluble. The reaction is:

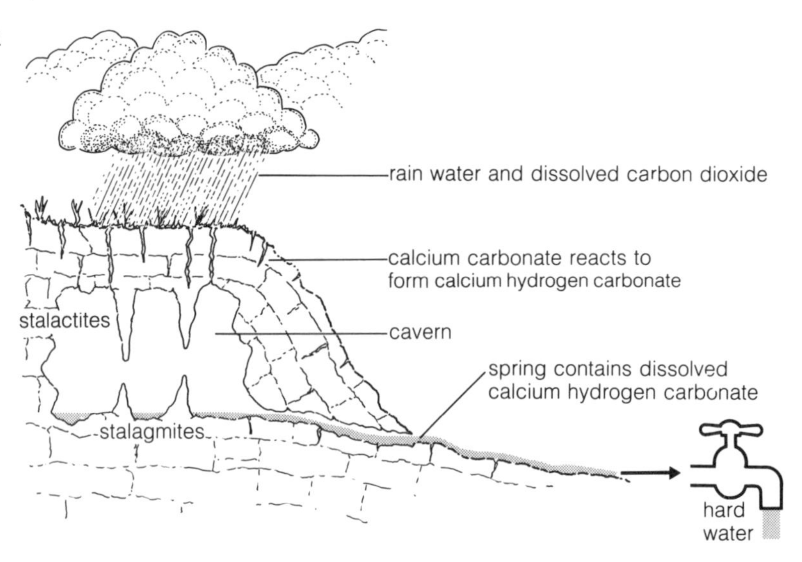

water + carbon dioxide + calcium carbonate ⟶ calcium hydrogen carbonate

$$H_2O(l) + CO_2(g) + CaCO_3(s) \longrightarrow Ca(HCO_3)_2(aq)$$

The other compounds that cause hardness are dissolved by rain from rocks such as **dolomite** and **gypsum**, and from the **soil**. Dolomite has the formula $CaCO_3.MgCO_3$. Gypsum has the formula $CaSO_4.2H_2O$.

Stalactites and stalagmites

Stalactites and stalagmites.

The diagram on the last page shows an underground cavern containing **stalactites** and **stalagmites**. Caverns like this are often found in limestone areas.

A stalactite is made of calcium carbonate, and hangs from the cavern roof. It begins when drops of water containing calcium hydrogen carbonate collect on the roof. Some of the water evaporates, losing carbon dioxide, and solid calcium carbonate is left behind:

calcium hydrogen carbonate ⟶ water + carbon dioxide + calcium carbonate

$$Ca(HCO_3)_2(aq) \longrightarrow H_2O(g) + CO_2(g) + CaCO_3(s)$$

This happens drop by drop, on the same part of the roof. Slowly, over thousands of years, a spike of calcium carbonate grows down. A stalagmite is formed in the same way, except that it grows up from the cavern floor. Stalactites and stalagmites are often found opposite each other. Can you explain why?

Temporary and permanent hardness

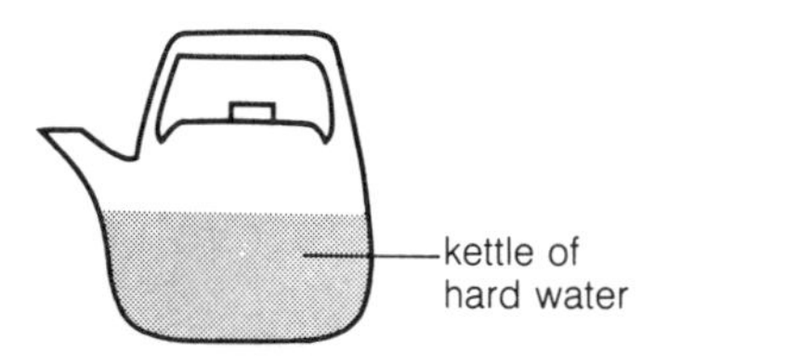

This water contains calcium hydrogen carbonate, so it is **hard**. When it is boiled, the calcium hydrogen carbonate . . .

scale forms on sides

. . . decomposes to calcium carbonate. This is insoluble. It collects on the sides of the kettle as a **fur** or **scale**.

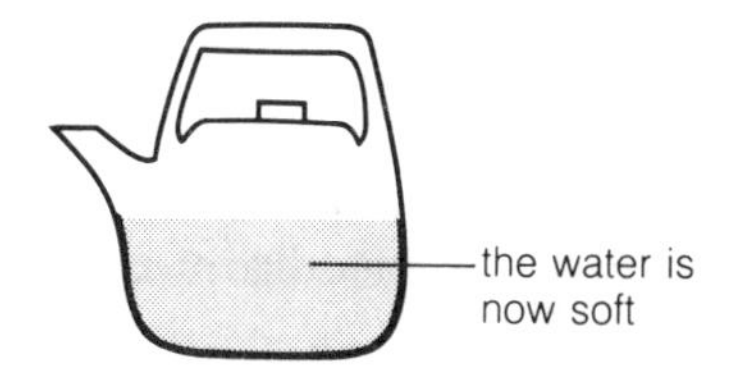

The result is that all the calcium hydrogen carbonate gets removed from the water. The water is now **soft**.

Hardness caused by calcium hydrogen carbonate is therefore called **temporary hardness**, because it can be removed by boiling the water. The reaction in the kettle is like the one for stalactites:

$$Ca(HCO_3)_2(aq) \longrightarrow H_2O(l) + CO_2(g) + CaCO_3(s)$$

Hardness caused by other calcium and magnesium compounds is called **permanent hardness**, because boiling does not affect it. But it *can* be removed in other ways, as you will see on the next page.

Questions

1 How can you tell whether tap water is hard or soft?
2 Explain why soap produces scum in hard water.
3 What are limestone and chalk made of?
4 Calcium carbonate is insoluble in pure water. Why does it dissolve in rain water?
5 How do you think a *stalagmite* gets formed? Write an equation for the reaction.
6 Name a substance that causes:
  a temporary hardness
  b permanent hardness

8.7 Making hard water soft

Hard water: good or bad?

Hard water does have some good points:

1 It has a pleasant taste, due to the dissolved compounds.
2 The calcium compounds in it are good for bones and teeth.
3 Doctors think it helps to prevent heart disease.

But it also has disadvantages:

1 Temporary hardness leaves a fur or scale in kettles, hot water pipes, boilers and radiators. This makes them less efficient, and can also cause blockage, so it has to be removed from time to time.
2 Hard water needs more soap than soft water, and leaves a messy scum that is difficult to wash out.

A pipe almost blocked by scale.

Soapless detergents The problem of scum is less serious now than it was thirty years ago. Now, soapless detergents are used for washing most things. Soapless detergents do *not* form a scum with calcium and magnesium compounds. Washing-up liquids, washing powders and most shampoos are soapless detergents.

Ways to soften hard water

Because of the problems it causes, hard water is often **softened** before use in factories, laundries and homes. Softening it means removing the calcium and magnesium compounds. Below are some ways to do that:

Boiling This removes temporary hardness, as you saw on page 109, by causing the calcium carbonate to precipitate.

Distilling Distilling the water gets rid of both sorts of hardness. The water is heated and boils off as steam, leaving all the dissolved compounds behind. The steam is then condensed to pure water. Distilled water is the softest you can get. It is also the most expensive, because of the fuel that the distillation needs.

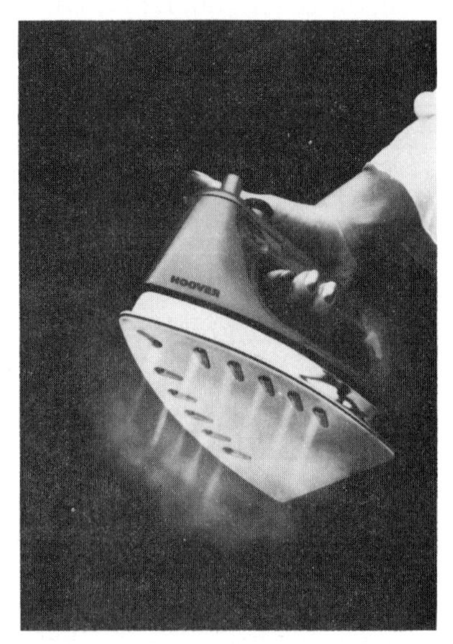

It is best to use distilled water in steam irons. Can you explain why?

Sodium carbonate (washing soda) When sodium carbonate is added to water, it removes both sorts of hardness, by precipitating calcium carbonate. For example, its reaction with calcium sulphate is:

sodium carbonate + calcium sulphate ⟶ calcium carbonate + sodium sulphate

$$Na_2CO_3(aq) + CaSO_4(aq) \longrightarrow CaCO_3(s) + Na_2SO_4(aq)$$

Bath salts usually contain sodium carbonate, to soften the bath water. The sodium sulphate that forms does not affect soap.

Ion exchangers Ion exchangers remove both sorts of hardness by removing all the calcium and magnesium ions.
An ion exchanger is a container full of small beads. These are made of special plastic called **ion exchange resin**, which has ions weakly attached to it—they could be sodium ions as shown on the right.

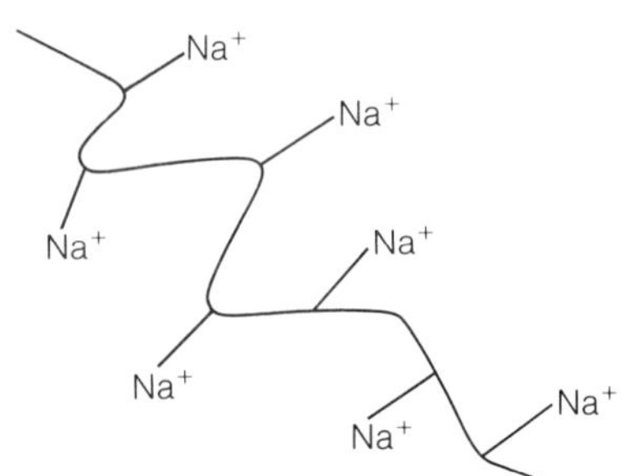

This is only part of a long molecule of resin (Each bead contains many molecules)

When hard water flows through the ion exchanger, the calcium and magnesium ions in it change places with the sodium ions, and attach themselves to the resin. The sodium ions dissolve in the water:

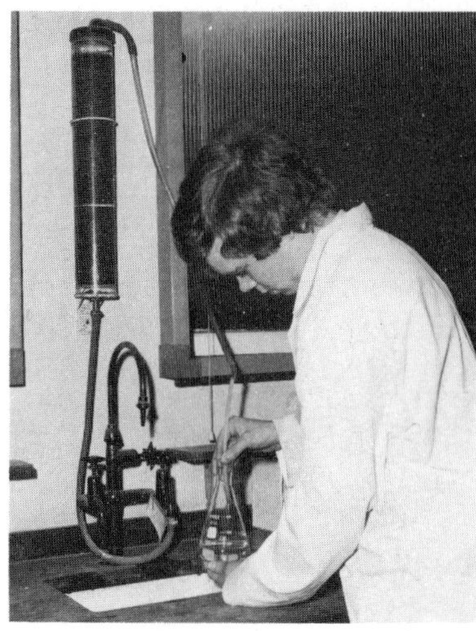
An ion exchanger in use.

After a time, all the sodium ions have gone, so no more hardness can be removed. The resin has to be **regenerated**. To do this, a concentrated solution of sodium chloride is poured in. The sodium ions push the calcium and magnesium ions off the resin, and the ion exchanger is ready for use again.

Comparing hardness

You can compare the hardness of different samples of water by finding how much soap each needs to give a good lather. Here are the steps:

1. Some water (say 20 cm^3) is measured into a conical flask.
2. A little soap solution is added from a burette, and the flask is carefully shaken.
3. More soap solution is added, a little at a time, until a lather lasting at least half a minute is obtained.
4. The volume of soap solution used is noted.

In the experiment in the photograph, water from the same tap was used for A, B, and C. For A it was left untreated, for B it was boiled first, and for C it was passed through an ion exchanger. D is distilled water. Here are the results:

Sample of water	Volume of soap solution used (cm^3)
A (untreated)	10
B (boiled)	7
C (ion exchanger)	0.5
D (distilled)	0.5

The experiment being carried out.

Do you agree that the tap water contains both types of hardness? Why do you think so?

Questions

1. Write down two advantages of hard water.
2. Explain how boiling removes temporary hardness. What problems can this method cause?
3. Why is sodium carbonate used in bath salts?
4. What is an ion exchanger? How does it work?
5. Explain how the experiment above shows that:
   a the tapwater contains both types of hardness
   b most of the hardness in it is permanent
   c the ion exchanger removes hardness just as effectively as distillation does.

8.8 Water pollution

Every day, hundreds of harmful substances find their way into our rivers. The result is **water pollution**. This diagram shows some of the things that cause water pollution:

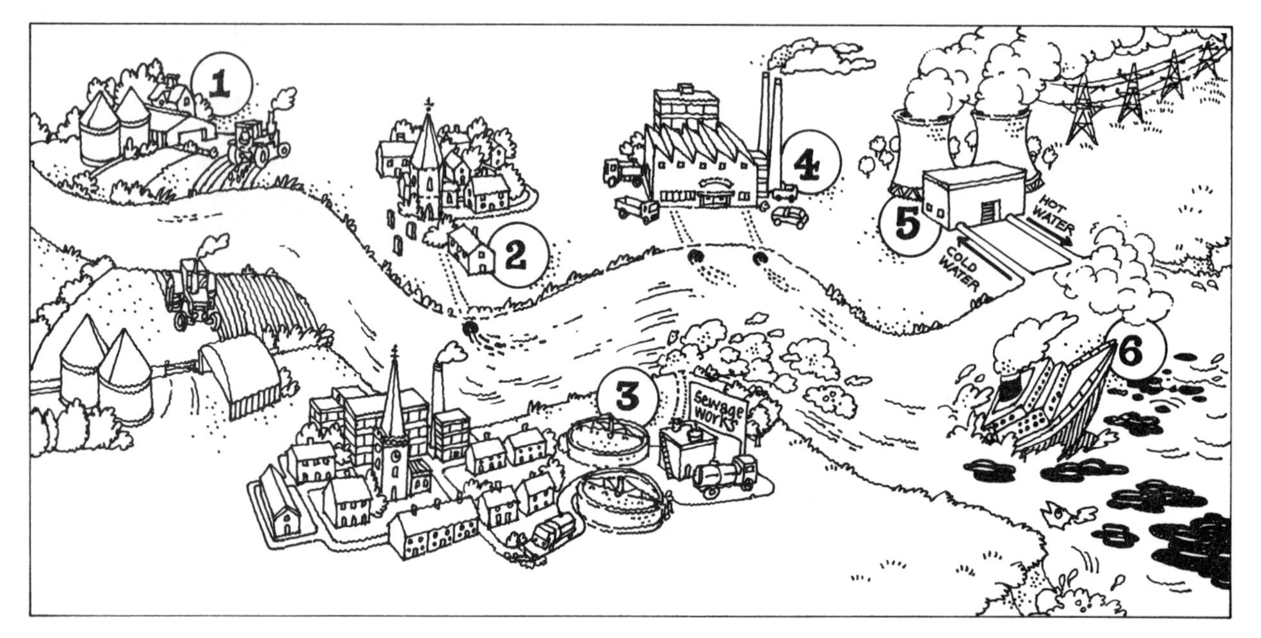

1 **Fertilisers.** On land, fertilisers are not harmful—they are put on the soil to help crops grow. However, if they drain into rivers, they also help tiny green water plants called **algae** to grow. These can grow so well that they cover the water surface. They cut off light from the plants below, and those plants then die. That is when the real problem begins. Bacteria feed on the dead plants, causing them to rot or **decompose**. During this process the bacteria use up the dissolved oxygen in the water. Soon, there is not enough oxygen for the fish and other living things, so they die from oxygen starvation.
Many fertilisers are **nitrates**. These can also harm humans, if enough of them get through the waterworks and into drinking water. They cause disease in young babies, and some scientists link them with stomach cancer.

2 **Untreated sewage.** Many villages have no sewage plants, so their sewage goes straight into nearby rivers. Farm manure often drains into rivers too. This untreated sewage contains bacteria that can cause disease in humans and other animals, so the water becomes dangerous to swim in or to drink from. Besides, other bacteria feed on the sewage, using up dissolved oxygen at the same time. The rest of the river life dies from oxygen starvation.

3 **Detergents.** Many different detergents go down the drain in our waste water. Some of them cannot be decomposed by bacteria, so they pass through the sewage works unchanged, and end up in the river. There they cause a lather, which spoils the look of the river. Even worse, they kill off the fish and other river life.

A 'dead' river. Pollution has killed most of the living things in it.

This fish has a cancer caused by pollution.

In Tokyo, people gather to protest about a company that dumps waste mercury compounds at sea.

Off Land's End, a British Navy ship sprays an oil slick with chemicals to break it up.

This problem is less serious since scientists developed **biodegradable detergents**. These *can* be broken down and made harmless by the bacteria in sewage plants and rivers.

4 **Waste from factories.** Waste liquid from some factories gets dumped straight into rivers. This waste may contain dangerous substances such as mercury, or its compounds. These do not kill fish, but get stored up in their bodies and eventually poison humans. Hair drops out and teeth rot. The victim becomes irritable and twitchy, and will die if not treated. Lead is another poisonous metal. It could leak out from factories that make car batteries, for example. But even more dangerous is the radioactive waste that may leak from nuclear waste processing plants, or nuclear power stations.

5 **Heat.** Factories and power stations use river water for cooling hot tanks and pipes. If the hot water drains back into the river it kills off fish. (Most fish cannot live above 30 °C.)
In some places, this problem is solved by spraying the hot water down cooling towers. When it reaches the bottom it is cold, and can be used for cooling again.

6 **Oil.** At sea, oil pollution is a hazard. If tankers run aground, or collide with other ships, oil can get spilled into the sea. There it kills sea birds, by poisoning them and by clogging up their wings so that they cannot fly to look for food. If it gets washed ashore the oil destroys all forms of life on rocks, in mud and on beaches. The bacteria in sea water *can* destroy oil. However, that is a very slow process. It is much faster to spray the oil spills with chemicals or to soak them up in special blankets.

A dying guillemot, its feathers covered in oil.

Questions

1 What are algae?
2 Explain why fertilisers in rivers can cause the death of fish.
3 Give two reasons why untreated sewage in rivers is harmful.
4 What is a biodegradable detergent?
5 Name three metals whose compounds can poison humans.
6 Spilled oil is considered a pollutant. Why is this? Describe some ways of clearing oil up, after a spill.

Questions on Chapter 8

1 Copy and complete the following paragraph:
Air is a .... of different gases. 99% of it consists of the two elements .... and .... One of these, ..., is needed for respiration, which is the process by which living things obtain the .... they need. The two elements above can be .... from liquid air by ...., because they have different .... Much of the .... obtained is used to make nitric acid and fertilisers. Some of the remaining 1% of air consists of two compounds, ... and .... One of these is important because it is taken in by plants, in the presence of ...., to form .... The rest of the air is made up of elements called the .... These are all members of Group ... of the Periodic Table.

2 Oxygen and nitrogen, the two main gases in air, are both slightly soluble in water. Using the apparatus below, a sample of water was boiled until 100 cm^3 of dissolved air had been collected.

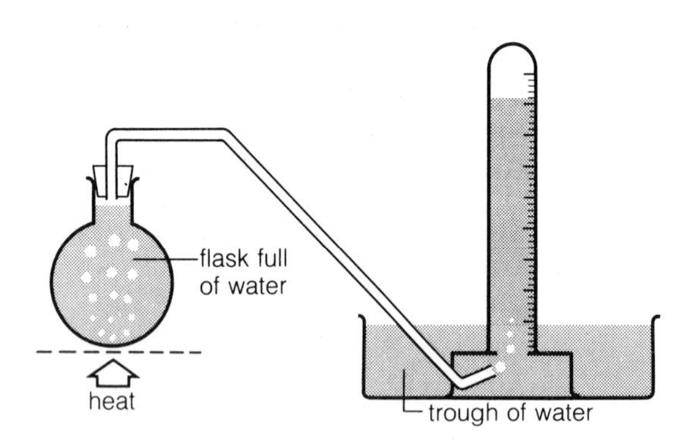

This air was then passed over heated copper. Its final volume was 67 cm^3.

a Is air more soluble or less soluble in water, as the temperature rises? How can you tell?
b The copper reacts with the oxygen in the dissolved air. Write an equation for the reaction.
c Draw a diagram of the apparatus used for passing air over heated copper.
d What volume of oxygen was present in 100 cm^3 of dissolved air?
e Calculate the approximate percentages of oxygen and nitrogen in dissolved air.
f What are the percentages of these gases in atmospheric air?
g Explain why the answers are different, for parts **e** and **f**.
h Which gas is more soluble in water, nitrogen or oxygen?
i What is the biological importance of air dissolved in water?

3 Air is a *mixture* of different gases.
a Which gas makes up approximately 78% of the air?
b Only one gas in the mixture will allow things to burn in it. Which gas is this?
c How are the gases in the mixture separated from each other, in industry?
d Which noble gas is present in the greatest amount in air?
e Which gas containing sulphur is a major cause of air pollution?
f Name two other gases which contribute to air pollution.
g Name one substance which is not a gas but which also pollutes the air.

4 Match each of the following ten equations with one of the descriptions below them.

i $C_7H_{16}(l) + 11O_2(g) \longrightarrow 7CO_2(g) + 8H_2O(l)$
heptane
(in petrol)

ii $S(s) + O_2(g) \longrightarrow SO_2(g)$

iii $C_6H_{12}O_6(s) + 6O_2(g) \longrightarrow 6CO_2(g) + 6H_2O(l)$
glucose

iv $2H_2(g) + O_2(g) \longrightarrow 2H_2O(g)$

v $6H_2O(g) + 6CO_2(g) \xrightarrow[\text{chlorophyll}]{\text{light}} C_6H_{12}O_6(s) + 6O_2(g)$

vi $2Cu(s) + O_2(g) \longrightarrow 2CuO(s)$

vii $CaCO_3(s) + H_2O(l) + CO_2(g) \longrightarrow Ca(HCO_3)_2(aq)$

viii $Ca(HCO_3)_2(aq) \xrightarrow{\text{heat}} CaCO_3(s) + CO_2(g) + H_2O(l)$

ix $CaSO_4(aq) + Na_2CO_3(aq) \longrightarrow CaCO_3(s) + Na_2SO_4(aq)$

x $2CH_4(s) + 3O_2(g) \longrightarrow 2CO(g) + 4H_2O(g)$

The descriptions are:
a removal of temporary hardness by boiling
b synthesis of water
c photosynthesis
d incomplete combustion of hydrocarbon
e oxidation of copper
f the dissolving of limestone by rain
g burning of sulphur
h respiration
i removal of hardness using bath salts
j burning of petrol

5 Write a word equation for each of the reactions in question 4.

Test	Sample A	B	C
1 Shaken with soap solution	poor lather	good lather	poor lather
2 Boiled first and then shaken with soap solution	good lather	good lather	poor lather
3 Some bath salts added, shaken with soap solution after filtering	good lather	good lather	good lather

6 Some samples of water were tested in the laboratory. The results are shown in the table above.

a Only one of the samples could have been pure water. Which one? Explain your answer.

b The other two samples were both from hard water areas.

i Which contained temporary hardness?

ii Which contained permanent hardness?

Explain how you were able to tell them apart.

c Name one substance that could cause the hardness in sample C.

7 A distilled water

B tap water

C calcium hydroxide solution

D sodium chloride solution

E unknown solution

Ten drops of soap solution were added to $10\,cm^3$ of each of the above liquids. After shaking for ½ minute, these heights of lather were obtained:

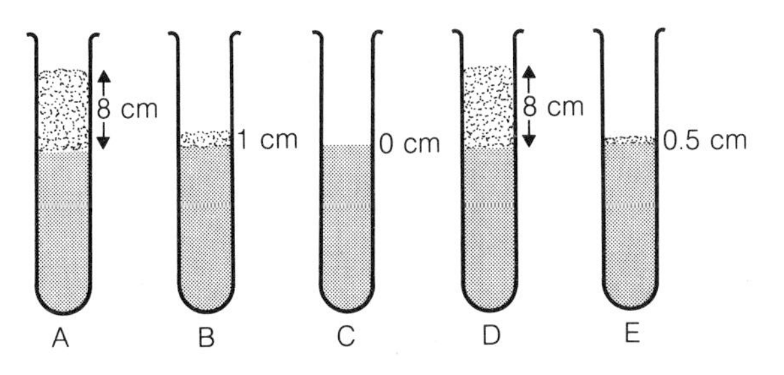

a Does the tap water come from a hard water or soft water area? Explain your answer?

b Does sodium chloride form a precipitate with the soap solution? Explain your answer.

c Does calcium hydroxide cause hardness? How can you tell from the experiment?

d What can you say about liquid E, the unknown solution?

e Another sample of solution E was boiled for 5 minutes, and then the test repeated. This time a lather of height 8 cm was obtained. Explain this result.

f What result would you expect for a further sample F, which is water from an ion exchanger?

g What results would be obtained, using 10 drops of a soapless detergent in place of the soap solution?

8 **a** Name the two elements from which water is made.

b Name a solid which is very soluble in water.

c Name a metal which reacts quickly with cold water

d Describe an experiment you could carry out in the laboratory to find out whether tap water contains any dissolved solids.

9 $25\,cm^3$ samples of water from four different areas were tested with soap solution, to see how much soap solution was needed for a lather that lasted at least half a minute. The experiment was repeated a second time using samples that had been boiled, and then a third time using samples that had been passed through an ion exchanger. The results are shown in the table below.

Sample	Volume of soap solution/cm^3 Untreated	Boiled	Passed through ion exchanger
A	14	1.9	1.9
B	16	16	1.8
C	25	20	1.9
D	1.8	1.8	1.8

a Which of the samples is the hardest water? Why do you think so?

b Which sample behaves like distilled water? Explain your choice.

c Decide whether the hardness is temporary, permanent or both, in:

i sample A

ii sample B

iii sample C

d Name a chemical which could be responsible for the hardness in:

i sample A

ii sample B

e Write an equation for the reaction in which the temporary hardness is removed, in samples A and C.

f Explain how an ion exchanger works, in removing hardness from water.

9.1 Acids and alkalis

Acids

One important group of chemicals is called **acids**:

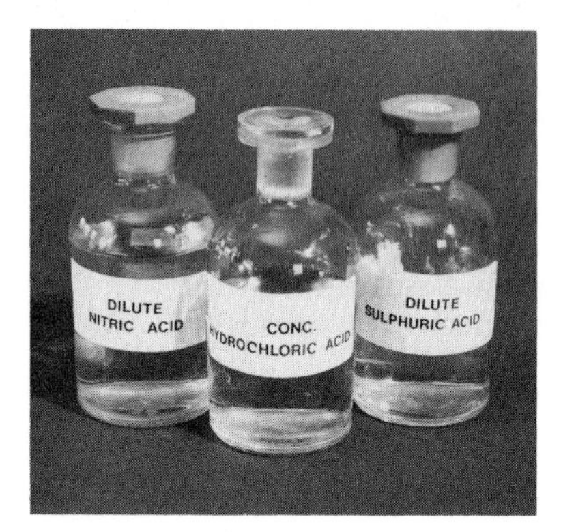

Here are some acids. They are all liquids. In fact they are **solutions** of pure compounds in water.

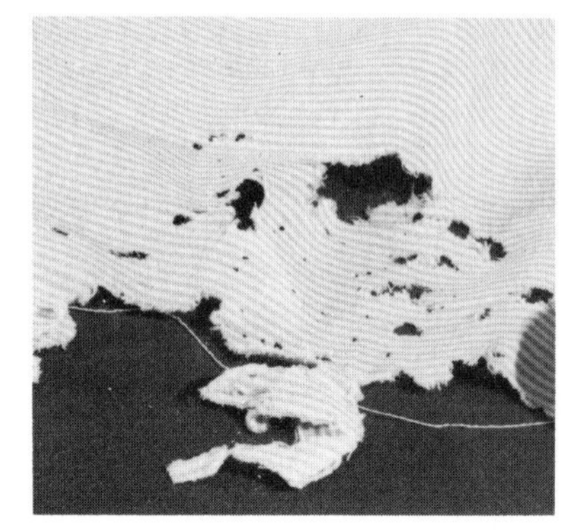

They must be handled carefully—especially the concentrated ones—for they are **corrosive**. They can eat away metals and skin—and cloth, as the holes in this shirt show.

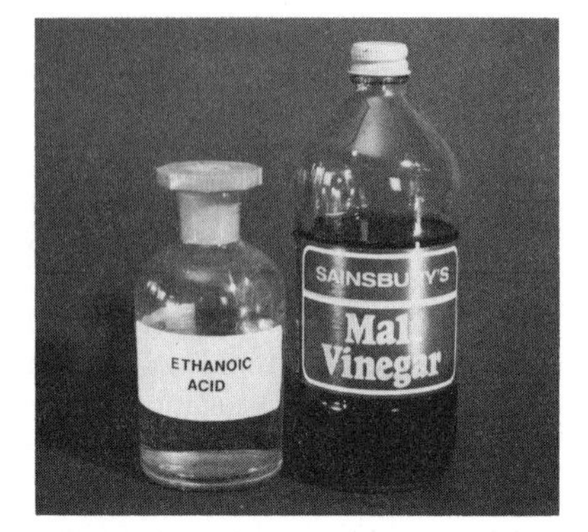

But there are some acids which are not so corrosive, even when concentrated. They are called **weak** acids. Ethanoic acid is one example. Vinegar contains some ethanoic acid.

You can tell if something is an acid, by its effect on **litmus**.
Litmus is a purple dye. It can be used as a solution, or on paper:

Litmus solution is purple. Litmus paper for testing acids is blue.

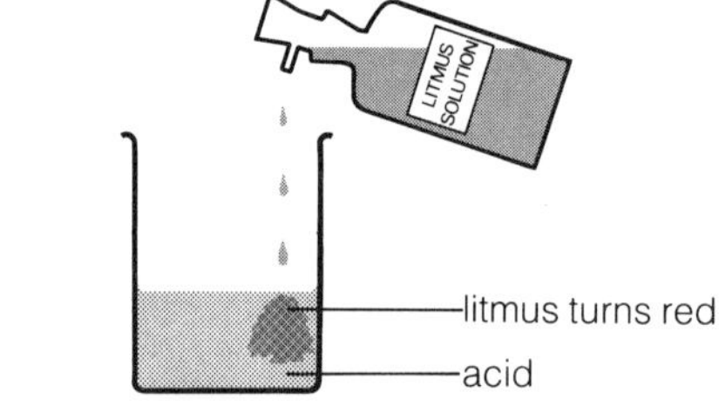

Acids turns litmus solution red.

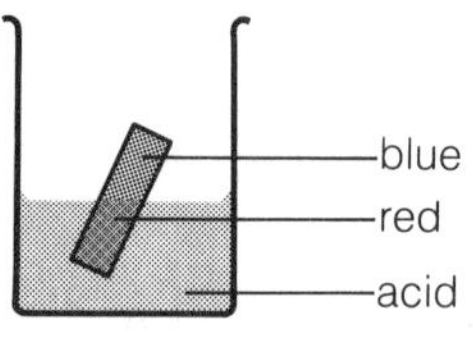

They turn blue litmus paper red too.

Some common acids The main ones are:

hydrochloric acid	$HCl\,(aq)$
sulphuric acid	$H_2SO_4\,(aq)$
nitric acid	$HNO_3\,(aq)$
ethanoic acid	$CH_3COOH\,(aq)$

But there are plenty of others. For example, lemon juice contains **citric acid**, ant and nettle stings contain **methanoic acid** and tea contains **tannic acid**.

Alkalis

There is another group of chemicals that also affect litmus, but in a different way to acids. They are the **alkalis**.
Alkalis turn litmus solution blue, and red litmus paper blue.
Like acids, they must be handled carefully, because they can burn skin.

Some common alkalis Most pure alkalis are solids. But they are usually used in laboratory as aqueous solutions. The main ones are:

sodium hydroxide	$NaOH\,(aq)$
potassium hydroxide	$KOH\,(aq)$
calcium hydroxide	$Ca(OH)_2\,(aq)$
ammonia	$NH_3\,(aq)$

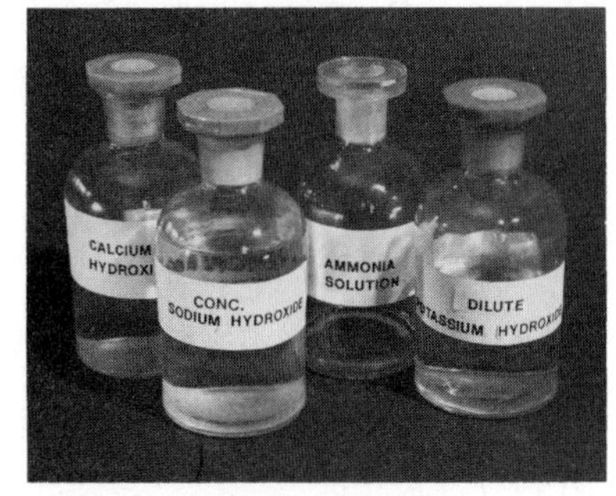

Common laboratory alkalis.

Neutral substances

Many substances do not affect the colour of litmus solution, so they are not acids or alkalis. They are **neutral**. Sodium chloride and sugar solutions are both neutral.

The pH scale

You saw on page 116 that some acids are weaker than others. It is the same with alkalis.
The strength of an acid or an alkali is shown using a scale of numbers called the **pH scale**. The numbers go from 0 to 14:

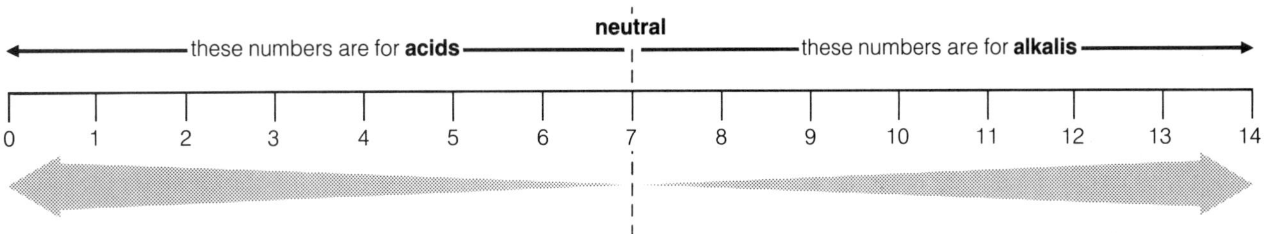

On this scale:
an acidic solution has a pH number less than 7
an alkaline solution has a pH number greater than 7
a neutral solution has a pH number of exactly 7.

You can find the pH of any solution by using **universal indicator**. Universal indicator is a mixture of dyes. Like litmus, it can be used as a solution, or as universal indicator paper. It goes a different colour at different pH values, as shown in this diagram:

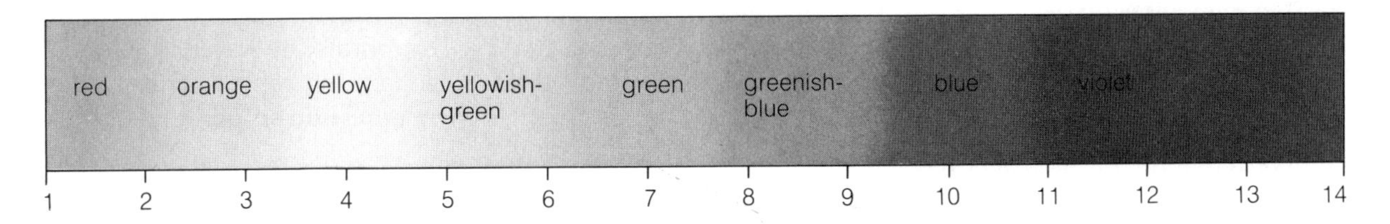

Questions

1. What does *corrosive* mean?
2. How would you test a substance, to see if it is an acid?
3. Write down the formula for:
   sulphuric acid nitric acid
   calcium hydroxide ammonia solution
4. What effect do alkalis have on litmus solution?
5. Say whether a solution is acidic, alkaline or neutral, if its pH number is:
   9 4 7 1 10 3
6. What colour would universal indicator show, in an aqueous solution of sugar? Why?

9.2 A closer look at acids

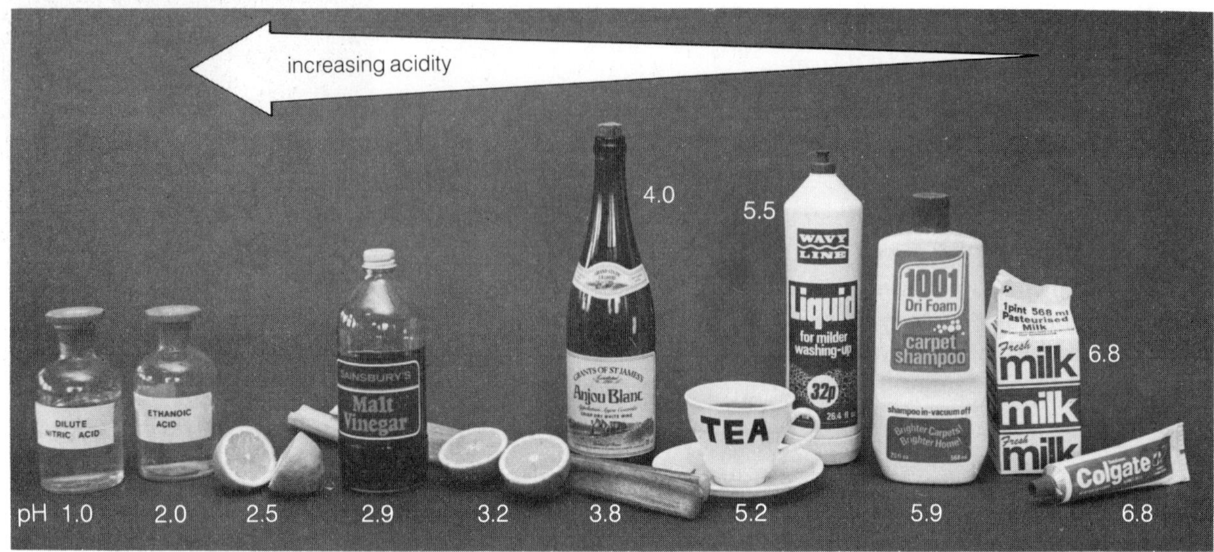

The liquids above are all **acidic**, as their pH numbers show.
Which are the most acidic? How can you tell?

The properties of acids

1. Acids have a sour taste. Think of the taste of vinegar.
   But *never* taste the laboratory acids, because they could burn you.
2. They turn litmus red.
3. They have pH numbers less than 7.
4. They usually react with **metals**, forming hydrogen and a **salt**:

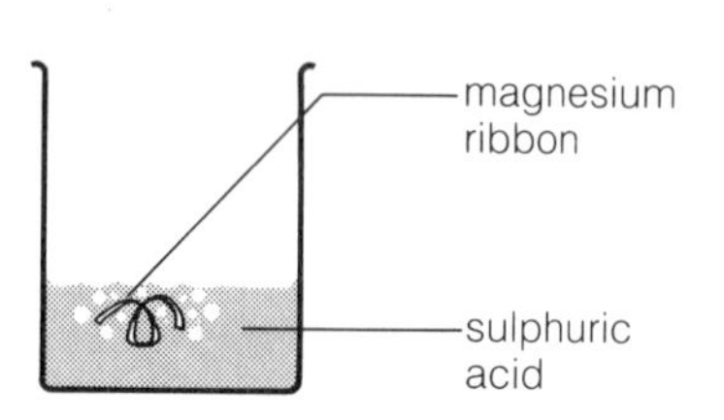

When magnesium is dropped into dilute sulphuric acid, hydrogen quickly bubbles off.

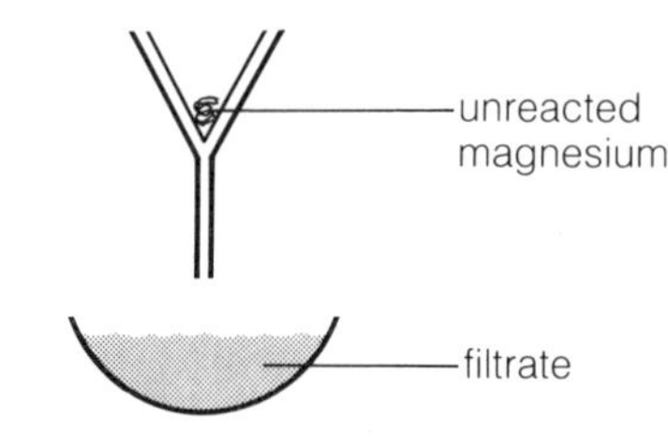

The bubbles stop when the reaction is over. The unreacted magnesium is then removed by filtering.

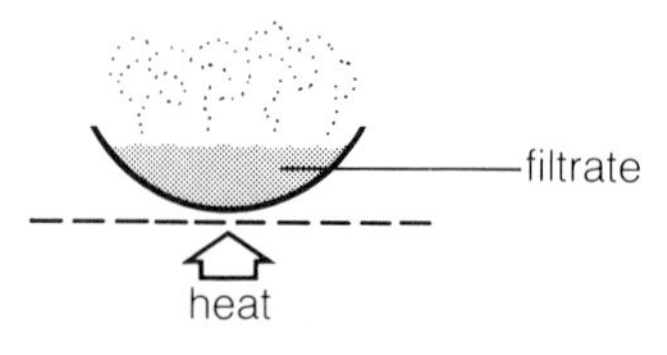

The filtrate is heated, to evaporate the water. A white solid is left behind. This solid is **magnesium sulphate**.

The equation for the reaction is:

magnesium + sulphuric acid ⟶ magnesium sulphate + hydrogen

$$Mg(s) + H_2SO_4(aq) \longrightarrow MgSO_4(aq) + H_2(g)$$

The magnesium has driven the hydrogen out of the acid, and taken its place. Magnesium sulphate is called a **salt**.
When a metal takes the place of hydrogen in an acid, the compound that forms is called a salt.

The salts of sulphuric acid are always called **sulphates**.
The salts of hydrochloric acid are called **chlorides**.
The salts of nitric acid are called **nitrates**.

5 Acids react with **carbonates**, forming a salt, water and carbon dioxide. Hydrochloric acid reacts with calcium carbonate like this:

calcium carbonate + hydrochloric acid ⟶ calcium chloride + water + carbon dioxide

$$CaCO_3(s) + 2HCl(aq) \longrightarrow CaCl_2(aq) + H_2O(l) + CO_2(g)$$

6 They react with **alkalis**, forming a salt and water. For example:

sodium hydroxide + nitric acid ⟶ sodium nitrate + water

$$NaOH(aq) + HNO_3(aq) \longrightarrow NaNO_3(aq) + H_2O(l)$$

7 They also react with **metal oxides**, forming salt and water:

zinc oxide + hydrochloric acid ⟶ zinc chloride + water

$$ZnO(s) + 2HCl(aq) \longrightarrow ZnCl_2(aq) + H_2O(l)$$

When an acid reacts with a carbonate, carbon dioxide fizzes off.

What causes acidity?

Acids have a lot in common, as you have just seen.
There must be something *in* them all, that makes them act alike.
That 'something' is **hydrogen ions.**
Acids contain hydrogen ions.
Acids are solutions of pure compounds in water. The pure compounds are molecular. But in water, the molecules break up to form ions. They *always* give hydrogen ions. For example, in hydrochloric acid:

$$HCl(aq) \longrightarrow H^+(aq) + Cl^-(aq)$$

The more H^+ ions there are in a solution, the more acidic it will be. In other words the more H^+ ions there are, the lower the pH number.

Dilute ethanoic acid reacts very slowly with magnesium, because it is a weak acid.

Strong and weak acids

Dilute hydrochloric acid reacts quickly with magnesium ribbon. The reaction could be over in minutes. But when ethanoic acid of the same concentration is used instead, the reaction is much slower. It could take all day.

The hydrochloric acid reacts faster *because it contains more hydrogen ions.*
In hydrochloric acid, *nearly all* the acid molecules break up to form ions. It is called a **strong** acid. But in ethanoic acid, only some of the acid molecules form ions, so it is called a **weak** acid.
In a strong acid, nearly all the acid molecules form ions. In a weak acid, only some of the acid molecules form ions.
So a strong acid *always* has a lower pH number than a weak acid of the same concentration.
Some strong and weak acids are shown in the box on the right.

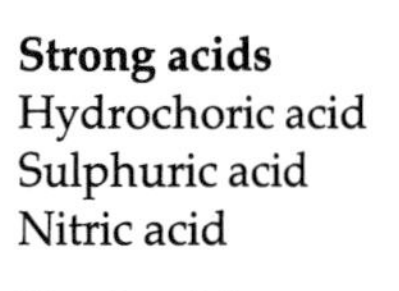

Strong acids
Hydrochoric acid
Sulphuric acid
Nitric acid

Weak acids
Ethanoic acid
Citric acid
Carbonic acid

Questions

1 Name the acid found in:
lemon juice wine tea vinegar

2 Write a word equation for the reaction of dilute sulphuric acid with: **a** zinc
b magnesium oxide **c** sodium carbonate

3 Bath salts contain sodium carbonate. Explain why they fizz when you put vinegar on them.

4 What causes the acidity, in acids?

5 Why is ethanoic acid called a weak acid?

6 Name two other weak acids, and two strong ones.

9.3 A closer look at alkalis

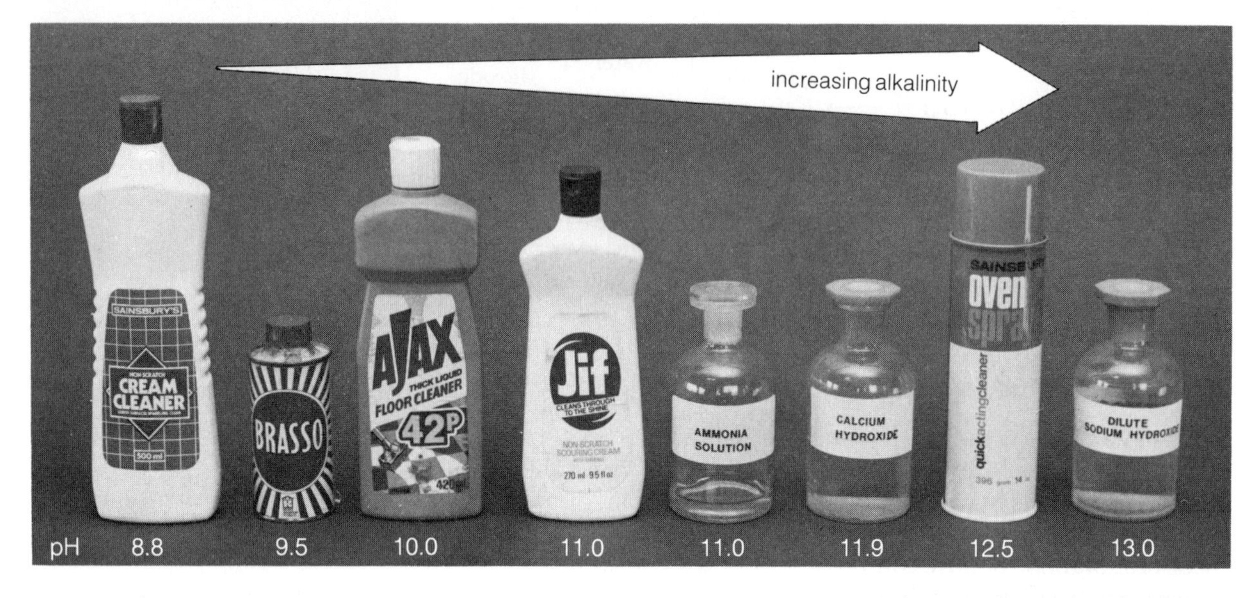

All the substances above are alkaline—you can tell by their pH numbers. Notice that several of them can be found in the kitchen! Many kitchen cleaners are alkaline because they contain ammonia or sodium hydroxide, which attack grease.

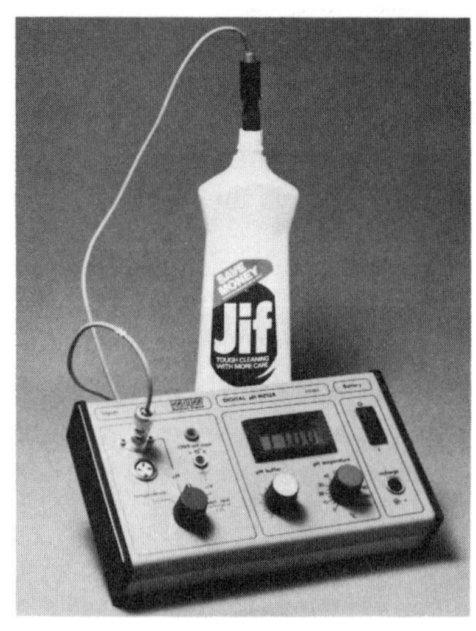

The pH number of a kitchen cleaner being measured with a pH meter.

What makes things alkaline?

You saw on page 119 that all acid solutions contain hydrogen ions, H^+. The alkalis also have something in common:
All alkaline solutions contain hydroxide ions, OH^-.
In sodium hydroxide solution, the ions are produced like this:

$NaOH\,(aq) \longrightarrow Na^+\,(aq) + OH^-\,(aq)$

In ammonia solution, ammonia molecules react with water molecules to form ions:

$NH_3\,(aq) + H_2O\,(l) \longrightarrow NH_4^+(aq) + OH^-\,(aq)$

The more OH^- ions there are in a solution, the more alkaline it will be. In other words, the more OH^- ions there are, the higher the pH number.

Strong and weak alkalis

Like acids, alkalis can also be strong or weak.
Sodium hydroxide is a **strong** alkali, because it exists almost completely as ions, in solution. Potassium hydroxide is also a strong alkali. But ammonia solution is a **weak** alkali because only some ammonia molecules form ions in solution.

Strong alkalis
Sodium hydroxide
Potassium hydroxide
Calcium hydroxide

Weak alkali
Ammonia

Properties of alkalis

1 Alkalis feel soapy to the touch. But it is dangerous to touch the laboratory alkalis (and some kitchen cleaners) because they can burn flesh.
2 Their solutions turn litmus blue.
3 Their solutions have pH numbers greater than 7.

4 All the alkalis except ammonia will react with **ammonium compounds**. They drive ammonia out of the compounds, as a gas. For example:

calcium hydroxide + ammonium chloride ⟶ calcium chloride + steam + ammonia

$$Ca(OH)_2\,(s) + 2NH_4Cl\,(s) \longrightarrow CaCl_2\,(s) + 2H_2O\,(g) + 2NH_3\,(g)$$

This reaction is used for making ammonia in the laboratory (page 156).

5 All alkalis react with **acids**, as you saw on page 119, producing a salt and water:

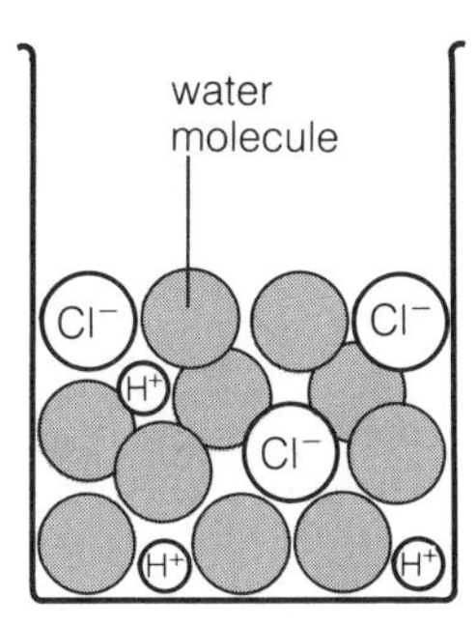

This is a solution of hydrochloric acid. It contains H^+ ions and Cl^- ions. It will turn litmus red.

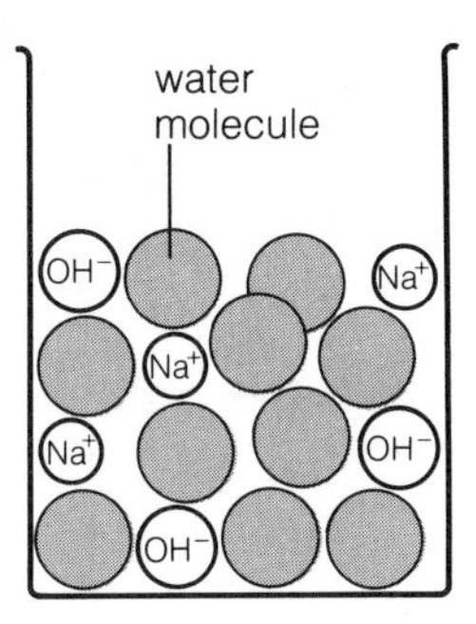

This is a solution of sodium hydroxide. It contains Na^+ ions and OH^- ions. It will turn litmus blue.

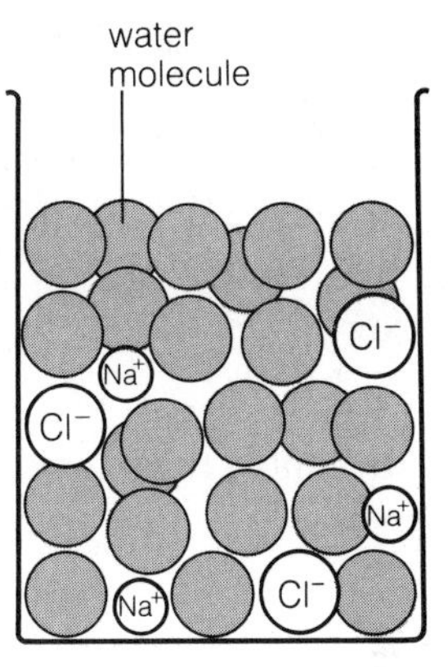

When the two solutions are mixed, the H^+ and OH^- ions join to form **water molecules**. The result is a neutral solution of sodium chloride, containing Na^+ and Cl^- ions. It has no effect on litmus.

The equation for this reaction is:

$$HCl\,(aq) + NaOH\,(aq) \longrightarrow NaCl\,(aq) + H_2O\,(l)$$

You could also write it showing only the ions that combine:

$$H^+\,(aq) + OH^-\,(aq) \longrightarrow H_2O\,(l)$$

The reaction is called a **neutralisation**. The alkali has **neutralised** the acid by removing its H^+ ions, and turning them into water.
During a neutralisation reaction, the H^+ ions of the acid are turned into water.
A neutralisation reaction is exothermic—it gives out heat. So the temperature of the solution rises a little.
You could obtain solid sodium chloride by heating the last solution above. The water evaporates leaving the solid behind.

Questions

1 Look at the substances shown at the top of page 120. Which ion do they all have in common?
2 What colour would litmus solution turn, if you mixed it with Ajax? Why?
3 A 0.1 M solution of potassium hydroxide has a higher pH number than a 0.1 M solution of ammonia. Why is this? Which will be a better conductor of electricity?
4 Write a word equation for the reaction between sodium hydroxide and ammonium chloride.
5 What happens to:
a the H^+ ions **b** the temperature
of the solution when an acid is neutralised?
6 Write a balanced equation for the reaction between sodium hydroxide and:
a sulphuric acid **b** nitric acid

9.4 Bases and neutralisation

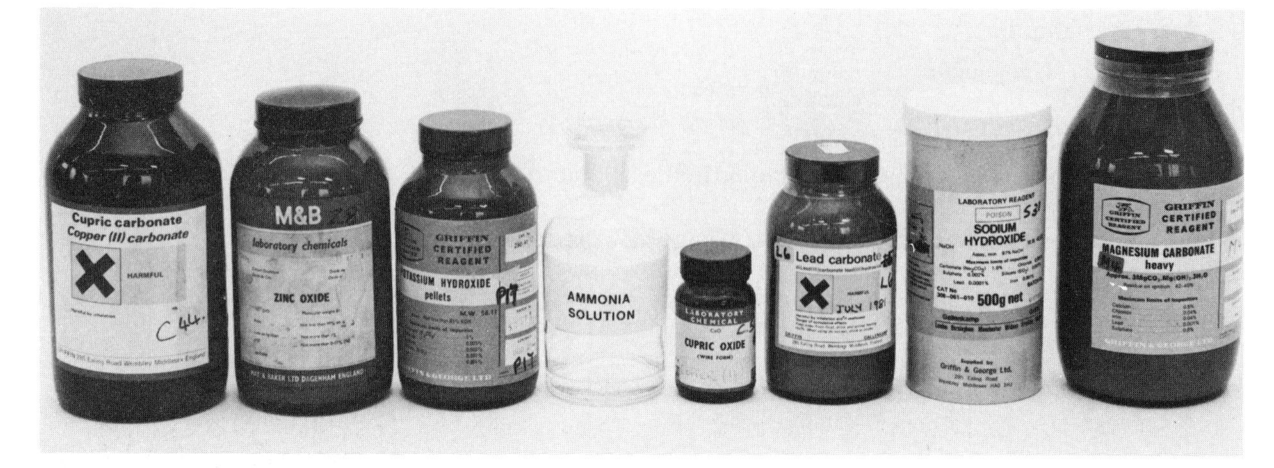

All these substances have something in common, as you will find out below.

Bases

An alkali can **neutralise** an acid, and destroy its acidity. It does this by removing the H^+ ions and converting them to water. But alkalis are not the only compounds that can neutralise acids:

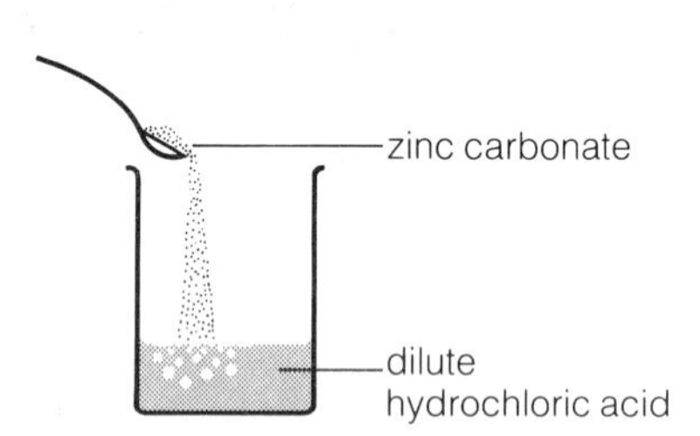

1 When zinc carbonate is added to dilute hydrochloric acid, it dissolves, and carbon dioxide bubbles off.

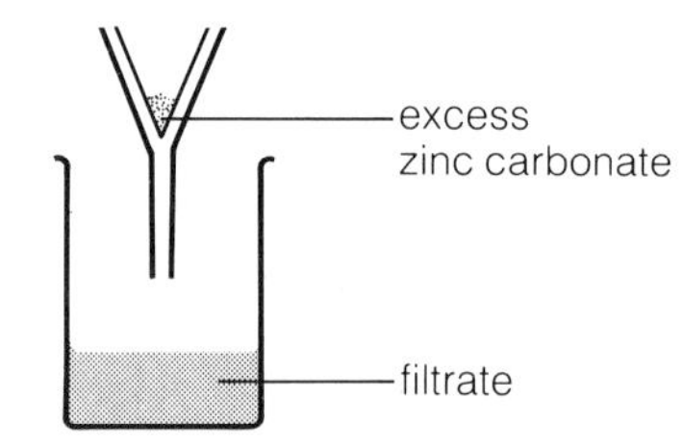

2 More is added until no more will dissolve, even with heating. The excess zinc carbonate is removed by filtering.

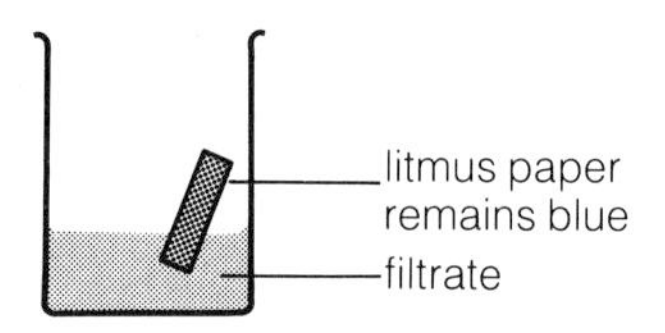

3 Then the filtrate is tested with blue litmus. There is no colour change. The acidity of the acid has been destroyed.

The equation for the reaction is:

$ZnCO_3\,(s) + 2\,HCl\,(aq) \longrightarrow ZnCl_2\,(aq) + H_2O\,(l) + CO_2\,(g)$

So the H^+ ions of the acid have been turned into water. The acid has been neutralised by the zinc carbonate.
In fact, acid can be neutralised by any of these compounds:

metal oxides
metal hydroxides
metal carbonates
metal hydrogen carbonates
ammonia solution

These compounds are all called **bases**.
Any compound that can neutralise an acid is called a base.

Alkalis are bases

The photograph at the top of the opposite page shows some bases. Notice that it includes alkalis—they can neutralise acids, so they are bases. They are bases that are soluble in water.
Alkalis are soluble bases.

Neutralisation always produces a salt

Neutralisation always produces a salt, as these general equations show:

acid + metal oxide ⟶ metal salt + water

acid + metal hydroxide ⟶ metal salt + water

acid + metal carbonate ⟶ metal salt + water + carbon dioxide

acid + metal hydrogen carbonate ⟶ metal salt + water + carbon dioxide

acid + ammonia solution ⟶ ammonium salt + water

Some neutralisations in everyday life

Insect stings When a bee stings, it injects an acidic liquid into the skin. The sting can be neutralised by rubbing on **calamine lotion**, which contains zinc carbonate, or **baking soda**, which is sodium hydrogen carbonate.
Wasp stings are alkaline, and can be neutralised with vinegar. Why? Ant stings and nettle stings contain methanoic acid. How would you treat them?

Bee stings are acidic.

Indigestion It may surprise you to know that you carry hydrochloric acid around in your stomach. It is a very dilute solution, and you need it for digesting food. But too much of it leads to **indigestion**, which can be very painful. To cure indigestion, you must neutralise the excess acid with a drink of sodium hydrogen carbonate solution (baking soda), or an indigestion tablet.

Soil treatment Most plants grow best when the pH of the soil is close to 7. If the soil is too acidic, or too alkaline, the plants grow badly or not at all.
Chemicals can be added to soil to adjust its pH. Most often it is too acid, so it is treated with **quicklime** (calcium oxide), **slaked lime** (calcium hydroxide), or **chalk** (calcium carbonate). These are all bases, and quite cheap.

Factory waste Liquid waste from factories often contains acid. If it reaches a river, the acid will kill fish and other river life. This can be prevented by adding slaked lime to the waste, to neutralise it.

Slaked lime being spread on fields where the soil is too acidic.

Questions

1 Look at the reaction shown on page 122. What salt does the filtrate contain? How would you obtain the dry salt from the filtrate.
2 What is a base? Name six bases.
3 What special property do alkalis have?
4 Which bases react with acids to give carbon dioxide?
5 Write a word equation for the reaction between:
  a dilute hydrochloric acid and copper(II) oxide
  b dilute sulphuric acid and potassium carbonate
6 Write a word equation for the reaction that takes place in your stomach, when you take baking soda to cure indigestion.

9.5 Making salts (I)

Making salts from acids

Acid + metal Zinc sulphate can be made by reacting dilute sulphuric acid with zinc:

$Zn\,(s) + H_2SO_4\,(aq) \longrightarrow ZnSO_4\,(aq) + H_2\,(g)$

These are the steps:

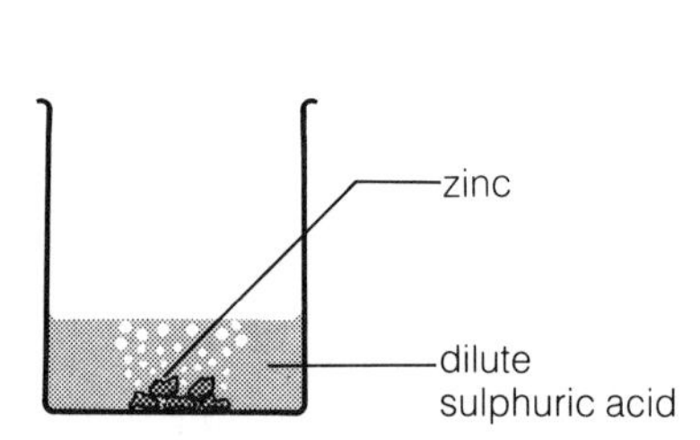

1 Some dilute sulphuric acid is put in a beaker, and zinc is added. The zinc begins to dissolve, and hydrogen bubbles off. The bubbles stop when all the acid has been used up.

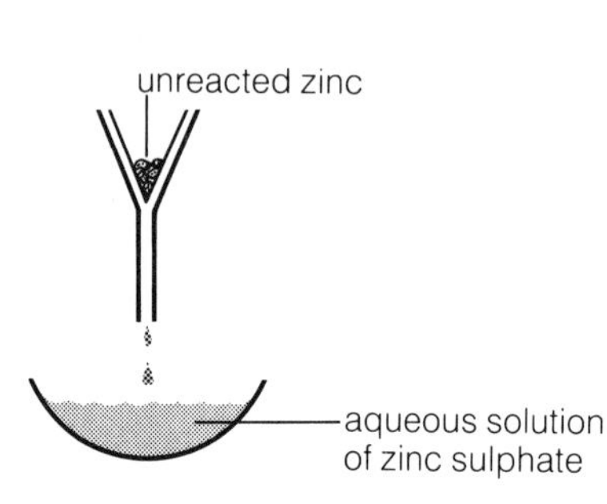

2 Some zinc is still left. It is removed by filtering, which leaves an aqueous solution of zinc sulphate.

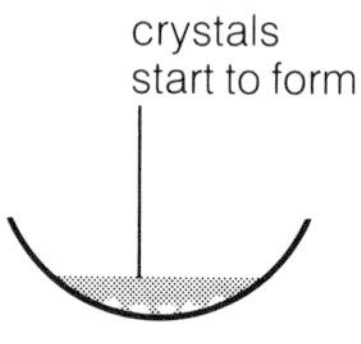

3 The solution is heated to evaporate some of the water. Then it is left to cool. Crystals of zinc sulphate start to form.

The method above is not suitable for *all* metals, or *all* acids. It is fine for magnesium, aluminium, zinc and iron. But the reactions of sodium, potassium and calcium with acid are dangerously violent. The reaction of lead is too slow, and copper, silver and gold do not react at all (page 134).

Acid + insoluble base Copper(II) oxide is an insoluble base. Although copper will not react with dilute sulphuric acid, copper(II) oxide will. The salt that forms is copper(II) sulphate:

$CuO\,(s) + H_2SO_4\,(aq) \longrightarrow CuSO_4\,(aq) + H_2O\,(l)$

The method is quite like the one above.

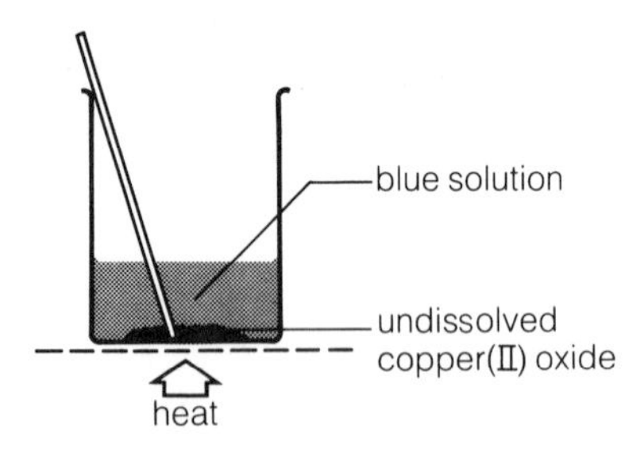

1 Some copper(II) oxide is added to dilute sulphuric acid. On warming it dissolves, and the solution turns blue. More is added until no more dissolves.

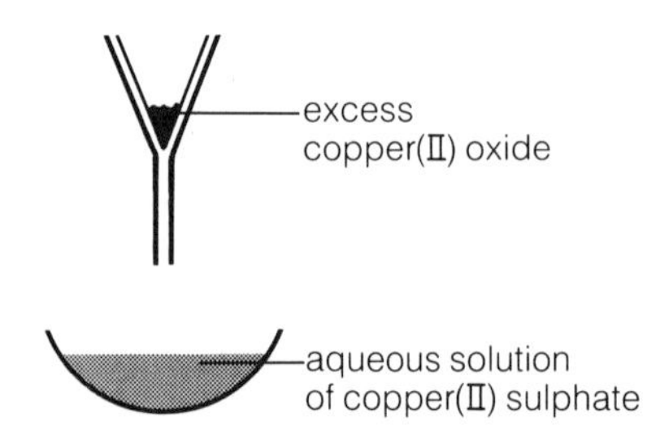

2 That means all the acid has been used up. The excess solid is removed by filtering. This leaves a blue solution of copper(II) sulphate in water.

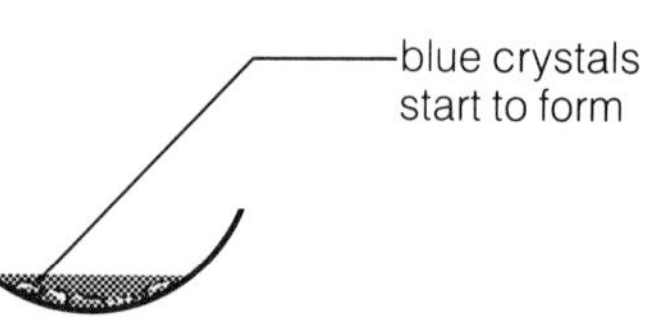

3 The solution is heated to evaporate some of the water. Then it is left to cool. Blue crystals of copper(II) sulphate start to form.

Acid + alkali (soluble base) The reaction of sodium with acids is very dangerous. So sodium salts are usually made by starting with sodium hydroxide. This reaction can be used to make sodium chloride:

$NaOH\,(aq) + HCl\,(aq) \longrightarrow NaCl\,(aq) + H_2O\,(l)$

Both reactants are soluble, and no gas is given off during the reaction. So it is difficult to know when the reaction is over. You have to use an **indicator**. Universal indicator or litmus could be used, but even better is **phenolphthalein**. This is pink in alkaline solution, but colourless in neutral and acid solutions:

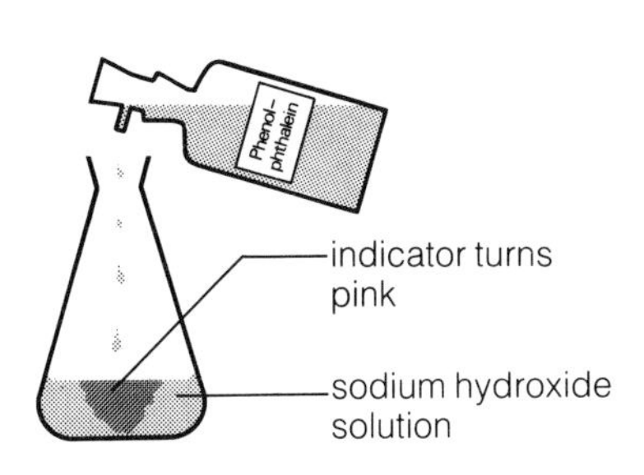

1 25 cm^3 of sodium hydroxide solution is measured into a flask, using a pipette. Then two drops of phenolphthalein are added. The indicator turns pink.

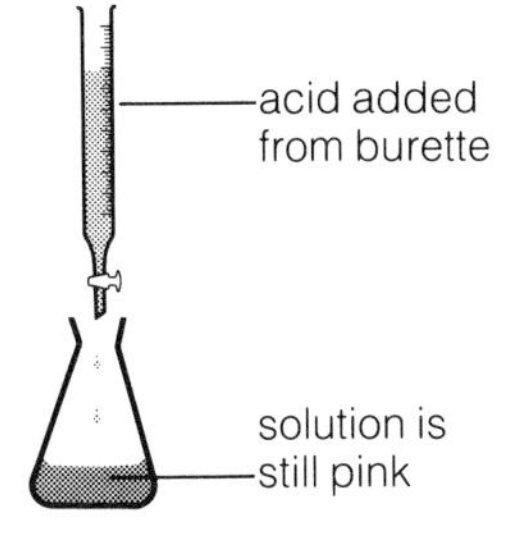

2 The acid is added from a burette, a little at a time. The flask is swirled to let the acid and alkali mix.

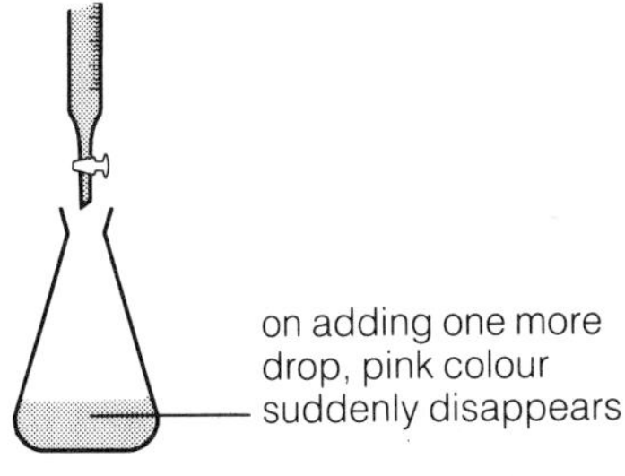

3 When all the alkali has been used up, the indicator suddenly turns colourless, showing that the solution is neutral. There is no need to add more acid.

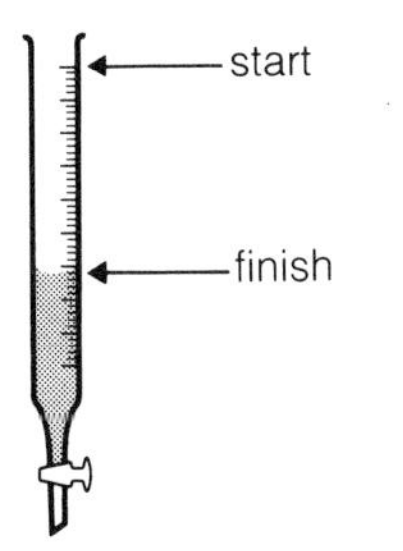

4 You can tell how much acid was added, using the scale on the burette. So now you know how much acid is needed to neutralise 25 cm^3 of alkali.

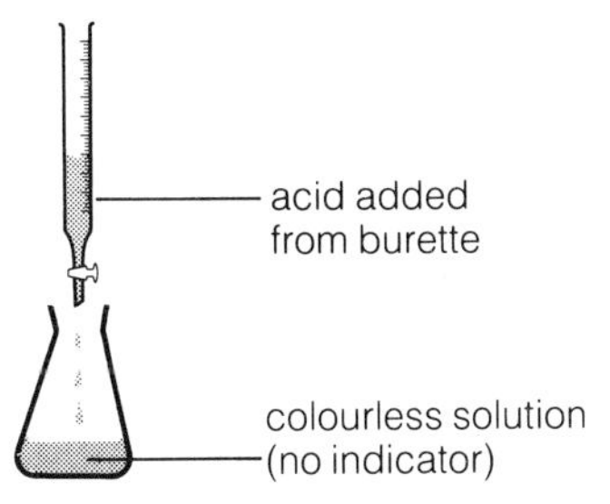

5 The reaction is carried out again, but this time there is no need for an indicator. 25 cm^3 of alkali is put in the flask, and the correct amount of acid added.

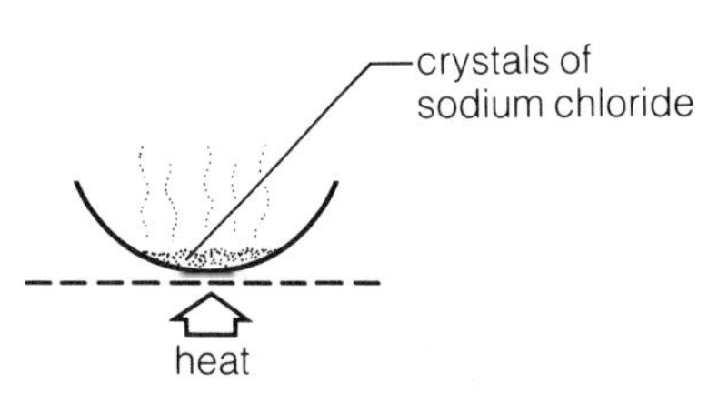

6 The solution from the flask is heated, to let the water evaporate. Dry crystals of sodium chloride are left behind.

In step 5 the reaction has to be carried out again, *without* indicator, because indicator would make the salt impure.
A similar method can be used for making potassium salts from potassium hydroxide, and ammonium salts from ammonia solution.

Questions

1 Name the acid and metal you would use for making:
a zinc chloride **b** magnesium sulphate

2 Why would you *not* make potassium chloride from potassium and hydrochloric acid?

3 How would you obtain lead(II) nitrate, starting with the insoluble compound lead(II) carbonate?

4 Write instructions for making potassium chloride, starting with solid potassium hydroxide.

9.6 Making salts (II)

Making insoluble salts

The salts made so far have all been soluble. They were obtained as crystals by evaporating solutions. But not all salts are soluble:

Soluble		Insoluble
All sodium, potassium, and ammonium salts		
All nitrates		
Chlorides . . .	*except*	silver and lead chloride
Sulphates . . .	*except*	calcium, barium and lead sulphate
Sodium, potassium, and ammonium carbonates . . .		but all other carbonates are insoluble

Insoluble salts can be made by **precipitation**.
For example, insoluble barium sulphate is precipitated when solutions of barium chloride and magnesium sulphate are mixed:

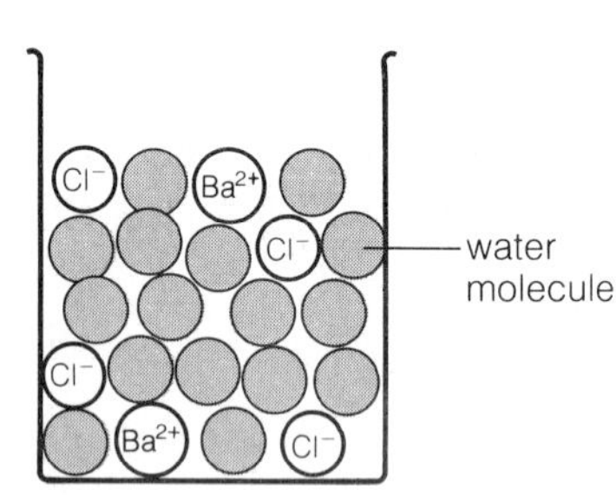

This is a solution of barium chloride, $BaCl_2$. It contains barium ions and chloride ions.

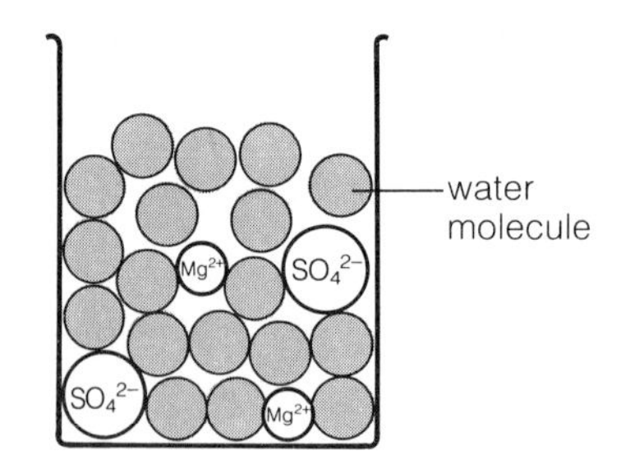

This is a solution of magnesium sulphate, $MgSO_4$. It contains magnesium ions and sulphate ions.

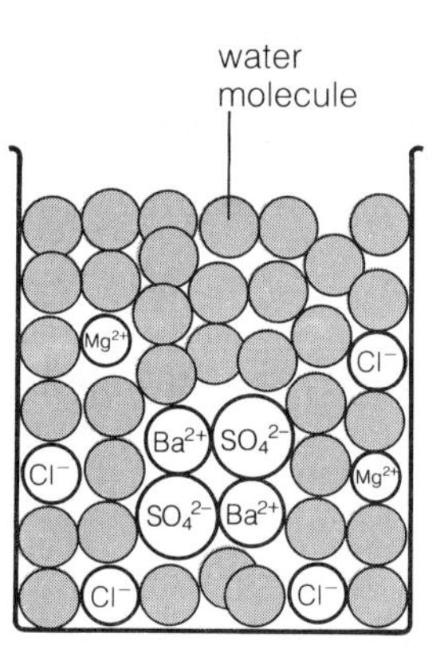

When the two solutions are mixed the barium ions and sulphate ions bond together, because they are strongly attracted to each other. Solid barium sulphate precipitates.

The equation for the reaction is:

$BaCl_2\,(aq) + MgSO_4\,(aq) \longrightarrow BaSO_4\,(s) + MgCl_2\,(aq)$

or you could write it in a shorter way as:

$Ba^{2+}\,(aq) + SO_4^{2-}\,(aq) \longrightarrow BaSO_4\,(s)$

These are the steps for obtaining the barium sulphate:

1. Solutions of barium chloride and magnesium sulphate are mixed. A white precipitate of barium sulphate forms at once.
2. The mixture is filtered. The barium sulphate gets trapped in the filter paper.
3. It is rinsed with distilled water.
4. Then it is put in a warm oven to dry.

The precipitation of barium sulphate.

Barium sulphate could also be made from barium nitrate and sodium sulphate, for example, since these are both soluble. As long as barium ions and sulphate ions are present, barium sulphate will be precipitated.
To precipitate an insoluble salt, you must mix a solution that contains its positive ions with a solution that contains its negative ions.

Making salts from their elements

Some salts contain just *two* elements. They can be made by direct combination of the elements.
An example is iron(III) chloride, $FeCl_3$. You could make it by heating iron in a stream of chlorine, as shown on the right. The reaction is exothermic. Once it begins, the bunsen can be turned off. Iron(III) chloride forms as a vapour, and solidifies in the cold collecting jar. Calcium chloride absorbs moisture, so it keeps the apparatus dry.

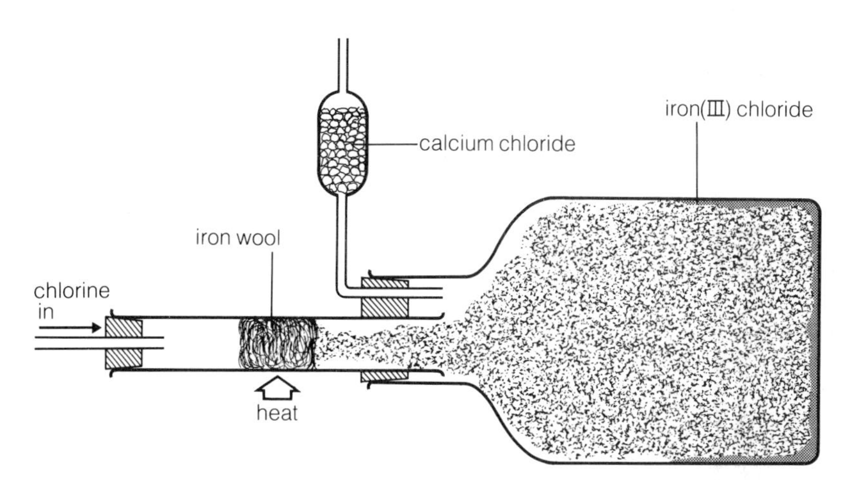

This method makes **anhydrous** iron(III) chloride. Anhydrous means *without water of crystallisation*. Anhydrous iron(III) chloride is dark brown. If left in air it absorbs moisture and turns yellow. It has become **hydrated** iron(III) chloride, with the formula $FeCl_3.6H_2O$.

Making salts in industry

Many salts occur naturally, and can be dug up out of the earth. Sodium chloride and magnesium sulphate are examples.
But others have to be made in factories, using methods like the ones on these pages. For example, huge amounts of nitrates and sulphates are needed every year as **fertilisers** (page 161). They are produced by neutralising nitric acid and sulphuric acid in giant tanks.
One important fertiliser is **ammonium nitrate**. It is made by neutralising nitric acid with ammonia solution:

$$HNO_3(aq) + NH_3(aq) \longrightarrow NH_4NO_3(aq)$$

The water is driven off in an evaporator, leaving molten ammonium nitrate. This is sprayed down a tall tower. By the time it reaches the bottom it has cooled into small pellets, that are easy to spread on soil.

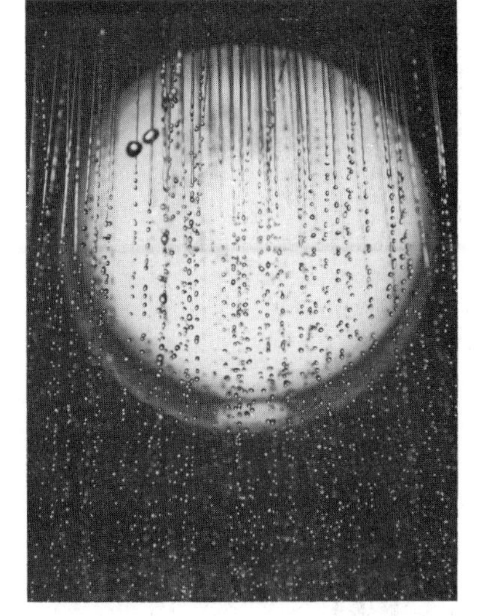

Molten ammonium nitrate being sprayed down a tower. It cools into pellets.

Questions

1 Choose two starting compounds you could use, to precipitate: **a** calcium sulphate
b zinc carbonate **c** lead chloride

2 Write a balanced equation for each reaction in question 1.

3 **a** The reaction between iron and chlorine is *exothermic*. What does that mean?
b What is the tube of calcium chloride for?

4 Potassium sulphate is a fertiliser. Suggest a way for making it in industry.

Questions on Chapter 9

1 Match each solution from list A with the correct formula from list B.

List A (solutions)	List B (formulae)
sodium hydroxide	$H_2SO_4\,(aq)$
hydrochloric acid	$HNO_3\,(aq)$
ammonia	$HCl\,(aq)$
calcium hydroxide	$CH_3COOH\,(aq)$
sulphuric acid	$NH_3\,(aq)$
nitric acid	$Ca(OH)_2\,(aq)$
ethanoic acid	$NaOH\,(aq)$

2 Five solutions A to E were tested with universal indicator solution, to find their pH. The results are shown below.

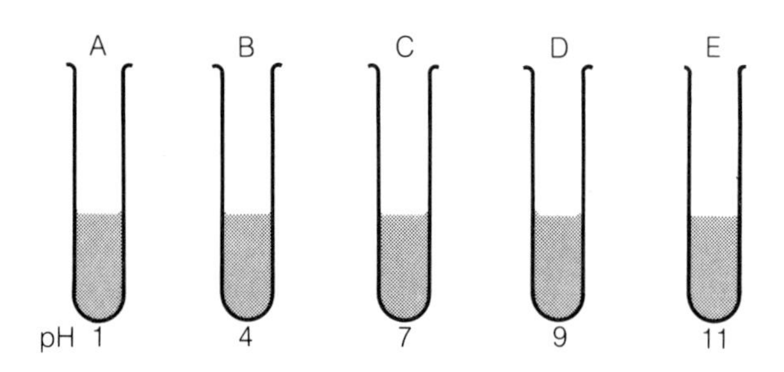

a What colour would each solution be?
b Which solution is:
i neutral?
ii strongly acidic?
iii weakly acidic?
iv strongly alkaline?
c The five solutions were known to be sodium chloride, sulphuric acid, ammonia solution, sodium hydroxide, and ethanoic acid. Now identify each of the solutions A to E.

3 Rewrite the following, choosing the correct word from each pair in brackets.
Acids are compounds which dissolve in water giving (hydrogen/hydroxide) ions. Sulphuric acid is one example. It is a (strong/weak) acid, which can be neutralised by (acids/alkalis) to form salts called (nitrates/sulphates).
Many (metals/non-metals) react with acids to give a gas called (hydrogen/carbon dioxide). Acids also react with (chlorides/carbonates) to give (chlorine/carbon dioxide).
Solutions of acids are (good/poor) conductors of electricity. They also effect indicators. For example, phenolphthalein turns (pink/colourless) in acids, while litmus turns (red/blue).
The strength of an acid is shown by its (concentration/pH) number. The (higher/lower) the number, the stronger the acid.

4 State whether the following properties belong to acids, or alkalis, or both.
a Sour taste
b pH values greater than 7
c Change the colour of litmus
d Soapy to touch
e Soluble in water
f May be strong or weak
g Neutralise bases
h Form ions in water
i Dangerous to handle
j Form salts with certain other chemicals
k Usually react with metals

5 This is a brief description of a neutralisation reaction.
"25 cm^3 of potassium hydroxide solution were placed in a flask and a few drops of phenolphthalein were added. Dilute hydrochloric acid was added until the indicator changed colour. It was found that 21 cm^3 of acid were used."
a Draw a labelled diagram of titration apparatus for this neutralisation.
b What piece of apparatus should be used to measure out accurately 25 cm^3 of sodium hydroxide solution?
c What colour was the solution in the flask at the start of the titration?
d What colour did it turn when the alkali had been neutralised?
e Was the acid more concentrated or less concentrated than the alkali? Explain your answer.
f Name the salt formed in this neutralisation.
g Write an equation for the reaction.
h How would you obtain *pure* crystals of the salt.

6 A and B are white powders. A is insoluble in water but B is soluble and its solution has a pH of 3.
A **mixture** of A and B bubbles or effervesces in water. A gas is given off and a clear solution forms.
a One of the white powders is an acid. Is it A or B?
b The other white powder is a carbonate. What gas is given off in the reaction?
c Although A is insoluble in water, a clear solution forms when the mixture of A and B is added to water. Explain why.

7 The chemical name for Aspirin is 2-acetyloxybenzoic acid.
This acid is soluble in hot water.
a How would you expect an aqueous solution of Aspirin to affect litmus paper?
b Do you think it is a strong acid or a weak one? Explain why you think so.
c What would you expect to see when baking soda is added to an aqueous solution of Aspirin?

Method of preparation	Reactants	Salt formed	Other products
a acid + carbonate	.... and	sodium chloride	water and ....
b acid + metal	.... and	iron(II) sulphate	....
c acid + ...	nitric acid and sodium hydroxide	....	....
d acid + base	.... and copper(II) oxide	copper(II) sulphate	....
e acid + ...	.... and	copper(II) sulphate	carbon dioxide and ...
f precipitation	silver nitrate and potassium chloride	....	....
g precipitation	lead nitrate and potassium iodide	....	....
h acid + alkali	.... and potassium hydroxide	potassium sulphate	water only
i direct combination	iron and sulphur	....	———
j direct combination	.... and	aluminium chloride	———

8 The table above is about the preparation of salts. Copy it and fill in the missing details.

9 Pink cobalt chloride is a **hydrated** salt. That means it contains **water of crystallisation**. Its formula is $COCl_2.6H_2O$.

Here are five other salts:

copper(II) sulphate $CuSO_4.5H_2O(aq)$
sodium chloride $NaCl$
zinc chloride $ZnCl_2.6H_2O$
sodium carbonate $Na_2CO_3.10H_2O$
ammonium nitrate NH_4NO_3

a Which of them are hydrated?
b Which of them are anhydrous (i.e. contain no water of crystallisation)?
c Which of them are coloured?
d Write down the formula of the compound known as **common salt.**
e Write down the formula of the salt which is often used as a fertiliser.
f Anhydrous copper(II) sulphate is a white powder. What is its formula?
g Hydrated copper(II) sulphate is blue. It turns white when it is heated. Explain why.
h i Anhydrous copper(II) sulphate could be used to test whether a liquid contained water. Explain how.
ii How could you *prove* that the liquid was *pure* water?
iii When some zinc chloride crystals are heated gently, a solution of zinc chloride forms. Suggest a reason why.

10 a Divide the following salts into two groups, *Soluble in water* and *Insoluble in water:*

sodium chloride
calcium carbonate
potassium chloride
barium sulphate
barium carbonate
silver chloride
sodium citrate
zinc chloride
sodium sulphate
copper(II) sulphate
lead sulphate
lead nitrate
sodium carbonate
ammonium carbonate

b Now write down two starting compounds that could be used to make each *insoluble* salt.

11 These diagrams show the stages in the preparation of copper(II) ethanoate, which is a salt of ethanoic acid.

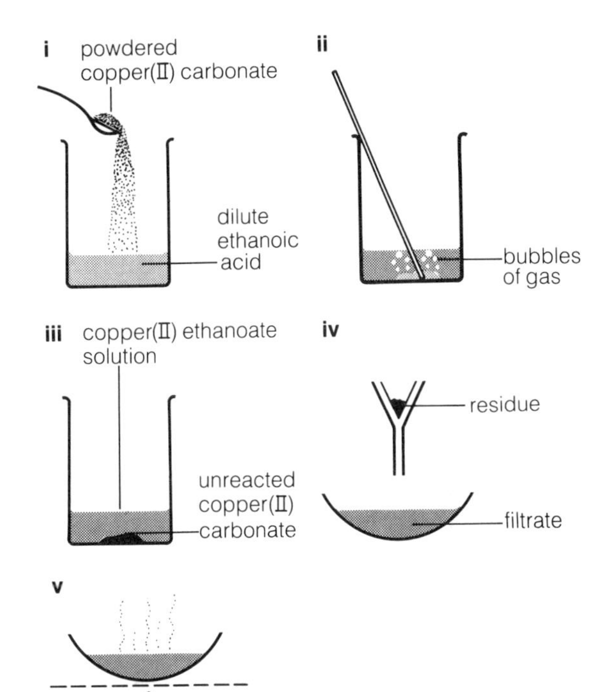

a Which gas is given off in stage **ii**?
b Write a word equation for the reaction in stage **ii**.
c How could you tell when the reaction is complete?
d Which reactant is completely used up in the reaction? Explain your answer.
e Why is copper(II) carbonate powder used, rather than lumps?
f Name the residue in stage **iv**.
g Write a list of instructions for carrying out this preparation, in the laboratory.
h Suggest another copper compound that could be used, instead of copper(II) carbonate, to make copper(II) ethanoate.

10.1 Metals and non-metals

On page 24, you saw that there are 105 different elements. Of these, 84 are **metals** and the rest are **non-metals**.
Some of the metals are listed in the box on the right.

Some of the metals:

aluminium, Al
calcium, Ca
copper, Cu
gold, Au
iron, Fe
lead, Pb
magnesium, Mg
potassium, K
silver, Ag
sodium, Na
tin, Sn
zinc, Zn

The properties of metals

Metals *usually* have these properties:

1. They are **strong**. They can hold heavy loads without breaking.
2. They are **malleable**. That means they can be hammered into different shapes without breaking.
3. They are **ductile**. That means they can be drawn out to make wires.
4. They are **sonorous**—they make a ringing noise when you strike them.
5. They are shiny.
6. They are good conductors of electricity and heat.
7. They have high melting and boiling points. (They are all solid at room temperature, except mercury.)
8. They have high densities. That means they feel 'heavy'. (The density of a substance is its mass per cubic centimetre.)
9. They react with oxygen to form oxides. For example, magnesium burns in the air like this:

   magnesium (silvery metal) + oxygen ⟶ magnesium oxide (white ash)

   Metal oxides are **bases**, which means they react with acids to form salts.
10. When metals form ions, the ions are positive. For example, in the reaction between magnesium and oxygen, magnesium ions (Mg^{2+}) and oxide ions (O^{2-}) are formed, as shown on page 43.

The last two properties above are called **chemical properties**, because they are about chemical changes in the metals. The other properties are **physical properties**.

Think of two reasons why metals are used to make gongs . . .

. . . and then three reasons why they are used for saucepans.

All metals are different

The properties on the last page are typical of metals. But not all metals have all of these properties. For example:

Iron has all the properties in the list. It is used for gates like these because it is both malleable and strong. It is used for anchors because of its high density. It melts at 1530 °C. But unlike most other metals, it is **magnetic**.

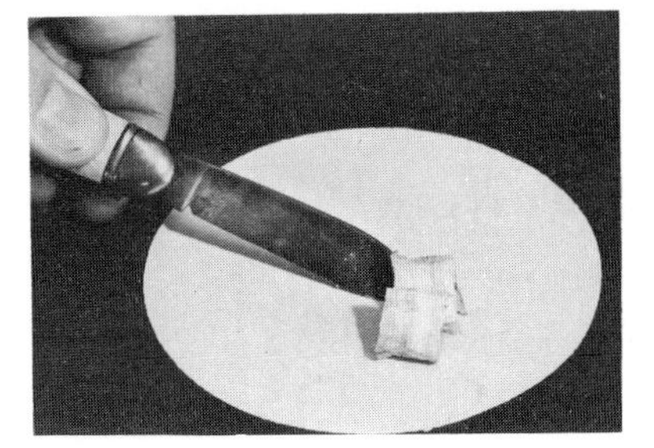

Sodium is quite different. It is so soft that it can be cut with a knife, and it melts at only 98 °C. It is so light that it floats on water, but it reacts immediately with the water forming a solution. No good for gates . . .

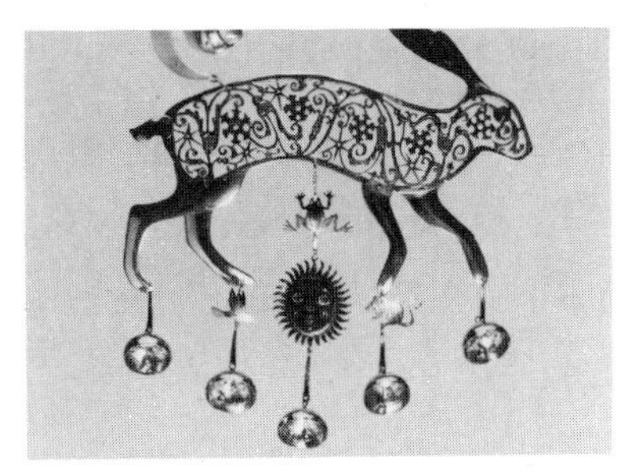

Gold melts at 1064 °C. Unlike most other metals it does not form an oxide—it is very unreactive. But it is malleable and ductile, and looks attractive. So it is used for making jewellery.

No two metals have exactly the same properties. You can find out more about the differences between them, on the next four pages.

Comparing metals with non-metals

Only 21 of the elements are non-metals. Some of them are listed on the right. Non-metals are quite different from metals. They *usually* have these properties:

1 They are not strong, or malleable, or ductile, or sonorous. In fact, when solid non-metals are hammered, they break up—they are **brittle**.
2 They have lower melting and boiling points than metals. (One of them is a liquid and eleven are gases, at room temperature.)
3 They are poor conductors of electricity. Graphite (carbon) is the only exception. They are also poor conductors of heat.
4 They have low densities.
5 Like metals, most of them react with oxygen to form oxides:

   sulphur + oxygen ⟶ sulphur dioxide

   But unlike metal oxides, these oxides are not bases. Many of them dissolve in water to give *acidic* solutions.
6 When they form ions, the ions are negative. Hydrogen is an exception—it forms the ion H^+.

Some of the non-metals:

bromine, Br
carbon, C
chlorine, Cl
helium, He
hydrogen, H
iodine, I
nitrogen, N
oxygen, O
sulphur, S

Questions

1 Make two lists, showing twenty *metals* and fifteen *non-metals*. Give their symbols too.
2 Try to think of a metal that is not malleable at room temperature.
3 Suggest reasons why:
   a silver is used for jewellery
   b copper is used for electrical wiring
4 For some uses, a highly sonorous metal is needed. Try to think of two examples.
5 Try to think of *two* reasons why:
   a mercury is used in thermometers
   b aluminium is used for beer cans
6 Look at the properties of non-metals, above. Which are *physical* properties? Which are *chemical*?

10.2 Metals and reactivity (I)

On the last page you saw that all metals are different.
The next few pages compare the way some metals react, to see how different they are.

The reaction between metals and oxygen

Look at the way sodium reacts with oxygen:

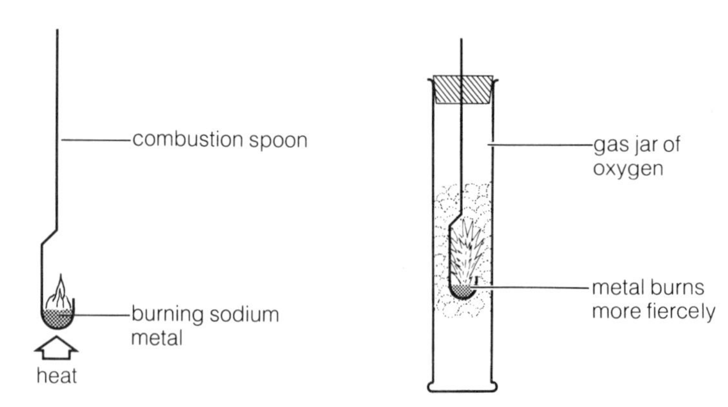

A small piece of sodium is put in a combustion spoon and heated over a bunsen flame. It melts quickly, and catches fire.

Then the spoon is plunged into a jar of oxygen. The metal burns even more fiercely, with a bright yellow flame.

Because it reacts with oxygen, sodium is stored under oil.

The steps above can be repeated for other metals. This table shows what happens:

Metal	Behaviour	Order of reactivity	Product
Sodium	Catches fire with only a little heating. Burns fiercely with a bright yellow flame	most reactive	Sodium peroxide, Na_2O_2, a pale yellow powder
Magnesium	Catches fire easily. Burns with a blinding white flame		Magnesium oxide, MgO a white powder
Iron	Does not burn, but the hot metal glows brightly in oxygen, and gives off yellow sparks		Iron oxide, Fe_3O_4 a black powder
Copper	Does not burn, but the hot metal becomes coated with a black substance		Copper oxide, CuO, a black powder
Gold	No reaction, no matter how much the metal is heated	least reactive	—

If a reaction takes place, the product is an oxide.
Sodium reacts the most vigorously with oxygen. It is the **most reactive** of the five metals. Gold does not react at all—it is the least reactive of them. The arrow in the table shows the **order of reactivity**.

The reaction between metals and water

Metals also show differences in the way they react with water. For example:

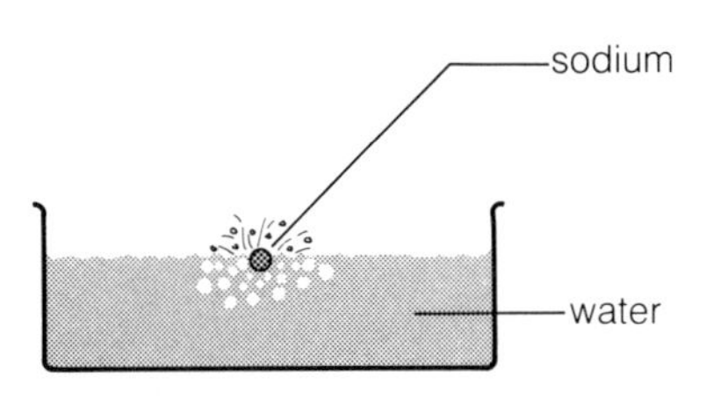

Sodium reacts violently with cold water, whizzing over the surface. Hydrogen gas and a clear solution of sodium hydroxide are formed.

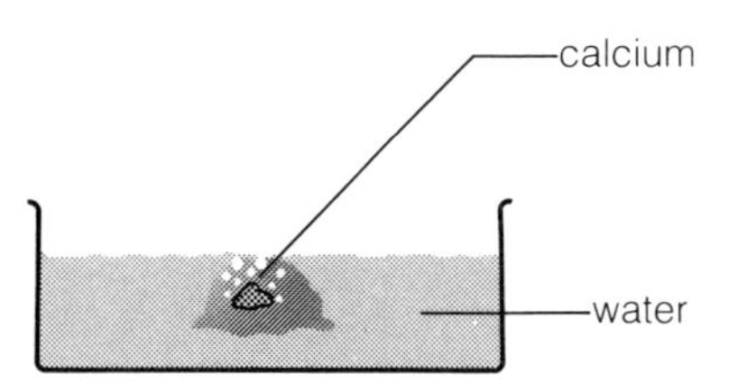

With calcium, the reaction is slower. Hydrogen bubbles off, and a cloudy solution of calcium hydroxide forms.

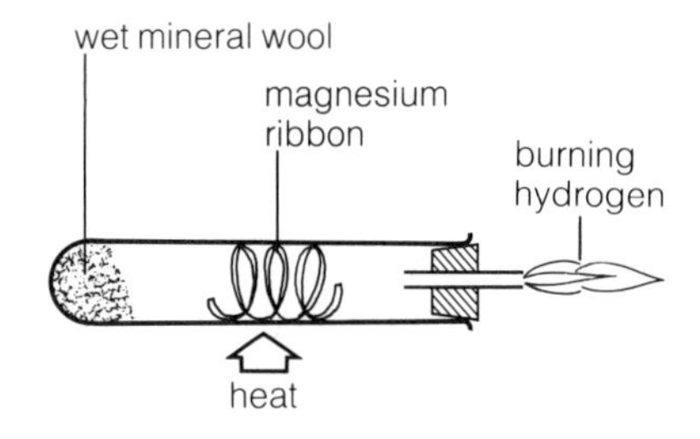

Magnesium reacts only very slowly with cold water. But it reacts vigorously when heated in steam. It glows brightly, and hydrogen and solid magnesium oxide are formed.

This table shows the results for other metals too:

Metals	Reaction	Order of reactivity	Products
Potassium	Very violent with cold water. Catches fire	most reactive	Hydrogen and a solution of potassium hydroxide, KOH
Sodium	Violent with cold water		Hydrogen and a solution of sodium hydroxide, $NaOH$
Calcium	Less violent with cold water		Hydrogen and calcium hydroxide, $Ca(OH)_2$, which is only slightly soluble
Magnesium	Very slow with cold water, but vigorous with steam		Hydrogen and solid magnesium oxide, MgO
Zinc	Quite slow with steam		Hydrogen and solid zinc oxide, ZnO
Iron	Slow with steam		Hydrogen and solid iron oxide, Fe_3O_4
Copper Gold	No reaction	least reactive	———

Notice that the first three metals in the list produce hydroxides. The others produce oxides, if they react.
Now compare this table with the one on the opposite page. Is sodium more reactive than iron each time? Is iron more reactive than copper each time?

Questions

1 Describe how magnesium and iron each react with oxygen. Write balanced equations for the reactions.
2 Which is more reactive, copper or iron?
3 Which is more reactive, sodium or zinc?
4 What gas is always produced, if a metal reacts with water?
5 Describe how magnesium reacts with steam. Write a balanced equation for the reaction.

10.3 Metals and reactivity (II)

Reaction with hydrochloric acid

On the last two pages, you saw that different metals react differently with oxygen and water. They also react differently with **acids.** Compare these results with hydrochloric acid:

Metal	Reaction with hydrochloric acid	Order of reactivity	Products
Magnesium	Vigorous	most reactive	Hydrogen and a solution of magnesium chloride, $MgCl_2$
Zinc	Quite slow		Hydrogen and a solution of zinc chloride, $ZnCl_2$
Iron	Slow		Hydrogen and a solution of iron(II) chloride, $FeCl_2$
Lead	Slow, and only if the acid is concentrated		Hydrogen and a solution of lead(II) chloride, $PbCl_2$
Copper Gold	No reaction, even with concentrated acid	least reactive	

Now compare this table with the last two tables. Is iron always more reactive than copper? Is magnesium always more reactive than iron?

Competing for oxygen

The reactions with oxygen, water, and hydrochloric acid show that iron is more reactive than copper. Now look at this experiment:

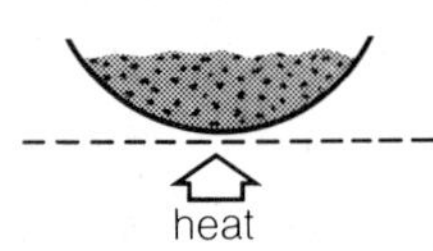

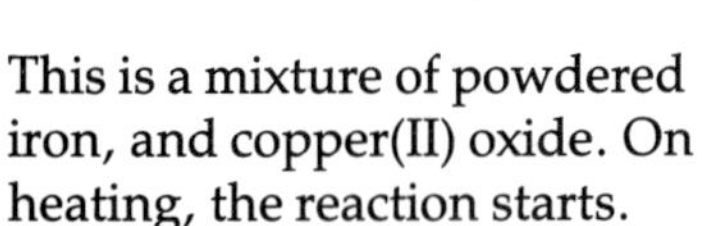

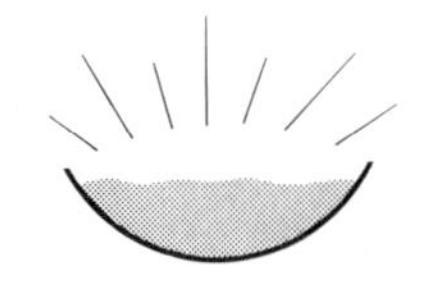

This is a mixture of powdered iron, and copper(II) oxide. On heating, the reaction starts.

The mixture glows, even after the bunsen is removed. Iron(II) oxide and copper are formed.

Here iron and copper are competing for oxygen. Iron wins:

$$\text{iron} + \text{copper(II) oxide} \longrightarrow \text{iron(II) oxide} + \text{copper}$$
$$Fe(s) + CuO(s) \longrightarrow FeO(s) + Cu(s)$$

By taking away the oxygen from copper, iron is acting as a **reducing agent** (page 72). Other metals behave in the same way when heated with the oxides of less reactive metals.

When a metal is heated with the oxide of a less reactive metal, it will remove the oxygen from it. The reaction is exothermic.

Using a competition reaction to repair railway lines. Aluminium and iron(III) oxide are being heated together, to give molten iron. This is run into gaps between rails. The process is called the Thermit process.

Displacement reactions

Let's look at another reaction involving iron and copper:

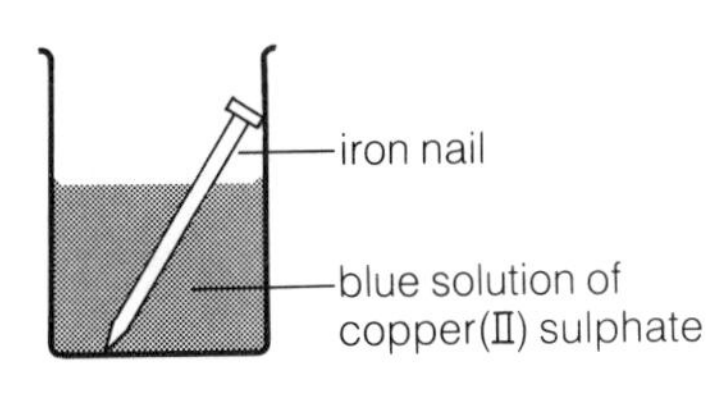

This time, an iron nail is placed in some blue copper(II) sulphate solution.

Soon a coat of copper appears on the nail. The solution turns pale green.

Here iron and copper are competing to be the compound in solution. Once again iron wins. It drives out or **displaces** copper from the copper(II) sulphate solution. Green iron(II) sulphate is formed:

$$\text{iron} + \text{copper(II) sulphate} \longrightarrow \text{iron(II) sulphate} + \text{copper}$$

$$\underset{}{Fe\,(s)} + \underset{\text{blue}}{CuSO_4\,(aq)} \longrightarrow \underset{\text{green}}{FeSO_4\,(aq)} + Cu\,(s)$$

Other metals displace less reactive metals in the same way.
A metal will always displace a less reactive metal from solutions of its compounds.

Now look at the way copper reacts with silver nitrate solution:

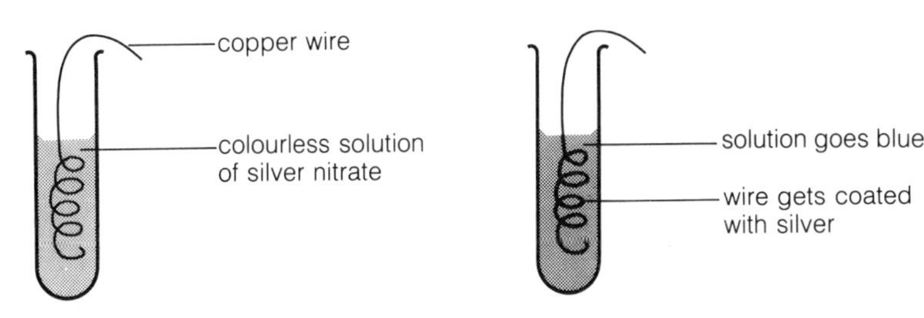

A coil of copper is placed in a solution of silver nitrate. The copper gradually dissolves . . .

. . . so the solution goes blue. At the same time, a coating of silver forms on the wire.

Crystals of silver on a copper wire. The copper has displaced the silver from silver nitrate solution.

Which metal is more reactive, copper or silver? (Check the reactivity series on the next page to see if you were right.)

Questions

The table on the opposite page will help you answer some of these questions.

1. Describe how iron and lead react with hydrochloric acid. Write balanced equations for the reactions.
2. Write a rule for the reaction of a metal with:
   - **a** oxides of other metals
   - **b** solutions of compounds of other metals
3. Will copper react with lead(II) oxide? Explain why.
4. Will iron react with lead(II) oxide? If *yes*, write a balanced equation for the reaction.
5. What would you *see* when zinc is added to copper(II) sulphate solution? (Zinc sulphate is colourless.)
6. Explain how the Thermit process works.
7. Tin does not react with iron(II) oxide. But it reduces lead(II) oxide to lead. Arrange tin, iron and lead in order of decreasing reactivity.

10.4 The reactivity series

By comparing their reactions with oxygen, water, acid, metal oxides, and solutions of metal salts, we can arrange the metals in order of reactivity. The list is called the **reactivity series**. Here it is:

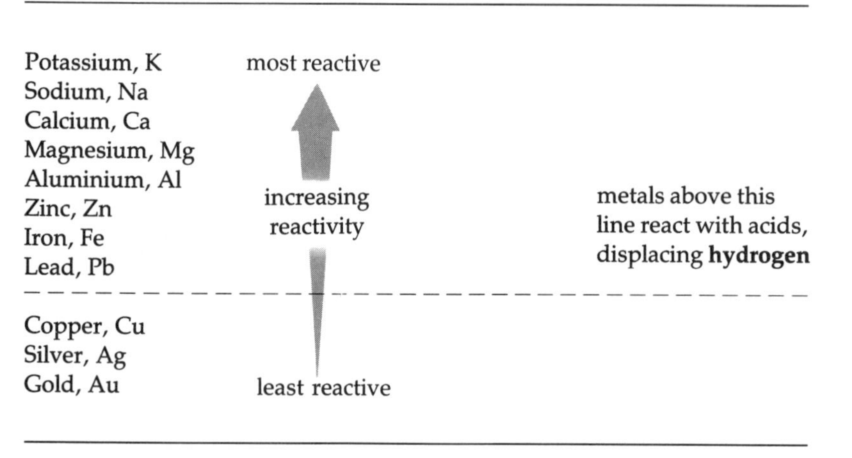

A metal's position in the reactivity series gives us a clue to its uses. Only unreactive metals are used to make coins.

Things to remember about the reactivity series

1 The more reactive the metal, the more it 'likes' to form compounds. So only copper, silver and gold are ever found as *elements* in the earth's crust. The other metals are *always* found as compounds.

2 The more reactive the metal, the more **stable** its compounds. A stable compound is difficult to break down or decompose, because the bonds holding it together are very strong. Compare what happens when these nitrates and carbonates are heated:

Metal	Effect of heat on carbonate	Effect of heat on nitrate
Potassium Sodium	no effect	decompose to nitrite and oxygen e.g. $2NaNO_3 \longrightarrow 2NaNO_2 + O_2$
Calcium Magnesium Zinc Iron Lead Copper	decompose to oxide and carbon dioxide $CuCO_3 \longrightarrow CuO + CO_2$	decompose further to oxide, nitrogen dioxide and oxygen $2Cu(NO_3)_2 \longrightarrow 2CuO + 4NO_2 + O_2$

compounds get more difficult to decompose

3 When a metal reacts, it gives up electrons to form ions. When magnesium reacts with oxygen it forms magnesium ions (Mg^{2+}). When copper reacts with oxygen it forms copper ions (Cu^{2+}). Magnesium is more reactive than copper because it gives up electrons more readily.
The more reactive the metal, the more readily it gives up electrons.

The reactivity series and cells

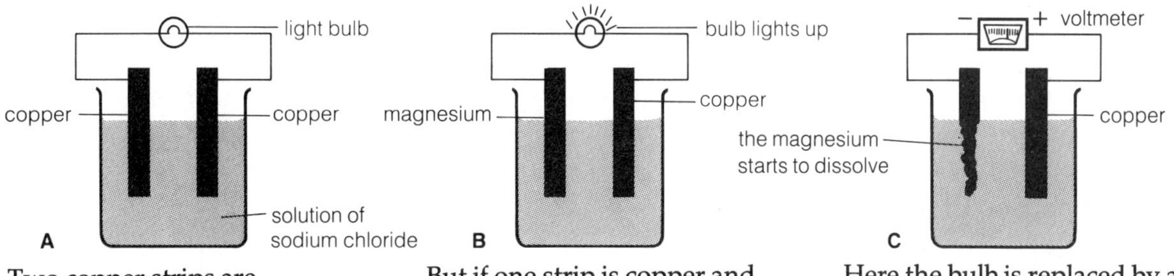

Two copper strips are connected to a light bulb and placed in a solution of sodium chloride. The solution is an **electrolyte**—it *can* conduct electricity. But nothing happens.

But if one strip is copper and the other magnesium, the bulb lights up. Electricity is being produced. Electrons are flowing through the wires even though there is no battery.

Here the bulb is replaced by a **voltmeter**. This measures the 'push' or **voltage** that makes the electrons flow. It is 2.7 volts. The needle shows the direction of the electron flow.

Where do the electrons come from? The answer is this:
Magnesium can give up electrons more readily than copper. So magnesium atoms give up electrons and go into solution as ions. The electrons flow along the wire to the copper strip.
This arrangement is called a **cell**. The magnesium strip is the **negative pole** of the cell. The copper strip is the **positive pole**.

A cell consists of two different metals and an electrolyte. In the cell, chemical energy produces electricity. The more reactive metal becomes the negative pole from which electrons flow.

The experiment can be repeated with other metals. As long as the strips are made of different metals, electrons will flow. But the voltage changes with the metals, as this table shows.

Metal strips	Volts
Copper and magnesium	2.70
Copper and iron	0.78
Lead and zinc	0.64
Lead and iron	0.32

Of these metals, copper and magnesium are furthest apart in the reactivity series. They give the highest voltage. Lead and iron are closest. Look at the voltage they give.
The further apart the metals are in the reactivity series, the higher the voltage of the cell.

The correct name for a torch battery is a 'dry cell'. It contains two different metals and an electrolyte. The electrolyte is a paste rather than a liquid, because a liquid would leak.

Questions

1 Why is sodium never found uncombined in nature?
2 Which will break down more easily on heating, magnesium nitrate or silver nitrate? Explain.
3 Why is magnesium more reactive than copper?
4 Explain why the bulb lights in experiment B above.
5 Why does the bulb not light in experiment A?
6 Will the bulb in B light if a sugar solution is used instead? Explain.
7 Which pair of metals gives the highest voltage? Why? iron, zinc; copper, iron; silver, magnesium
8 In question 7, which metal in each pair becomes the negative pole?

10.5 Metals in the earth's crust

We get some metals from the sea, but most from the earth's **crust**. The earth's crust is the outer layer of the earth. It is only about 6 km thick under the oceans, and about 70 km thick under the highest mountain. But no-one has yet managed to dig a hole through it, to reach the next layer!

Iron is a very important metal. We use about nine times more iron than all the other metals put together. It is made into steel and used for large things like the Forth bridge (above) as well as small things like needles.

The composition of the earth's crust

The earth's crust is a huge mixture of many different compounds. It also contains *elements* such as sulphur, copper, silver, platinum and gold. These occur uncombined or **native** because they are unreactive. If you could dig up the earth's crust and break down all the compounds, you would find it is almost half oxygen! Its composition is:

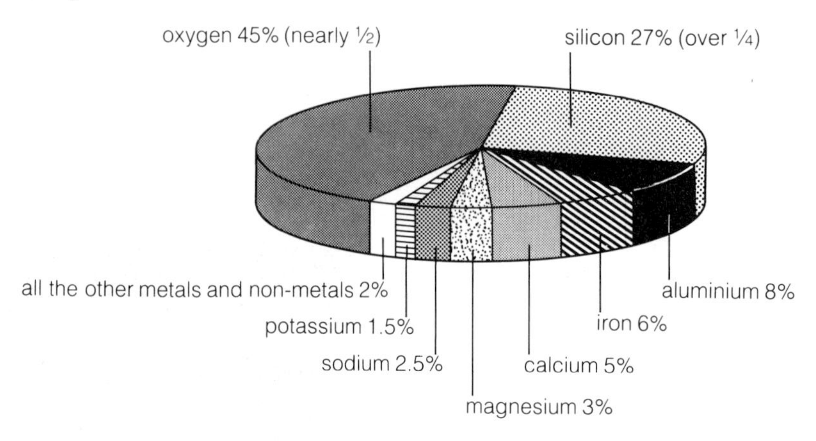

Nearly $\frac{3}{4}$ of the earth's crust is made of just two non-metals, oxygen and silicon. These are found combined in compounds such as silicon dioxide (**silica** or **sand**). Oxygen also occurs in compounds such as aluminium oxide, iron oxide and calcium carbonate.
The rest of the crust consists mainly of just six metals. Aluminium is the most abundant of these, and iron is next. None of the six is found uncombined, because they are all reactive metals, and reactive metals 'like' to form compounds.

Some metals are scarce

Look again at the pie chart above. It shows that there are just six plentiful metals. All the other metals *together* make up less than one-fiftieth of the earth's crust, so they are not plentiful.
If a metal makes up less than one-thousandth of the earth's crust, it is called **scarce**. Here are some of the scarce metals:

copper	zinc	lead	tin
mercury	silver	gold	platinum

The scarce metals are expensive to buy. Even so, we are using them up very quickly, and many people are worried that they will soon run out.

A wedding kimono woven from platinum, a scarce metal. It is worth £100,000.

Metal ores

The rocks in the earth's crust are a mixture of many different compounds. But some contain a larger amount of just one metal compound, or one metal, and it may be worth digging these up to extract the metal. Rocks from which metals are obtained are called **ores**. Below are some examples:

This is a chunk of **rock salt**, the main ore of sodium. It is mostly **sodium chloride**. Nearly all countries have some rock salt. The largest deposits are in Russia and the United States.

This is a piece of **bauxite**, the main ore of aluminium. It is mostly **aluminium oxide**. Huge amounts of it are found in Australia and Jamaica. But there is none in Britain.

Since gold is unreactive, it is found as the free element. This is a lump of nearly pure gold. Most of the world's gold comes from South Africa.

To mine or not to mine?

Before starting to mine an ore, the mining company must decide whether it is economical. It must find the answers to questions like these:

1 How much ore is there?
2 How much metal will we get from it?
3 Are there any special problems about getting the ore out?
4 How much will it cost to mine the ore and extract the metal from it? (The cost will include buildings, mining equipment, extraction plant equipment, transport, fuel, chemicals, and wages.)
5 How much will we be able to sell the metal for?
6 So will we make a profit if we go ahead?

The answers to these questions will change from year to year. If the cost of fuel drops, and the selling price of a metal rises, even a low quality or **low grade** ore may become worth mining.

The mining company also has to think about the local community. People may worry that the landscape will be ruined by dust and pits and scars. They may also worry that the extraction plant will cause pollution. On the other hand they may welcome the new jobs that mining will bring.

A landscape spoiled by mining. These days, many mining companies take care to clear and restore the land when a mine is exhausted.

Questions

1 Which is the main *element* in the earth's crust?
2 Which is the most common *metal* in the earth's crust? Which is the second?
3 Gold occurs *native* in the earth's crust. Explain.
4 Is it true that the reactive metals are plentiful in the earth's crust?
5 What is a *scarce* metal? Name four.
6 One metal is used more than all the others put together. Which one?
7 What is a metal ore?
8 Name the main ore of: **a** sodium **b** aluminium What is the main compound in each ore?

10.6 Extracting and recycling metals

Ways of extracting metals

On the last page you saw that metals are obtained from **ores**. An ore is usually a compound of the metal, mixed with impurities. When the ore has been dug up, it must be decomposed in some way, to get the metal. This is called **extracting** the metal.
The method of extraction depends on how reactive the metal is.
The more reactive the metal, the more difficult its compounds are to decompose.
Electrolysis is the most powerful way to decompose a metal compound. But it needs a lot of electricity, and that makes it expensive. So it is used only for the more reactive metals, as this table shows.

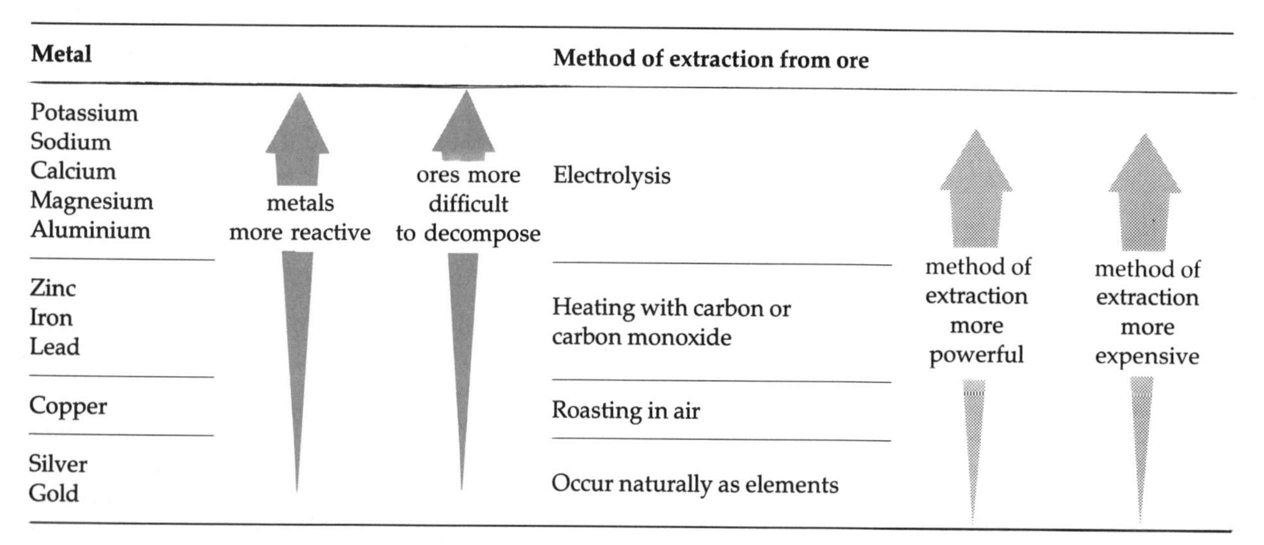

Metal	Method of extraction from ore
Potassium Sodium Calcium Magnesium Aluminium	Electrolysis
Zinc Iron Lead	Heating with carbon or carbon monoxide
Copper	Roasting in air
Silver Gold	Occur naturally as elements

Examples of the different methods of extraction

1 **Electrolysis**. It is used for extracting sodium from rock salt. The rock salt is first melted in giant steel tanks:

sodium chloride ⟶ sodium + chlorine

$$2NaCl(l) \longrightarrow 2Na(l) + Cl_2(g)$$

Electrolysis is also used for extracting aluminium (page 144).

2 **Heating with carbon monoxide**. This is used for extracting iron from iron ore in the blast furnace (page 146):

iron(III) oxide + carbon monoxide ⟶ iron + carbon dioxide

$$Fe_2O_3(s) + 3CO(g) \longrightarrow 2Fe(l) + 3CO_2(g)$$

3 **Roasting in air**. Some copper is found native. But most occurs as copper(I) sulphide, in an ore called **copper pyrites**. The copper is extracted by roasting the sulphide in air:

copper(I) sulphide + oxygen ⟶ copper + sulphur dioxide

$$Cu_2S(s) + O_2(g) \longrightarrow 2Cu(l) + SO_2(g)$$

Until a hundred years ago, aluminium was rarely used because it was difficult to extract. Then in 1886 the modern process of electrolysis was developed. The statue of Eros in London's Piccadilly Circus was made from aluminium in 1893.

The recycling of metals

Metals are a **non-renewable** resource. In other words, when you dig up an ore, no new ore forms to replace it. So the supply will eventually run out. Look at this table:

Metal	Amount used up each year (million tonnes)	How many years before the metal runs out
Iron	800	114
Aluminium	12	350
Copper	8	45
Zinc	4.5	68
Lead	4	23
Tin	0.25	18

These are only rough figures. But they show how careful we need to be. It makes sense to **recycle** metals—that is, to melt down used metals and use them again, rather than throw them away.

A metal company will recycle used metal only if it is economical. The company has to work out the cost of collecting the scrap, transporting it, melting it down, getting rid of impurities, and paying the workers' wages. In future, as metals get scarcer and more expensive, recycling will become a more important process.

The electrolysis of aluminium is very expensive, because of all the electricity it uses. Aluminium beer cans can be melted down for only $\frac{1}{20}$ of the cost.

Scrap iron can be separated from other metals using a magnet.

Every year people in Britain use and throw away 10000 million tins. The metal in them is worth £15000000. But unfortunately it is not easy to separate the layer of tin from the steel can.

Questions

1. What does the *extraction* of a metal mean?
2. For the extraction of sodium, name:
   the ore used the method used
   Why must the ore be melted first?
3. Aluminium is more common than iron in the earth's crust, but more *expensive* too. Suggest why.
4. Why is electrolysis needed for extracting calcium? Why is it *not* needed for extracting copper?
5. Metals are a non-renewable resource. It makes sense to recycle them. Explain the underlined words.
6. Which property of iron can be used to separate it from other kinds of scrap metal?

10.7 Making use of metals

Pure metals and alloys

The way a metal is used depends on its **properties**:

Pure aluminium can be rolled into very thin sheets, which are quite strong but easily cut. So it is used for milk tops and cooking foil.

Pure lead is soft, and bends easily without being heated. It also resists corrosion. So it is used to protect brickwork around chimneys and projecting windows.

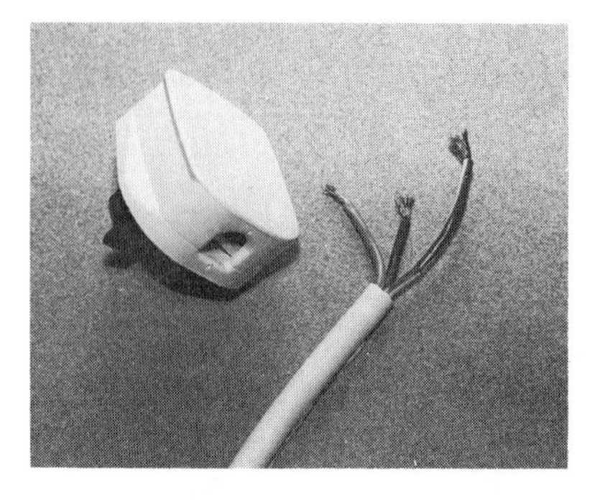

Pure copper is easily drawn into wires, and is an excellent conductor of electricity. So it is used for electrical wiring.

Sometimes a metal is most useful when it is pure. For example, copper is not nearly such a good conductor, when it contains impurities.
But many metals are more useful when they are *not* pure. Iron is the most widely-used metal of all, and it is almost never used pure:

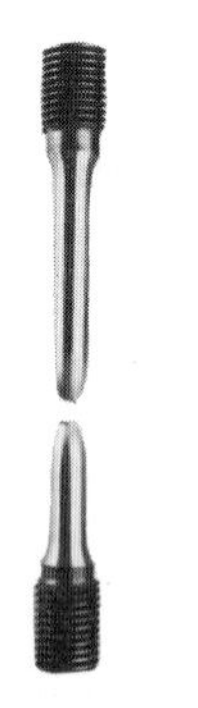

Pure iron is no good for building things, because it is too soft and stretches easily. Besides, it rusts easily too.

But when a little carbon (0.5%) is mixed with it, it becomes much harder and stronger. It is now called **mild steel**, and is used for buildings, bridges, ships and car bodies.

When nickel and chromium are mixed with it, it becomes harder and no longer rusts. It is called stainless steel, and is used for car parts, kitchen sinks, and all the tableware shown above.

You can see that the properties of the iron have been changed, by mixing other substances with it. The mixtures are called **alloys**.
The properties of any metal can be changed, by mixing other substances with it. The mixtures are called alloys.
The added substances are usually metals, but sometimes non-metals like carbon or silicon. An alloy is usually made by melting the main metal and then dissolving the other substances in it.

Uses of pure metals

This table summarises some uses of pure metals:

Metal	Uses	Properties that make it suitable
Sodium	A coolant in nuclear reactors	Conducts heat well. Melts at only 98 °C so the hot metal will flow along pipes.
Aluminium	Overhead electricity cables Saucepans	A good conductor of electricity (not as good as copper, but cheaper and much lighter). Conducts heat well and is non-toxic (non-poisonous).
Zinc	Coating iron, to give **galvanized** iron	Protects the iron from rusting.
Tin	Coating steel cans or 'tins'	Unreactive and non-toxic. Protects the steel from rusting.
Mercury	Thermometers	Liquid at room temperature. Expands on heating. Easy to see, and does not wet the sides of tubes.

Notice that sodium is not used for making things—it is too reactive.
But in nuclear reactors it is well protected from any reactions.

Uses of alloys

There are thousands of different alloys. Here are just a few!

Alloy	Made from	Special properties	Uses
Cupronickel	75% copper 25% nickel	Hard wearing, attractive silver colour	'Silver' coins
Stainless steel	70% iron 20% chromium 10% nickel	Does not rust	Car parts, kitchen sinks, cutlery
Duralium	96% aluminium 4% copper	Light, and stronger than aluminium	Aircraft parts
Brass	70% copper 30% zinc	Harder than copper, does not corrode	Musical instruments
Bronze	95% copper 5% tin	Harder than brass, does not corrode, sonorous	Statues, ornaments, church bells
Solder	70% tin 30% lead	Low melting point	Joining wires and pipes

You can find out more about the alloys of iron, on page 147.

Questions

1 Why is iron more useful, when it is mixed with a little carbon?
2 What are alloys? How are they made?
3 Explain why tin is used to coat food tins.
4 Name an alloy that:
a has a low melting point **b** never rusts
5 Which metals are used to make:
stainless steel? duralium? bronze?

10.8 More about aluminium

From rocks to rockets

Aluminium is the most abundant metal in the earth's crust.
Its main ore is **bauxite**, which is aluminium oxide mixed with impurities like sand and iron oxide. The impurities make it reddish brown.

These are the steps in obtaining aluminium:

1 First, geologists test rocks to find out how much bauxite there is, and whether it is worth mining. If the tests are satisfactory, mining begins.

2 Bauxite usually lies near the surface, so it is easy to dig up. This is a bauxite mine in Jamaica. Everything gets coated with red-brown bauxite dust.

3 From the mine, the ore is taken to a bauxite plant, where it is treated to remove the impurities. The result is white **aluminium oxide**, or **alumina**.

4 The alumina is taken to another plant for electrolysis. Much of the Jamaican alumina is shipped to plants like this one, in Canada or the USA . . .

5 . . .where electricity is cheaper. There it is electrolysed to give aluminium. The metal is made into sheets and blocks, and sold to other industries.

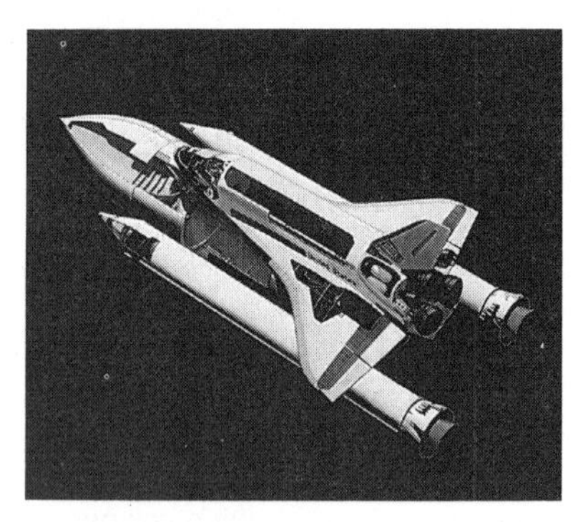

6 It is used to make beer cans, cooking foil, saucepans, racing bikes, T.V. aerials, electricity cables, ships, aeroplanes and even space rockets.

A closer look at the electrolysis

In step 5 above you saw that aluminium is obtained from alumina by electrolysis.
The electrolysis is carried out in a huge steel tank. The tank is lined with graphite, which acts as the cathode. Huge blocks of graphite hang in the middle of the tank, and act as anodes.

Pure alumina melts at 2045 °C. It would be expensive, and dangerous, to keep the tank at that temperature. Instead, the alumina is dissolved in molten **cryolite**, for the electrolysis. (Cryolite is another aluminium compound, with a much lower melting point.)

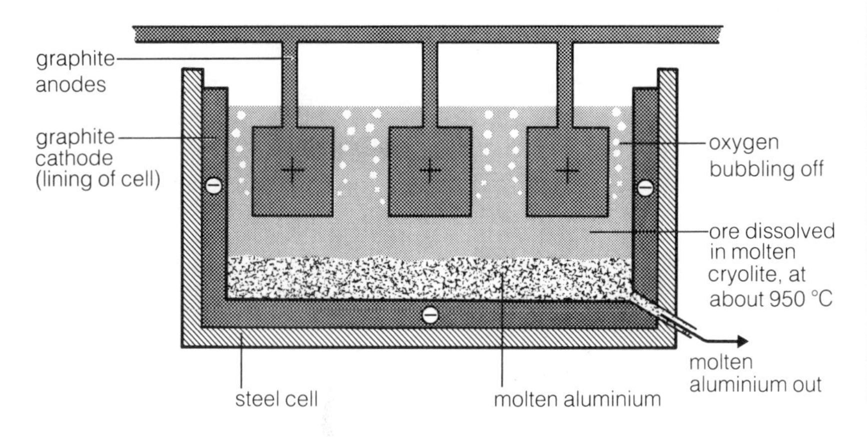

A carbon anode on its way to the electrolysis tank.

When the alumina dissolves, its aluminium ions and oxide ions become free to move.

At the cathode The aluminium ions receive electrons:

$4\,Al^{3+} + 12\,e^- \longrightarrow 4\,Al$

The aluminium atoms collect together, and drop to the bottom of the cell as molten metal. This is run off at intervals.

At the anodes The oxygen ions give up electrons:

$6\,O^{2-} \longrightarrow 3\,O_2 + 12\,e^-$

Oxygen gas bubbles off. But unfortunately it attacks the graphite anodes and eats them away, so they must be replaced from time to time.

Some properties of aluminium

1. Aluminium is a bluish-silver, shiny metal.
2. Unlike most metals, it has a low density—it is 'light'.
3. It is a good conductor of heat and electricity.
4. It is malleable and ductile.
5. It is non-toxic.
6. It is not very strong when pure, but it can be made stronger by mixing it with other metals to form alloys (page 142).

These properties lead to the wide range of uses for aluminium, given in step 6 on the opposite page.

This underground train is made of aluminium, strengthened with small amounts of other metals.

Questions

1. Copy and complete: The chief ore of aluminium is called .... It is first purified to give .... or ...., which has the formula .... Then this is .... to give aluminium.
2. Draw the cell for the electrolysis of aluminium.
3. Why do the aluminium ions move to the cathode?
4. What happens at the cathode?
5. Why must the anodes be replaced from time to time?
6. List six uses of aluminium. For each, say what properties of the metal make it suitable.

10.9 More about iron

The extraction of iron

A stockpile of iron ore.

Iron is the second most abundant metal in the earth's crust. To extract it, three substances are needed:

1 **Iron ore**. The chief ore of iron is called **haematite**. It is mainly iron oxide, Fe_2O_3, mixed with sand.
2 **Limestone**. This is a common rock. It is mainly calcium carbonate, $CaCO_3$.
3 **Coke**. This is made from coal, and is nearly pure carbon.

These three substances are mixed together, to give a mixture called **charge**. The charge is heated in a tall oven called a **blast furnace**. Several reactions take place, and finally liquid iron is produced.

In the blast furnace A blast furnace is like a giant chimney, at least 30 m tall. It is made of steel, and lined with fireproof bricks. The charge is added through the top. Hot air is blasted in through the bottom, making the charge glow white hot. These reactions take place:

1 The coke reacts with oxygen in the air, giving **carbon dioxide**:

$C(s) + O_2(g) \longrightarrow CO_2(g)$

2 The limestone decomposes to **calcium oxide** and **carbon dioxide**:

$CaCO_3(s) \longrightarrow CaO(s) + CO_2(g)$

3 The carbon dioxide reacts with more coke, giving **carbon monoxide**:

$C(s) + CO_2(g) \longrightarrow 2CO(g)$

4 This reacts with iron oxide in the ore, giving liquid **iron**:

$Fe_2O_3(s) + 3CO(g) \longrightarrow 2Fe(l) + 3CO_2(g)$

The iron trickles to the bottom of the furnace.

5 Calcium oxide from step 2 reacts with sand in the ore, to form **calcium silicate** or **slag**:

$CaO(s) + SiO_2(s) \longrightarrow CaSiO_3(l)$

The slag runs down the furnace and floats on the iron.
The slag and iron are drained from the bottom of the furnace. When the slag solidifies it is sold, mostly for roadbuilding. Only *some* of the iron is left to solidify, in moulds. The rest is taken away, while still hot, and turned into steel.

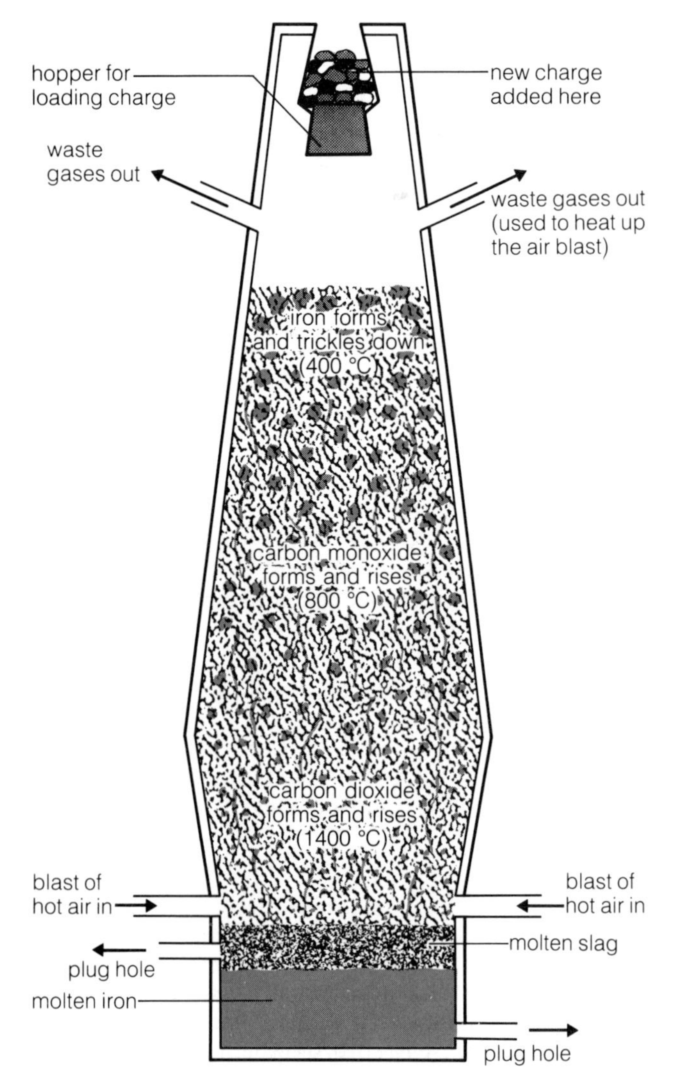

Cast iron

The iron from the blast furnace usually contains a lot of carbon (up to 4 %) as well as other impurities. You saw that some of the iron is allowed to solidify, in moulds or **casts**. It is called **cast iron**. The carbon makes it very hard, but also brittle—it snaps under strain. So these days is only used for things like gas cylinders, railings and storage tanks, that are not likely to get bent during use.

The famous bridge at Ironbridge is made of cast iron. Opened in 1781, it was the first iron bridge in the world. Modern bridges are made of steel.

Steel

Most of the iron from the blast furnace is turned into **steel**. Steel is an alloy of iron. There are many different types, and each has different properties. This is how they are made:

1. **First, unwanted impurities are removed from the iron.** This is done in an **oxygen furnace**. The molten metal is poured into the furnace, as shown on the right. A lot of scrap iron is added too. Then some calcium oxide is added, and a jet of oxygen is turned on. The calcium oxide reacts with some of the impurities, forming a slag that can be skimmed off. Oxygen reacts with the others, and they burn away.
   For some steels, *all* the impurities are removed from the iron. But many steels are just iron plus a small amount of carbon—enough to make the metal hard, but not brittle. So the carbon content has to be checked continually. When it is correct, the oxygen is turned off.
2. **Then, other elements may be added.** As you saw on page 142, different elements affect iron in different ways. The added elements are carefully measured out, to give steels of exactly the required properties.

Molten iron being poured into an oxygen furnace.

There are thousands of different steels. Below are just three of them:

Name	Contains			Special property	Uses
Mild steel	99.5 % Fe,	0.5 % C		Hard but easily worked	Buildings, car bodies
Hard steel	99 % Fe,	1 % C		Very hard	Blades for cutting tools
Duriron	84 % Fe,	1 % C,	15 % Si	Not affected by acid	Tanks and pipes in chemical factories

Questions

1. Name the raw materials for extracting iron.
2. Write an equation for the reaction that gives iron.
3. The calcium carbonate helps to purify the iron. Explain how, with equations.
4. Name a waste gas from the furnace.
5. The slag and waste gas are both useful. How?
6. What is cast iron? Why is it brittle?
7. Explain how mild steel is made.
8. What makes *hard steel* harder than mild steel?
9. Explain why most iron is turned into steel.

10.10 Corrosion

When a metal is attacked by air, water or other substances in its surroundings, the metal is said to **corrode**:

Damp air quickly attacks potassium, turning it into a pool of potassium hydroxide solution.

Iron also corrodes in damp air. So do most steels. The result is **rust** and the process is called **rusting**. In time, this car will fall apart.

But gold is unreactive and never corrodes. This gold mask of King Tutankhamoun was buried in his tomb for over 3000 years, and is still as good as new.

In general, the more reactive a metal is, the more readily it corrodes.

The corrosion of iron and steel

The corrosion of iron and steel is called **rusting**. The iron is **oxidised**. It needs both air and water, as these tests show.

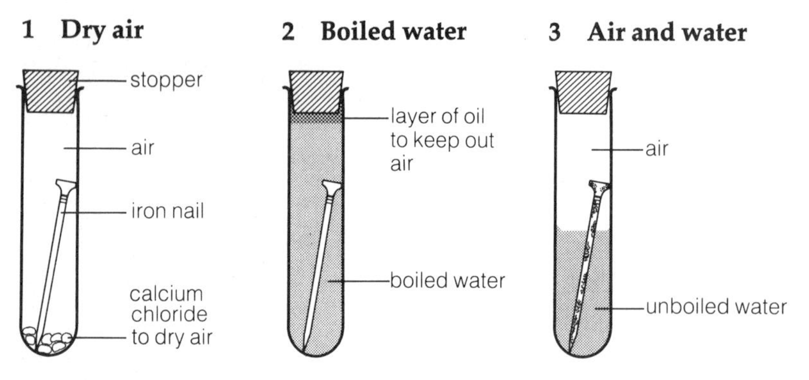

Nails 1 and 2 do not rust. Nail 3 does rust.

How to stop rust. When iron is made into stainless steel, it does *not* rust. But stainless steel is too expensive to use in large amounts. So other methods of rust prevention are needed.
Below are some of the methods. They mostly involve *coating the metal with something, to keep out air and water*:

1 **Paint.** Steel bridges and railings are usually painted. Paints that contain lead or zinc are mostly used, because these are especially good at preventing rust. For example, 'red lead' paints contain an oxide of lead, Pb_3O_4.
2 **Grease.** Tools, and machine parts, are coated with grease or oil.

Steel girders for building bridges are first sprayed with special paint.

3 **Plastic.** Steel is coated with plastic for use in garden chairs, bicycle baskets and dish racks. Plastic is cheap and can be made to look attractive.
4 **Galvanizing.** Iron for sheds and dustbins is usually coated with zinc. It is called **galvanized iron**.
5 **Tin plating.** Baked beans come in 'tins' which are made from steel coated on both sides with a fine layer of tin. Tin is used because it is unreactive, and non-toxic. It is deposited on the steel by electrolysis, in a process called **tin plating**.
6 **Chromium plating.** Chromium is used to coat steel with a shiny protective layer, for example for car bumpers. Like tin the chromium is deposited by electrolysis.
7 **Sacrificial protection.** Magnesium is more reactive than iron. When a bar of magnesium is attached to the side of a steel ship, it corrodes instead of the steel. When it is nearly eaten away it can be replaced by a fresh bar. This is called **sacrificial protection**, because the magnesium is sacrificed to protect the steel. Zinc can be used in the same way.

Steel cans for food are plated with tin.

Does aluminium corrode?

Aluminium is more reactive than iron, so you might expect it to corrode faster in damp air. In fact, clean aluminium starts corroding immediately, but the reaction quickly stops. These diagrams show why.

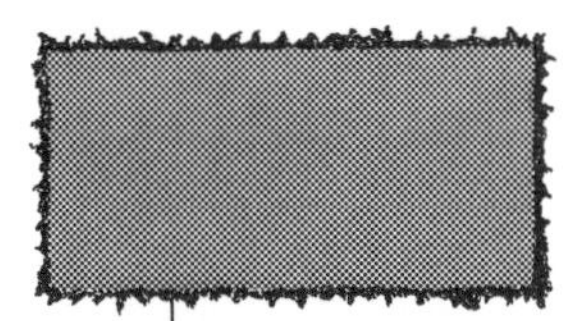

rust flakes

When iron corrodes, rust forms in tiny flakes. Damp air can get past the flakes, to attack the metal below. In time, it rusts all the way through.

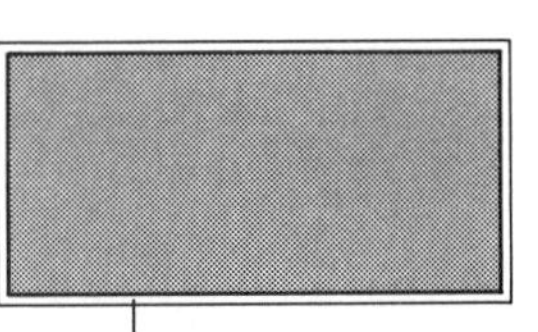

coat of aluminium oxide

But when aluminium reacts with air, an even coat of aluminium oxide forms. This seals the metal surface and protects it from further attack.

The layer of aluminium oxide can be made thicker by electrolysis, to give even more protection. This process is called **anodising**. The aluminium is used as the anode of a cell in which dilute sulphuric acid is electrolysed. Oxygen forms at the anode and reacts with the aluminium, so the layer of oxide grows.

Anodised aluminium is used for cookers, fridges, cooking utensils, saucepans, window frames and sometimes for wall panels on buildings. The oxide layer can easily be dyed to give bright colours.

Anodised aluminium is used for door and window frames.

Questions

1 What is *corrosion*?
2 What two substances cause rusting?
3 Iron that is tin-plated does not rust. Why?
4 In one method of rust prevention, the iron is not coated with anything. Which method is this?
5 Explain why magnesium can prevent the rusting of iron. Why would zinc do instead?
6 Why does aluminium not corrode right through?
7 How can a layer of aluminium oxide be made thicker? What is the process called?

Questions on Chapter 10

1 Read the following passage about the physical properties of metals.
Elements are divided into metals and non-metals. All metals are electrical conductors. Many of them have a high density and they are usually ductile and malleable. All these properties influence the way the metals are used. Some metals are sonorous and this leads to special uses for them.

a Explain the meaning of the words underlined.
b Copper is ductile. How is this property useful in everyday life?
c Aluminium is hammered and bent to make large structures for use in ships and aeroplanes. What property is important in the shaping of this metal?
d Name one metal that has a *low* density.
e Some metals are cast into bells. What property must the chosen metals have?
f Add the correct word: *Metals are good conductors of .... and electricity.*
g Name one other physical property of metals and give two examples of how this property is useful.

2 **a** Write a short passage, like that in question 1, about the physical properties of non-metals.
b Give one way in which the *chemical* properties of non-metals and metals differ.

3 This is a list of metals in order of their chemical reactivity:
sodium (most reactive)
calcium
magnesium
zinc
iron
lead
copper
silver (least reactive)

a Which element is stored in oil?
b Which elements will react with *cold* water?
c Choose one metal that will not react with cold water but will react with steam. Draw a diagram of suitable apparatus to demonstrate this reaction. (You must show how the steam is produced.)
d Name the gas given off in **b** and **c**.
e Name another reagent that reacts with metals to give the same gas.
f Which of the metals will *not* react with oxygen when heated?
g How does iron react when heated in oxygen?
h How would you expect
i lead **ii** calcium
to react when heated in oxygen?
(Hint: look at the table on page 132 and the full reactivity series on page 136.)

4 Look again at the list of metals in question 3. Because zinc is more reactive than iron, it will remove the oxygen from iron(III) oxide, on heating.
a Write a word equation for the reaction.
b Decide whether these chemicals will react together, when heated:
i magnesium + lead(II) oxide
ii copper + lead(II) oxide
iii magnesium + copper(II) oxide
c For those which react:
i describe what you would *see*
ii write a word equation
d What name is given to reactions of this type?

5 When magnesium powder is added to copper(II) sulphate solution, solid copper forms. This change occurs because magnesium is more reactive than copper.
a Write a word equation for the reaction.
b Use the list of metals in question 3 to decide whether these will react together:
i iron + copper(II) sulphate solution
ii silver + calcium nitrate solution
iii zinc + lead(II) nitrate solution
c For those which react:
i describe what you would *see*
ii write the word equation
d What name is given to reactions of this type?

6 Choose one metal to fit each description below. (You must choose a different one each time.) Write down the name of the metal. Then write a balanced equation for the reaction that takes place.
a A metal which burns in oxygen.
b A metal which reacts with oxygen without burning.
c A metal which reacts gently with dilute hydrochloric acid.
d A metal which floats on water and reacts vigorously with it.

7 Strips of copper foil and magnesium ribbon were cleaned with sandpaper and then connected as shown below. The bulb lit up.

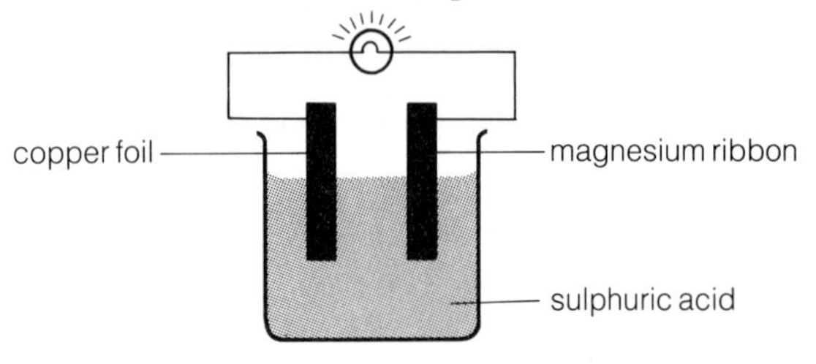

a Why were the metals first cleaned with sandpaper?
b Name the electrolyte used.

c Explain why the bulb lit up.
d Which is the more reactive of the two metals?
e Which of the two metals releases electrons into the circuit?
f What could replace the bulb to measure the 'push' that makes the electrons flow?
g What energy change takes place while the bulb is lit?
h What name is usually given to this type of arrangement?
i Give two good reasons why this particular arrangement could not be used commercially as a battery.

8 Only a few elements are found uncombined, in the earth's crust. Gold is one example. The rest occur as compounds, and have to be extracted from their ores. This is usually carried out by heating with carbon, or by electrolysis.
Some information about the extraction of four different metals is shown below.

Metal	Formula of main ore	Method of extraction
Iron	Fe_2O_3	Heating with carbon
Aluminium	$Al_2O_3.2H_2O$	Electrolysis
Copper	Cu_2S	Roasting the ore
Sodium	NaCl	Electrolysis

a Give the chemical name of each ore.
b Arrange the four metals in order of reactivity.
c How are the more reactive metals extracted from their ores?
d i How is the least reactive metal extracted from its ore?
ii Why can this method not be used for the more reactive metals?
e Iron and aluminium ores are relatively cheap, but aluminium metal is a lot more expensive than iron metal. Why is this?
f Which of the methods would you use to extract:
i potassium?
ii lead?
iii magnesium?
(Hint: look at the reactivity series on page 136.)
g Gold is a metal found native in the earth's crust. Explain what *native* means.
h Where should gold go, in your list for **b**?
i Name another metal which occurs native.

9 Explain why the following metals are suitable for the given uses. (There should be more than one reason in each case.)
a Aluminium for window frames
b Iron for bridges
c Copper for electrical wiring
d Lead for roofing
e Zinc for coating steel

10 Many metals are more useful when mixed with other elements than when they are pure.
a What name is given to the mixtures?
b What metals are found in these mixtures?
brass, solder, stainless steel
c Describe the useful properties of the mixtures in **b**.

11 Bauxite is an important raw material. It is the hydrated oxide of a certain metal. The metal is extracted from the oxide by electrolysis.
a Which metal is extracted from bauxite?
b The compound cryolite is also needed for the extraction. Why is this?
c What are the electrodes made of?
d i At which electrode is the metal obtained?
ii Write an equation for the reaction that takes place at this electrode.
e i What product is released at the other electrode?
ii This product reacts with the electrode itself. What problem does that cause?
f Give three uses of the metal obtained from bauxite.
g To improve its resistance to corrosion, the metal is often anodized. How is this carried out? What happens to the surface of the metal?

12 a Draw a diagram of the blast furnace. Try it first without looking back at the diagram on page 146. Show clearly on your diagram:
i where air is 'blasted' into the furnace
ii where the molten iron is removed
iii where the second liquid is removed.
b i Name the three raw materials added at the top of the furnace.
ii What is the purpose of each material?
c i What is the name of the second liquid that is removed from the bottom of the furnace?
ii When it solidifies, does it have any uses? If so, name one.
d i Name a waste gas that comes out at the top of the furnace.
ii Does this gas have a use? If so, what?
e Write an equation for the chemical reaction which produces the iron.
f Most of the iron that is obtained from the blast furnace is used to make steel. What element other than iron is present in most steel.

13 *aluminium gold iron tin magnesium mild steel calcium stainless steel*
a In the above list of metals and alloys, only four are resistant to corrosion.
i Which are they?
ii Explain why each is resistant to corrosion.
b Which of the other metals or alloys will corrode most quickly? Explain your answer.

11.1 Hydrogen

Hydrogen is the lightest of all elements. On earth, it occurs in many compounds, such as water and methane. But the free element is so light that it has escaped far above the earth, into the outer atmosphere.
Outside earth, hydrogen is the most common element in the universe. The sun is a white-hot ball of gas, more than a million kilometres across, and largely consisting of hydrogen. Its heat and light are produced by nuclear reactions in which hydrogen is converted to helium.

Sunshine – thanks to hydrogen!

How hydrogen is made

In the laboratory Most metals react with acids, to give hydrogen:

metal + acid ⟶ metal salt + hydrogen

This is a good way to make the gas in the laboratory. Usually, zinc and dilute hydrochloric acid are used. The equation is:

zinc + hydrochloric acid ⟶ zinc chloride + hydrogen

$$Zn(s) + 2HCl(aq) \longrightarrow ZnCl_2(aq) + H_2(g)$$

The apparatus is shown below. Note how the hydrogen is collected. The method works because hydrogen is lighter than air.

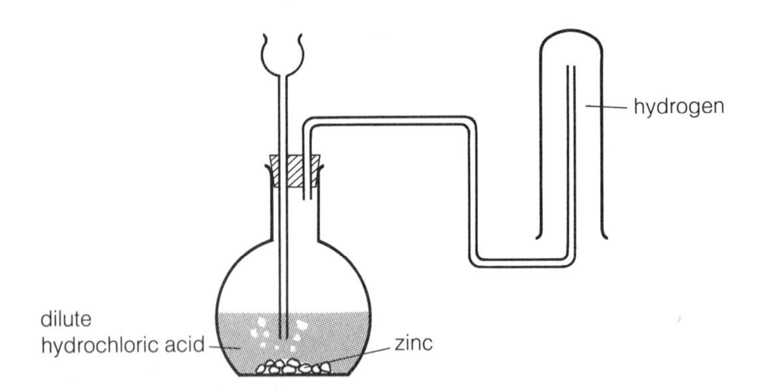

In industry Most hydrogen is made by reacting methane with steam. (In Britain, methane occurs naturally as North Sea gas. Some is also obtained from crude oil.) The methane and steam are mixed, and passed over catalysts. The reaction is:

$$CH_4(g) + 2H_2O(g) \xrightarrow{\text{catalysts}} CO_2(g) + 4H_2(g)$$

The carbon dioxide is removed by dissolving it in water—under pressure, since it is not very soluble. Hydrogen is left behind.

Hydrogen was once used to fill airships but in 1937 an airship called 'The Hindenburg' was destroyed, and 35 passengers killed, when its hydrogen exploded.

The properties of hydrogen

1 It is the lightest of all gases. It is about 20 times lighter than air.
2 It is colourless and has no smell.
3 It is almost insoluble in water.

4 It combines with oxygen to form water:

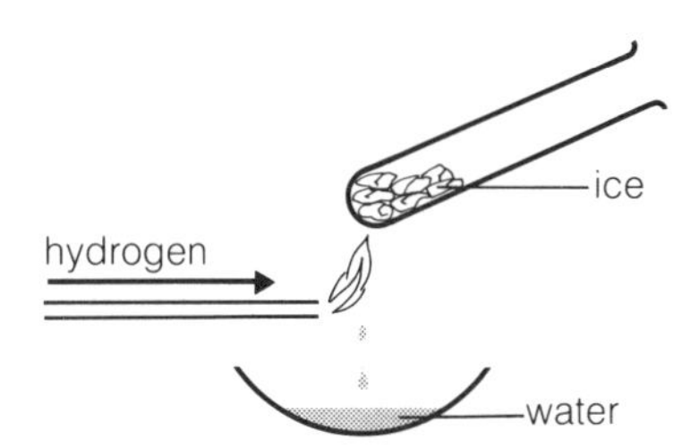

A jet of hydrogen burns in air or oxygen with a blue flame. It can be condensed to water on a cold surface.

A mixture of the two gases explodes, when it is lit. The reaction obviously gives out a lot of energy.

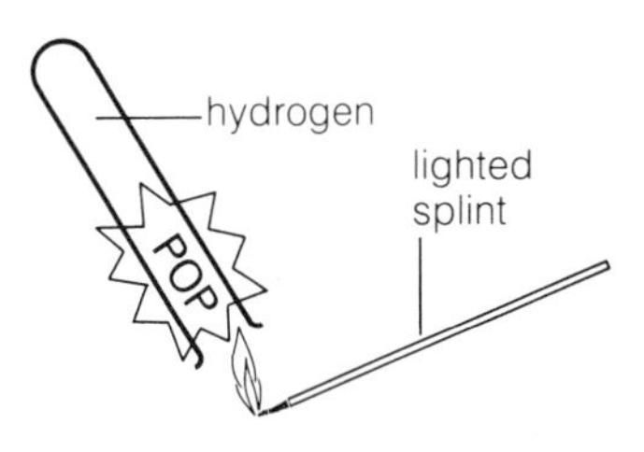

When a lighted splint is held to a test tube of hydrogen, the gas burns with a squeaky pop. This is used as a test for hydrogen.

In each case, the reaction is:

$2H_2(g) + O_2(g) \longrightarrow 2H_2O(g)$

This reaction gives out so much energy that it is used to drive space rockets. The hydrogen and oxygen are stored in tanks in the rockets, in liquid form. Care is needed to make sure they do not leak out and mix accidentally, because that would cause an explosion. In future, cars may run on hydrogen instead of petrol. Since hydrogen forms water when it burns, this would help to cut down pollution.

5 Hydrogen acts as a reducing agent, by removing oxygen. For example, copper(II) oxide is reduced to copper, by heating it in a stream of hydrogen. The reaction is:

$CuO(s) + H_2(g) \longrightarrow Cu(s) + H_2O(g)$

Lead(II) oxide is reduced to lead, in the same way:

$PbO(s) + H_2(g) \longrightarrow Pb(s) + H_2O(g)$

You could carry out these reactions using the apparatus below:

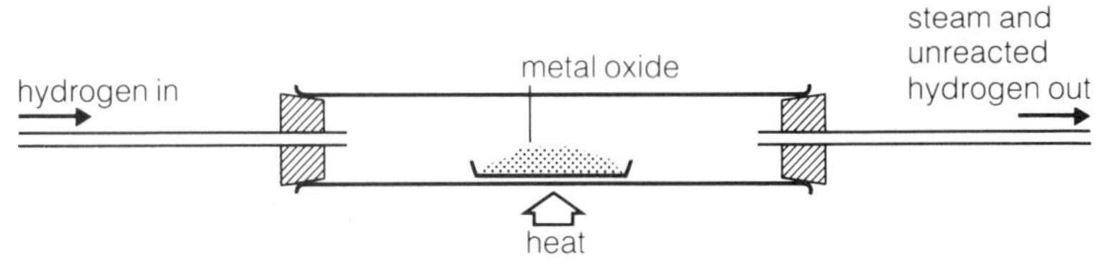

Notice that the hydrogen is oxidised to water, each time.

Uses of hydrogen

Some is used as fuel. Since hydrogen forms water when it burns, rather than harmful gases, this would cut down pollution.
Some is reacted with vegetable oils, to make margarine.
Some is used to make solvents such as methanol (CH_3OH).

Hydrogen is the fuel for this car of the future. In the engine the hydrogen burns in oxygen, giving out energy.

Questions

1. How is hydrogen usually made in the laboratory? Draw a labelled diagram of the apparatus.
2. Name the raw materials for making the gas in industry.
3. Give three *chemical* properties of hydrogen.
4. Complete the equation:
   $FeO(s) + H_2(g) \longrightarrow$ ....
5. List *six* uses of hydrogen.

11.2 Nitrogen

Nitrogen is all around us—it makes up nearly $\frac{4}{5}$ of the air. These are some of its properties:

1 It is a colourless gas, with no smell.
2 It is only slightly soluble in water.
3 It is very unreactive, compared with oxygen. But it does react with some substances:
 (i) it combines with hydrogen, in the presence of a catalyst, to form ammonia, NH_3.
 (ii) it combines with oxygen, at high temperatures, to form oxides such as nitrogen monoxide, NO, and nitrogen dioxide, NO_2. This happens in car engines (page 102). It also happens during flashes of lightning, when the air gets heated up. Oxides of nitrogen are usually harmful, and cause air pollution. But they are helpful in one way, as you will see below.

three shared pairs of electrons

N N

The bonding in nitrogen. Since three pairs of electrons are shared, the bond is a triple bond. It can be shown as N≡N.

The importance of nitrogen

Nitrogen is essential for living things. It is needed to make compounds called **proteins**, that are part of every plant and animal. Living things cannot use nitrogen directly from the air, to make proteins. Instead, plants obtain it from **nitrates** in the soil, and animals obtain it by eating plants. This is what happens:

1 **First, nitrogen from the air is converted to nitrates in the soil.** The change is brought about in three ways:

By lightning. During a lightning flash, oxides of nitrogen form in the air. They dissolve in rain, making an acidic solution. This reacts with compounds in the soil, to give nitrates.

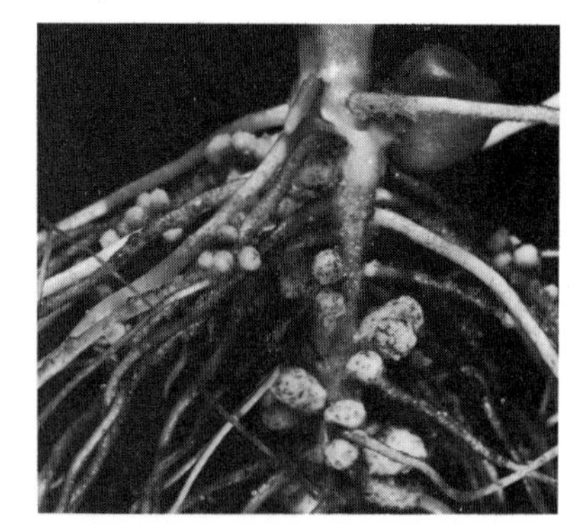

By bacteria. Some bacteria can change nitrogen into nitrates. They live in soil, or in swellings called **root nodules** on the roots of plants such as clover, peas and beans.

By man. In industry nitrogen is turned into **fertilisers.** These are compounds such as potassium nitrate and ammonium nitrate. Farmers then spread the fertilisers on the soil.

2 **Next, plants use these nitrates to make proteins.** The nitrates dissolve in water, and the plants take in the solutions through their roots.
3 **Then animals eat the plants.** They digest the plant proteins, and turn them into the proteins their bodies need. Some animals eat other animals, and obtain protein that way. We humans do both:

we obtain plant protein from cereals and vegetables, and animal protein from meat, fish, eggs, milk and cheese.

4 **Waste protein gets changed back to nitrates.** In soil, there are bacteria which feed on the protein in dead plants and animals, and in animal manure. They change it back to nitrates, which can be used again by plants.

5 **Some nitrates get changed back to nitrogen.** Some bacteria attack nitrates and change them to nitrogen, which goes back into the air. These are called **denitrifying bacteria** and they usually live in heavy wet soils.

Denitrifying bacteria live in heavy wet soil.

The nitrogen cycle

You saw above that nitrogen circulates between the air, the soil and living things. This process is called the nitrogen cycle. Here is a drawing of it:

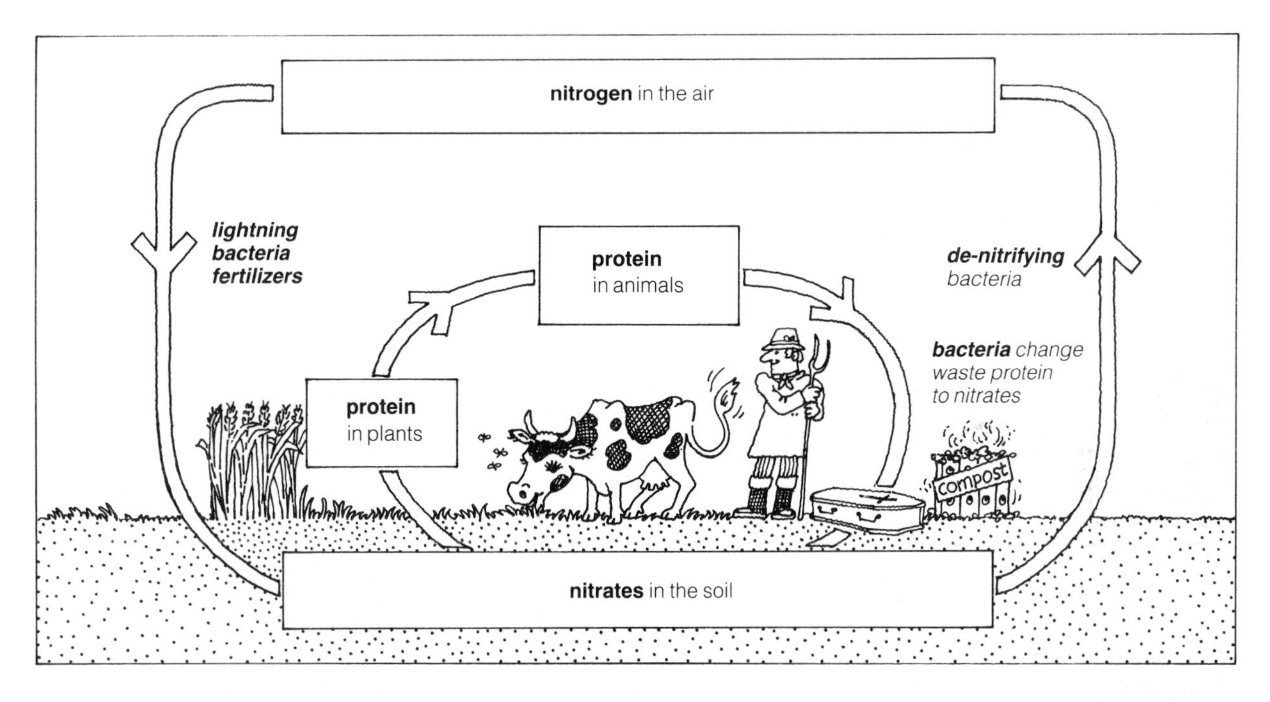

Other uses of nitrogen

Nitrogen is essential for life, but it is also useful in many other ways. That is why about 2500 tonnes of it are produced in Britain every day, by fractional distillation of liquid air. The gas is used for filling spaces in food packages and oil tanks. Liquid nitrogen is very cold, so is used for freezing food. Other uses are given on page 101.

Questions

1 Draw a molecule of nitrogen. The bond in it is called a triple bond. Can you suggest why?
2 In what form do plants take in nitrogen?
3 Give two *natural* ways in which nitrogen is converted to nitrates in the soil.
4 Explain how waste protein is converted to nitrates.
5 What do denitrifying bacteria do?
6 Make a copy of the nitrogen cycle.
7 List *five* uses of nitrogen, in industry.
8 List *five* nitrogen compounds, and their formulae.

11.3 Ammonia

Ammonia is a gas with the formula NH_3.
You could make it in the laboratory by heating any ammonium compound with any alkali. The alkali drives out or displaces ammonia from the compound. For example:

ammonium chloride + calcium hydroxide ⟶ calcium chloride + steam + ammonia

$$2NH_4Cl(s) + Ca(OH)_2(s) \longrightarrow CaCl_2(s) + 2H_2O(g) + 2NH_3(g)$$

In fact this can be used to test for an ammonium compound. If a compound gives off ammonia when it is heated with an alkali, it must be an ammonium compound.

The properties of ammonia

1 Ammonia is a colourless gas with a strong choking smell.
2 It is less dense than air.
3 It is easily liquefied, either by cooling to −33 °C or by compressing. This makes it easy to transport in tanks and cylinders.
4 It is very soluble in water. The fountain experiment shows this:

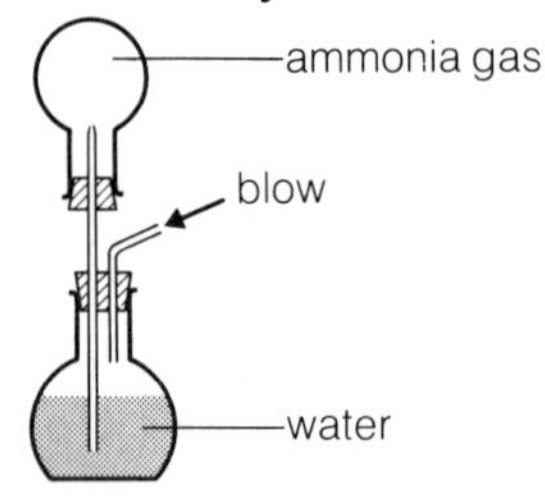

By blowing into the bent tube, you can make water rise up the straight tube.

When the first drops of water reach the top of the tube, they dissolve nearly all the ammonia in the flask.

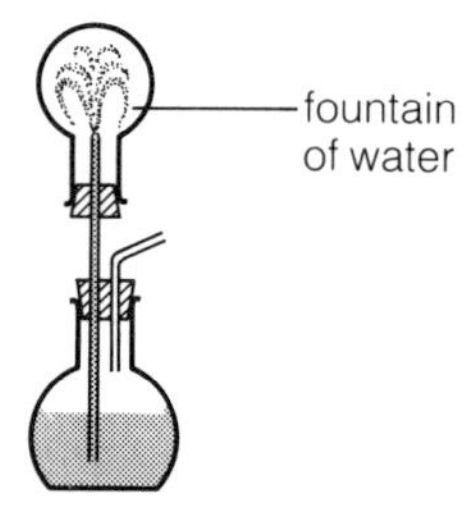

To fill the space left by the gas, water rushes up the tube and bursts out like a fountain.

5 It reacts with hydrogen chloride gas to form a white smoke:

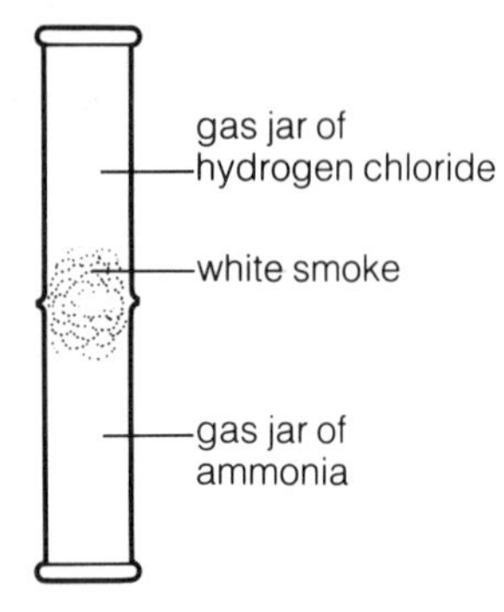

The white smoke is made of tiny particles of solid ammonium chloride:

$$NH_3(g) + HCl(g) \longrightarrow NH_4Cl(s)$$

This reaction can be used to test whether a gas is ammonia.

You can test for ammonia, using a glass rod dipped in concentrated hydrochloric acid. The white fumes prove that the gas coming from the beaker is ammonia.

The properties of ammonia solution

A solution of ammonia in water has these properties:

Ammonium nitrate, a fertiliser made by reacting ammonia with nitric acid.

1 It turns red litmus blue—it is **alkaline**. That means it contains hydroxide ions. Some of the ammonia has reacted with the water, forming ammonium ions and hydroxide ions:

$NH_3\,(aq) + H_2O\,(l) \longrightarrow NH_4^+\,(aq) + OH^-\,(aq)$

Since only some of its molecules form ions, ammonia is a **weak** alkali.

2 Since ammonia solution is alkaline, it reacts with acids to form salts. For example, it reacts with nitric acid to form ammonium nitrate:

$NH_3\,(aq) + HNO_3\,(aq) \longrightarrow NH_4NO_3\,(aq)$

Ammonium nitrate is an important fertiliser.

3 It reacts with a solution of iron(III) chloride to form an orange-brown precipitate:

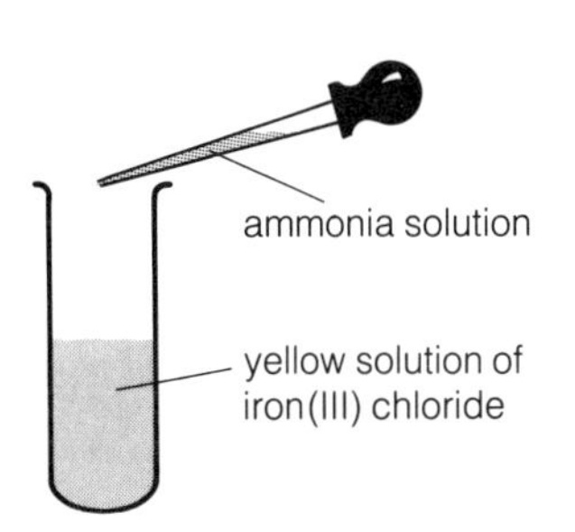

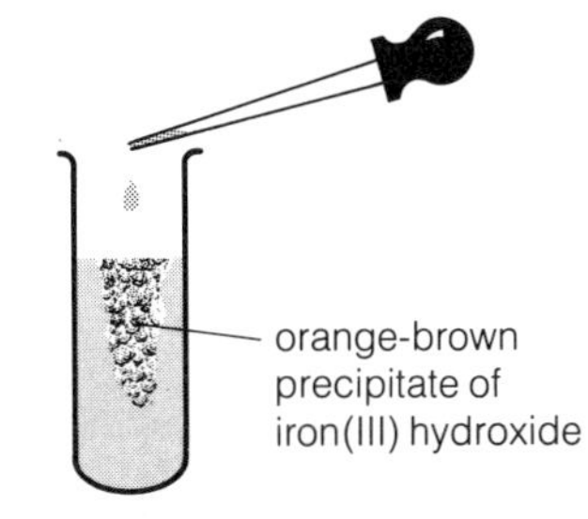

Iron(III) chloride solution contains Fe^{3+} and Cl^- ions. Ammonia solution contains NH_4^+ and OH^- ions.

The Fe^{3+} ions have joined up with the OH^- ions to form insoluble iron(III) hydroxide:
$Fe^{3+}(aq) + 3OH^-(aq) \longrightarrow Fe(OH)_3(s)$

In the same way, ammonia solution reacts with solutions of zinc salts, aluminium salts, iron(II) salts and copper(II) salts, to form insoluble metal hydroxides.

4 It can be used to test for copper(II) compounds:
(i) If a solution contains a copper(II) compound, it will react with ammonia solution to give a blue precipitate of copper(II) hydroxide.
(ii) When more ammonia solution is added, the precipitate will dissolve again, giving a deep blue solution.
The blue solution proves that a copper(II) compound was present. The deep blue colour is due to ions with the formula $[Cu(NH_3)_4]^{2+}$.

These cleaners contain ammonia. Like other alkalis, ammonia attacks grease. Since it is a weak alkali it is safe to use in household cleaners.

Questions

1 Explain how you could test an unknown compound to see if it was an ammonium compound.
2 Write down three physical properties of ammonia.
3 Describe how you would test for ammonia gas, using concentrated hydrochloric acid and a glass rod.
4 Write down two properties of ammonia solution that show it is alkaline.
5 Why is ammonia used in household cleaners?
6 Name the precipitate that forms when ammonia solution is added to zinc chloride solution.

11.4 Ammonia and nitric acid in industry

Making ammonia in industry

Part of the ICI ammonia plant at Billingham in Cleveland.

In industry, ammonia is made from nitrogen and hydrogen. The first step is to obtain these gases.

(i) Hydrogen is made from methane (North Sea gas) and steam, as shown on page 152:

$$CH_4(g) + 2H_2O(g) \xrightarrow{\text{catalysts}} CO_2(g) + 4H_2(g)$$

(ii) Nitrogen is obtained by burning hydrogen in air. Air is mostly nitrogen and oxygen, with small amounts of other gases. Only the oxygen reacts with hydrogen, forming steam:

$$2H_2(g) + O_2(g) \longrightarrow 2H_2O(g)$$

When the steam condenses, the gas that remains is mainly nitrogen.

The reaction between them Nitrogen is unreactive. To make it react with hydrogen, a process called the **Haber process** is used:

1. The two gases are mixed. The mixture is cleaned or **scrubbed**, to get rid of any impurities.
2. Next it is compressed. This pushes the gas molecules closer together.
3. Then it goes to the **convertor**. This is a round tank containing beds of hot iron. The iron is a catalyst for this reaction:

   $$N_2(g) + 3H_2(g) \rightleftharpoons 2NH_3(g)$$

   The reaction is **reversible** (page 73). So it does not go to completion. A mixture of nitrogen, hydrogen and ammonia leaves the convertor.
4. This mixture is cooled until the ammonia condenses. Then the nitrogen and hydrogen are pumped back to the catalyst, for another chance to react.
5. The ammonia is run into tanks and stored as a liquid, under pressure.

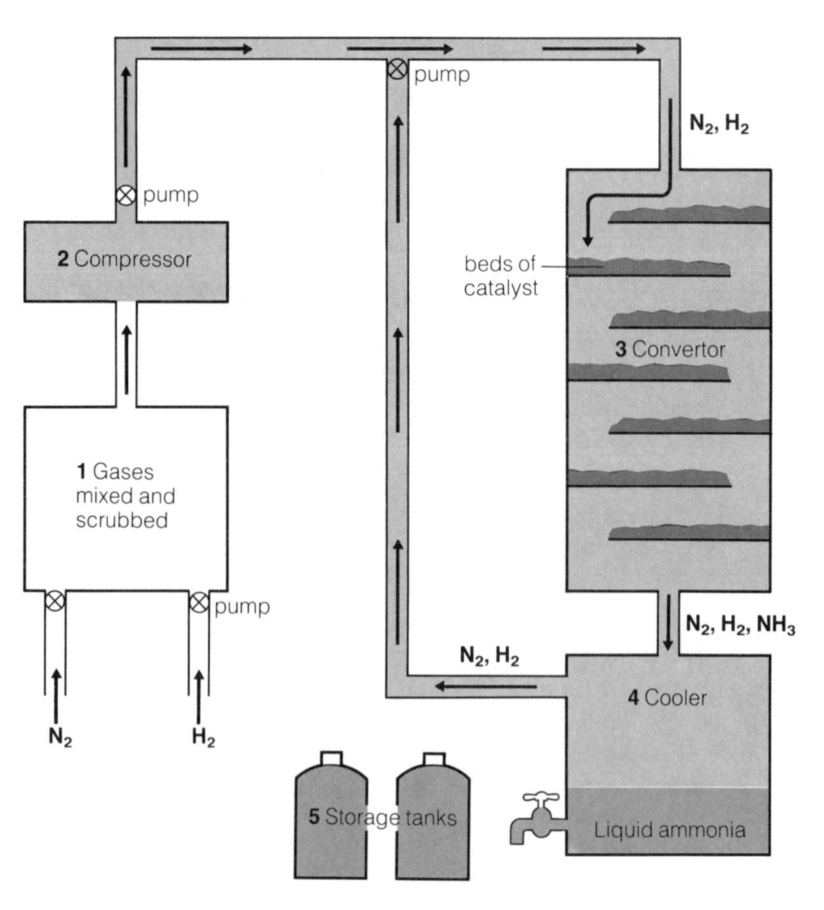

Uses of ammonia

Most of the ammonia from the Haber process is used to make **fertilisers**, such as ammonium nitrate and ammonium sulphate. Some is used to make household cleaners, dyes, explosives and nylon. A lot is used to make nitric acid, as on the next page.

Making nitric acid in industry

A lot of the ammonia from the Haber process is used to make **nitric acid**. The raw materials for nitric acid are **ammonia, air** and **water**. This flow chart shows the stages in the process:

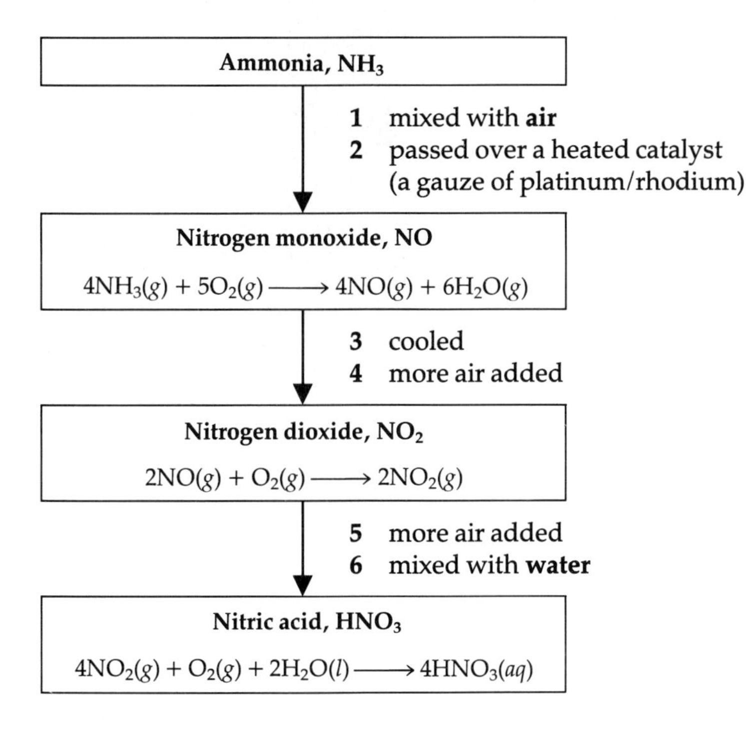

Part of the ICI nitric acid plant at Billingham. It is built close to ICI's ammonia plant. Can you explain why?

The overall result is that ammonia is oxidised to nitric acid. Chemical engineers must make sure that no nitrogen monoxide or nitrogen dioxide can escape from the plant, since these gases cause acid rain. Besides, if any nitric acid goes down the drain, it will end up in the river, killing fish and other river life.

Uses of nitric acid

Most of the nitric acid is used to make **fertilisers.**
Some is used to make explosives such as trinitrotoluene (TNT).
Some is used in making nylon and terylene.
Some is used in making drugs.

Questions

1 Ammonia is made from nitrogen and hydrogen.
- **a** How are the nitrogen and hydrogen obtained?
- **b** What is the process for making ammonia called?
- **c** What catalyst is used? What does it do?
- **d** Write an equation for the reaction.
- **e** The reaction is *reversible*. What does that mean?

2 **a** Name the raw materials for making nitric acid.
b Write word equations for the three reactions that take place, during its manufacture.
c The result of the reactions is that ammonia is *oxidised* to nitric acid. What does that mean?
d What is the main use of nitric acid?

11.5 Fertilisers

We need more food

Every year, the world's population gets larger, as this graph shows:

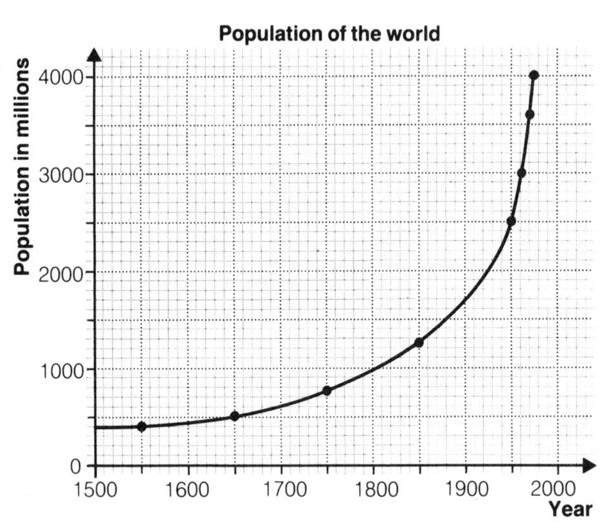

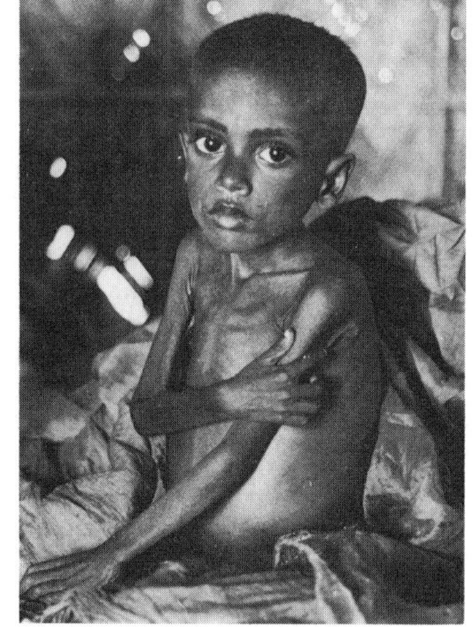

Every day, over 40 000 children die in poor countries. Starvation is mostly to blame.

Just now there are around four thousand million people on earth, and at least one in every eight of them is starving. By the year 2000, the problem will be even more serious.

We depend on plants for all our food. Bread comes from wheat, and even baked beans start off in fields. Animals need grass and other crops, in order to give milk and meat. So one way to tackle the hunger problem is to grow more plants. But they will only grow properly if the soil is in good condition.

What plants need from soil

Plants need **water** from the soil. They also need different **elements**. The three they need most are **nitrogen, potassium** and **phosphorus**.

Nitrogen is essential for proteins, which make strong stems and healthy leaves.

Potassium helps a plant to survive frost and to resist disease.

Phosphorus is needed to make roots grow and seeds form.

Plants also need smaller amounts of **calcium, magnesium, sodium,** and **sulphur,** and tiny amounts of **copper, iron, zinc, manganese,** and **boron**. They obtain all these elements from compounds in the soil, which they take in as solutions, through their roots.

Fertilisers

Plants use up nitrogen, phosphorus and potassium more quickly than the other elements. Small amounts return to the soil in animal manure, and when dead plants and animals decay. Some nitrogen is also put back by lightning and by bacteria (page 154). But not enough. Soon, the soil is too poor for new crops. So the farmer must replace the missing elements by adding **fertilisers**. Fertilisers are chemicals like these:

ammonium nitrate NH_4NO_3
calcium nitrate $Ca(NO_3)_2$
sodium nitrate $NaNO_3$
urea $CO(NH_2)_2$
ammonium sulphate $(NH_4)_2SO_4$
potassium sulphate K_2SO_4
ammonium phosphate $(NH_4)_3PO_4$
ammonia solution $NH_3(aq)$

You can see that ammonium phosphate will put back nitrogen and phosphorus in the soil. What about potassium sulphate? Or urea?

Choosing a fertiliser The fertilisers on sale are usually a mixture of chemicals. They are made up to suit different soils and crops. For example, some contain more nitrogen than others.
A farmer can have his soil tested, to find which elements it lacks. Then he can buy a fertiliser to suit it.
For example, suppose he wants to grow grass, for feeding cattle. Grass needs a lot of nitrogen. So, if his soil lacks nitrogen, the first bag of fertiliser above would give him better value for money.

Problems with fertilisers Many fertilisers are **nitrates**, as you can see from the list above. Nitrates are soluble in water, so they are taken in easily by plants. But they are also washed out of the soil easily by rain. This is wasteful. Besides, when the rainwater drains into rivers, the nitrates cause pollution (page 112).
Urea also contains nitrogen, but it is almost insoluble. That means it acts slowly on plants, but is not washed away.

The fertiliser mixture in this bag contains 20% nitrogen, 8% phosphorus and 14% potassium . . .

. . . And this one contains 10% nitrogen, 25% phosphorus and 15% potassium.

A machine spreading a slurry of animal manure. It contains the natural fertiliser urea, among others.

Cattle farmers use nitrogen fertilisers to help their grass grow.

Questions

1 Name the three main elements that plants need, and say why they are important.
2 How do plants obtain the elements they need?
3 What are fertilisers, and why are they used?
4 Name three fertilisers that would put nitrogen back in the soil.
5 Name one *natural* fertiliser, found in manure.
6 Fertilisers can cause water pollution. Explain why.

11.6 A look at a fertiliser factory

Fertilisers are made all year round, but sales are seasonal. A pellet could lie buried in a 20 metre heap for several months. It must be able to survive storage without losing shape or sticking to other pellets.

What the factory makes

A modern fertiliser factory will make two main kinds of fertiliser:

1 **straight N fertiliser** for farmers who want only nitrogen. In Britain this is usually in the form of ammonium nitrate, NH_4NO_3.
2 **NPK compound fertilisers** for farmers who want nitrogen, phosphorus, and potassium. These are usually a mixture of ammonium nitrate, ammonium phosphate, and potassium chloride.

The factory may make other fertilisers too, but in very much smaller amounts.

The production process

This diagram shows the raw materials the factory needs to make 1000 tonnes of straight and 1000 tonnes of compound fertiliser a day, and the steps in the production process; t/day stands for *tonnes per day*.

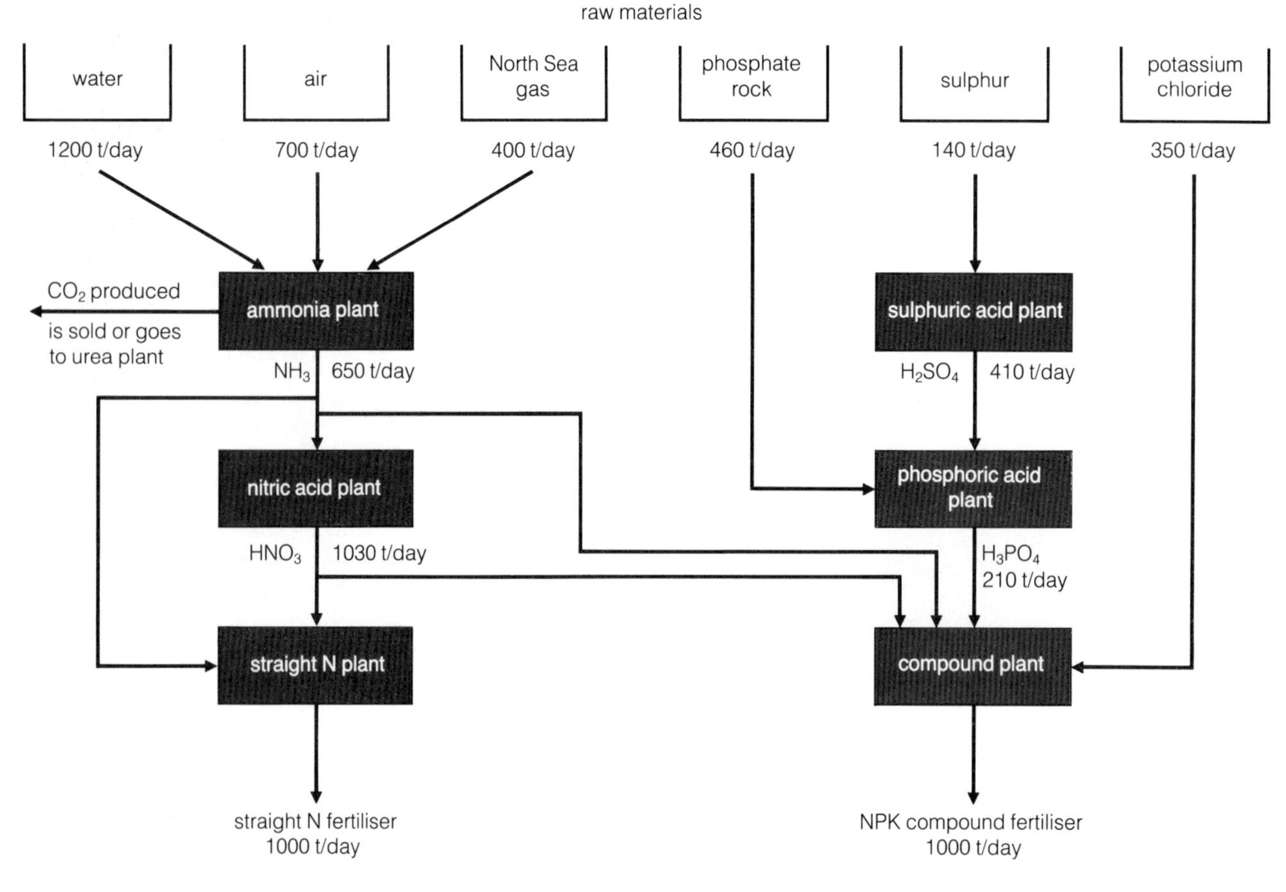

You can see that the factory is not just a single unit. It is six separate plants built close together on the same site. Each plant is controlled so that it is making the right amount of a substance at the right time.

Mollie Travis, personnel officer: 'Of course any new factory has to staff up well in advance of production. We were running training courses for local workers six months before we opened.

Michael Nelson, chemical engineer at the same factory: 'My job is mostly to do with the environment. We continually monitor the waste gases and liquids leaving the plant to make sure they're safe.'

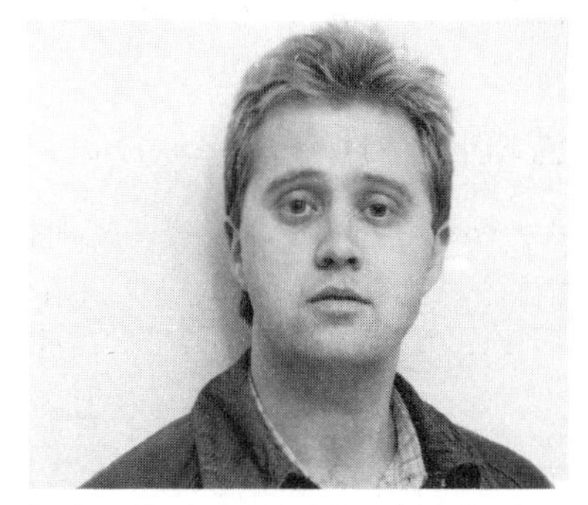

Graham Taylor, technician: 'Me? I've just come off the night shift. A fertiliser factory like this one keeps going 24 hours a day, every day. It's got to, to be economical.

Choosing the factory site

A company considering a new site for a fertiliser factory needs to think about things like these:

1. **Closeness to raw materials.** For example, potassium chloride is mined near Whitby in Yorkshire. Should we build near there?
2. **Closeness to ports.** We need to import sulphur and phosphate. We want to export fertiliser. Should we be close to a port?
3. **The road network.** We will be delivering fertiliser all over the country. Are we close to major roads?
4. **The water supply.** We need hundreds of tonnes of water a day as a raw material, and for cooling. Is it available here?
5. **The gas supply.** We need North Sea gas as a raw material, and as a fuel. Any problems getting it here?
6. **The workforce.** Can we get the kinds of workers we need locally? Will we be able to attract key staff from other areas?
7. **The cost of land.** Is there somewhere cheaper we should go?
8. **Government grants.** Are there special grants for starting a factory in this area?
9. **Objections from the local people.** They may worry that the factory will pollute the air, or the river, or make too much noise, or spoil a view. Will we be able to set their minds at rest?

The company is not likely to find a perfect site. It will have to balance all these factors and choose the site that fits best.

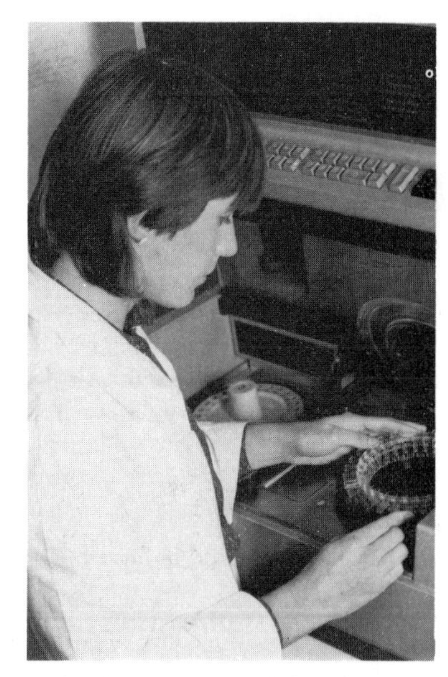

Emily Benson, chemist: 'In fact there are strict legal requirements for the amount of N, P and K in fertilisers. We have machines that check the contents automatically. But we do lab checks too.'

Questions

You may need to look back at pages 158 and 159 to answer some of these questions.

1. What are the two main types of fertiliser a factory is likely to make? Which chemicals do these usually contain?
2. List the raw materials the factory needs.
3. List the acids the factory makes.
4. What is the sulphur used for?
5. North Sea gas is used as a raw material. For which process? Write an equation for the reaction.
6. Write word equations for the reactions involved in making nitric acid from ammonia.
7. List all the ways a fertiliser factory might cause pollution.
8. When choosing a factory site, which *four* factors do you think are most important? Explain your choice.

Questions on Chapter 11

1 The reaction that takes place when hydrogen is passed over iron(II) oxide is:

$FeO(s) + H_2(g) \longrightarrow Fe(s) + H_2O(g)$

a Complete the sentences below.
In the reaction the iron(II) oxide is .... to iron and the hydrogen is .... to water. In the reaction, hydrogen is acting as a ....

b Draw a diagram of the apparatus that could be used to demonstrate this reaction in the laboratory.

c Suggest why carbon monoxide and not hydrogen is used in industry, to extract iron from iron oxide.

d Give one use of hydrogen, in industry.

2 This paragraph is about the element nitrogen. Rewrite it, choosing the correct item from each pair in brackets.
Air is a (mixture/compound) which contains ($\frac{1}{5}$/$\frac{4}{5}$) nitrogen. The symbol for nitrogen is (N/N_2) and the gas is made of (molecules/atoms) represented by the formula (N/N_2). It is therefore a (monatomic/diatomic) gas. Nitrogen is needed by plants to make (proteins/sugars). Most plants are unable to take nitrogen directly from the (water/air), so some 'fixing' of the gas is required. Some is fixed by (rain/lightning) and some is fixed by (leaves/bacteria), but most is fixed artificially by making (pollutants/fertilisers). For these, the nitrogen is first reacted with (hydrogen/oxygen) to give the (gas/solid) ammonia. This is turned into (sulphuric/nitric) acid, which is then used to make fertilisers such as ammonium nitrate.

3 Say where the following words should fit in the diagram below. For example, **a** is **air**. You will need to use some words twice.

nitrogen	*air*	*ammonia*
nitrates	*proteins*	*bacteria*

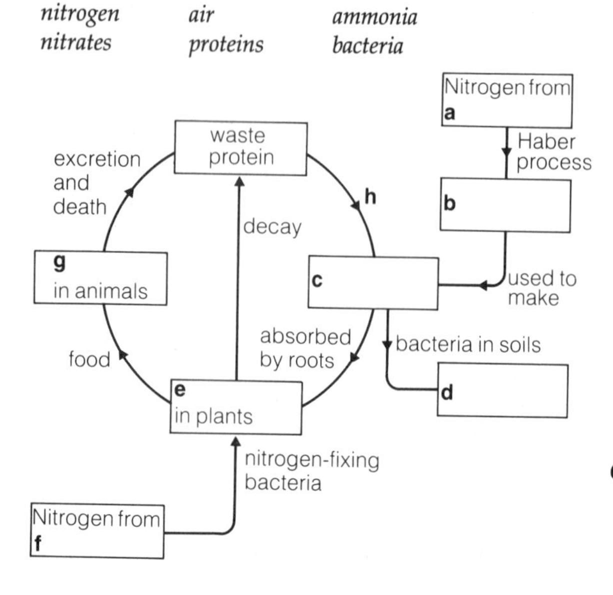

4 This apparatus can be used to prepare *safely* a solution of ammonia in water:

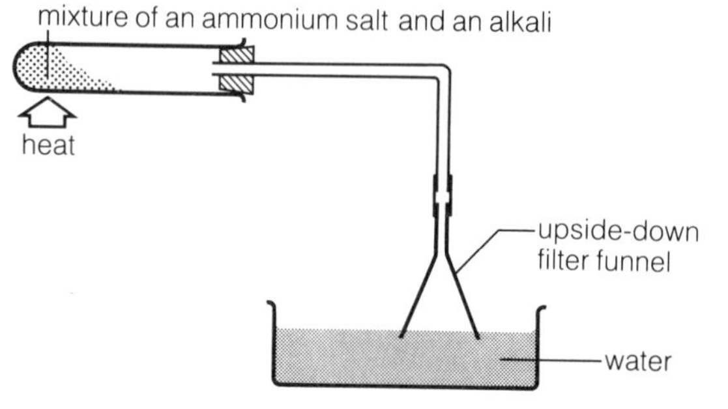

Ammonia gas is given off when the test tube is heated.

a Name two compounds which could be heated together in the test tube.

b Write an equation for the reaction.

c Why would it be dangerous to dip the glass tube into the water, *without* using the filter funnel? (Hint: think about the fountain experiment.)

d If a few drops of litmus solution are added to the water in the trough, what colour change would be seen during the experiment?

e If the water is replaced by a dilute solution of nitric acid, what salt would be formed in the trough?

5 The diagram shows apparatus for an experiment to make ammonia. X and Y are two gas syringes, connected to a combustion tube A. At the start, syringe X contained 75 cm^3 of hydrogen and syringe Y contained 25 cm^3 of nitrogen.

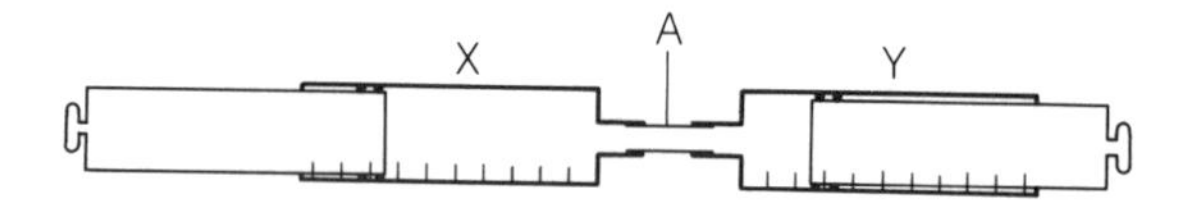

a Copy out the diagram, labelling the gases.

b How would you make the gases mix?

c The reaction needs a catalyst. Why?

d Name a suitable catalyst.

e Where should the catalyst be placed? Add the catalyst to your diagram.

f Where should the apparatus be heated? Show this on your diagram.

g How would you show that some ammonia had been obtained, at the end of the experiment?

6 Look at question 2. Make up a similar question about the properties of ammonia. Use brackets to provide alternatives from which another pupil could choose.

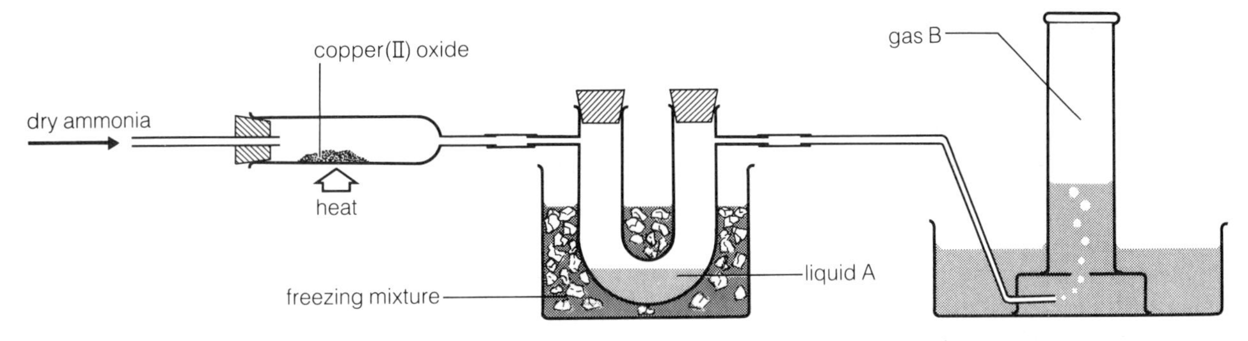

7 Using the above apparatus, dry ammonia is passed over heated copper(II) oxide. The gases given off are passed through a cooled U-tube. A liquid (A) forms in the U-tube and a colourless gas (B) collects in the gas jar.

a The copper(II) oxide is reduced to copper. What would you see as the gas passes over the heated copper oxide?

b Why is the U-tube surrounded by a freezing mixture?

c The liquid A is found to turn blue cobalt chloride paper pink and to have a boiling point of 100 °C. Identify liquid A.

d Is gas B soluble or insoluble in water?

e Identify gas B. (Hint: look at the other chemicals in this reaction.)

f Write a word equation for the reaction.

g The copper oxide is *reduced* by the ammonia. Explain what this means.

h How will the mass of the heated tube and contents change, during the reaction?

8 The manufacture of ammonia and nitric acid are both very important industrial processes.

A AMMONIA

a Name the raw materials used.

b Which two gases react together?

c Why are the two gases scrubbed?

d Why is the mixture passed over iron?

e What happens to the *unreacted* nitrogen and hydrogen?

f Why is the manufactured ammonia stored at high pressure?

B NITRIC ACID

a Name the raw materials used.

b Which chemicals react together to form nitric acid?

c What would happen if the gauze containing platinum and rhodium was removed?

d Why must the chemical plant be constantly checked for leaks?

9 Write equations for the chemical reactions in question 8.

10 Write a paragraph explaining why ammonia and nitric acid are very important chemicals.

11 Ammonium compounds and nitrates are of great importance as fertilisers.

a Why do these compounds help plant growth?

b Name one *natural* fertiliser.

c Name two compounds containing nitrogen which are manufactured for use as fertilisers. Write the chemical formulae for these compounds.

d Name two elements other than nitrogen which plants need, and explain their importance to the plants.

e Why are some fertilisers not suitable for quick-growing vegetables like lettuce?

f Some fertilisers are acidic. What is usually added to soils to correct the level of acidity?

g Land which is intensively farmed needs regular applications of fertiliser. Explain why.

h Fertilisers obviously have advantages. But many people are worried about the increasing use of fertilisers, especially nitrates, by farmers. Can you suggest why?

12 Ammonia gas is bubbled into copper(II) sulphate solution. At first a blue precipitate forms. This then turns to a deep blue solution.

a What is the blue precipitate? (Hint: remember that ammonia solution contains hydroxide ions.)

b Explain why the colour deepens as more ammonia gas is added.

c Is the deep blue compound soluble or insoluble in water?

d This reaction of ammonia is a useful one. Explain why.

13 Nitric acid acts both as an oxidising agent and as an acid. Say which way it is behaving when it:

a is neutralised by sodium hydroxide.

b reacts with a carbonate giving carbon dioxide gas.

c reacts with copper releasing nitrogen dioxide gas.

d reacts with copper oxide to form copper nitrate.

e reacts with dry sawdust, causing it to burst into flames.

f reacts with magnesium, giving a dark brown gas.

g turns litmus solution red

12.1 Oxygen

Making oxygen in the laboratory

The diagram on the right shows one way to make oxygen in the laboratory.
Hydrogen peroxide is a clear colourless liquid, with the formula H_2O_2. It decomposes to water and oxygen, like this:

$2H_2O_2(aq) \longrightarrow 2H_2O(l) + O_2(g)$

The reaction is very slow, so black manganese(IV) oxide is added as a catalyst.
Oxygen is not very soluble in water. How can you tell, from the apparatus?

The properties of oxygen

1. It is a clear, colourless gas, with no smell.
2. It is only slightly soluble in water.
3. It is very reactive. It reacts with a great many substances, to produce **oxides**, and the reactions usually give out a lot of energy. For us, its two most important reactions are **respiration** and the **combustion of fuels**.

Respiration This is the process that keeps us alive. During respiration, oxygen reacts with glucose in our bodies. The reaction produces carbon dioxide, water, and the energy we need:

glucose + oxygen $\longrightarrow$ carbon dioxide + water + energy

The glucose comes from digested food, and the oxygen from air:

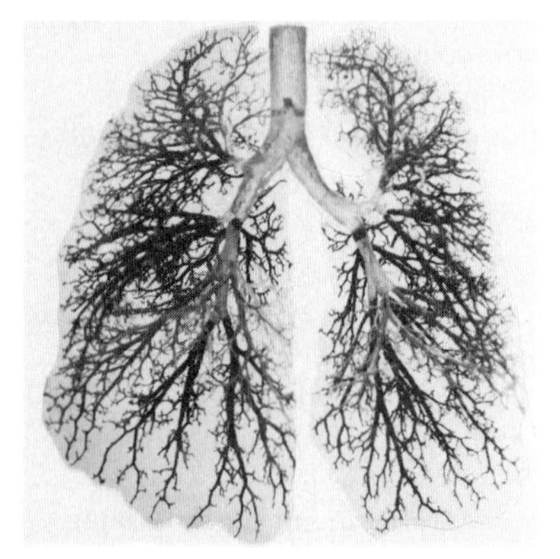

When we breathe in, air travels to our lungs. It passes along tiny tubes to the surface of the lungs. There the oxygen diffuses through the surface, into the blood.

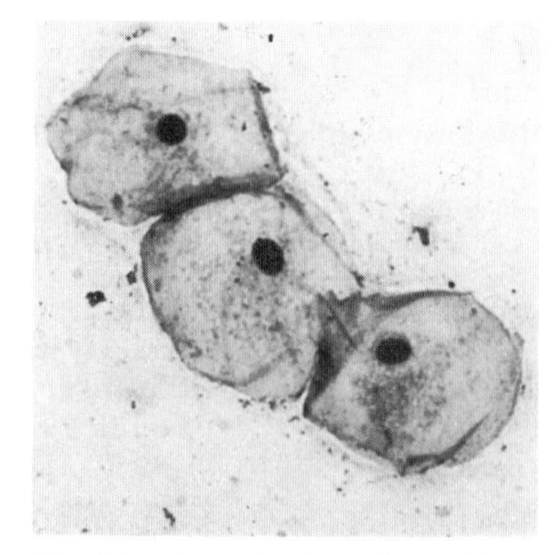

The blood carries it to the cells, along with glucose from digested food. There are millions of cells in the body. Respiration takes place in each one.

The energy from respiration keeps our hearts and muscles working. It also keeps us warm.

The carbon dioxide and water pass from the cells back into the blood. The blood carries them to the lungs, and we breathe them out. Respiration goes on in the cells of *all* living things, not just humans. Fish use the oxygen dissolved in water, which they take in through their gills. Plants use oxygen from the air, and take it in through tiny holes in their leaves.

Combustion of fuels Fuels are substances we burn to get energy—usually in the form of heat. The burning needs oxygen:

North Sea gas is a fuel. It is mainly **methane**. It is pumped to towns and cities from gas wells in the North Sea.

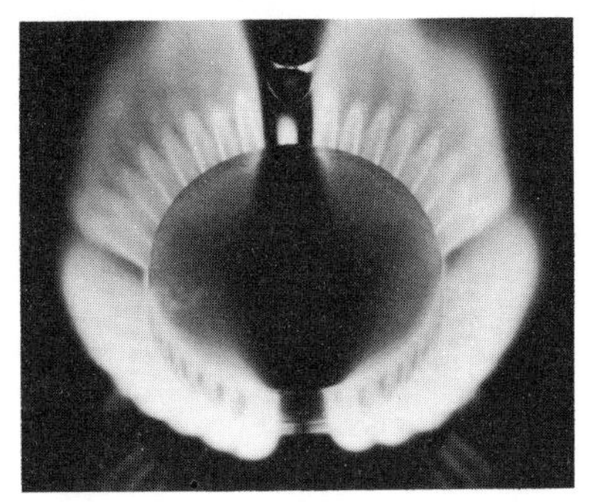

In the pipes of a gas cooker, the methane mixes with air. When the mixture is lit, the methane reacts with oxygen in the air.

The reaction gives out energy as heat and light. The heat is used to cook food.

Methane is also burned in gas fires, and gas central heating systems. The energy from the reaction is used to warm houses and heat water:

methane + oxygen ⟶ carbon dioxide + water + energy

$CH_4(g) + 2O_2(g) \longrightarrow CO_2(g) + 2H_2O(g) + \text{energy}$

Petrol, oil, coal and wood are also used as fuels. Each is a mixture of compounds that contain carbon and hydrogen. When they burn in plenty of oxygen, they all produce carbon dioxide, water and energy.

If there is only a limited amount of oxygen, **carbon monoxide** (CO) is produced instead of carbon dioxide. This is a problem, because carbon monoxide is a deadly poisonous gas.

Test for oxygen Things burn much faster in pure oxygen than in air. The reason is that the oxygen in air is **diluted** by nitrogen and other gases. This gives us a way to test an unknown gas, to see if it is oxygen:

1 A wooden splint is lit. Then the flame is blown out. The splint keeps on glowing, because the wood is reacting with the oxygen in air.
2 The glowing splint is plunged into the unknown gas.
3 If the gas is oxygen, the splint immediately bursts into flame.

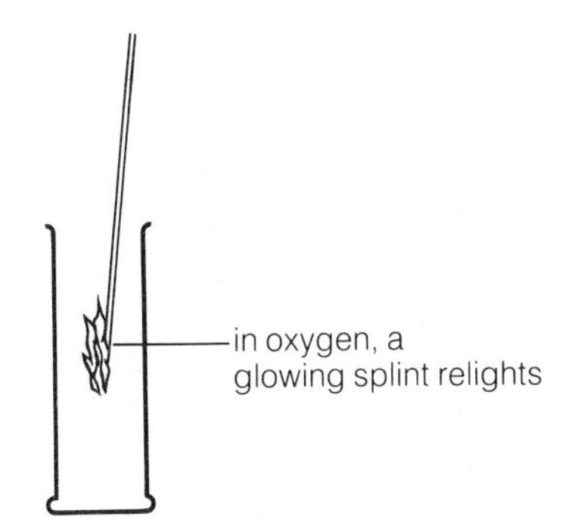

The test for oxygen.

Questions

1 Draw a labelled diagram, showing how you would make oxygen from hydrogen peroxide.
2 This method produces 'damp' oxygen. Why? Can you suggest a way to dry it?
3 What is respiration? Where does it take place?
4 How do fish obtain oxygen for respiration?
5 What is a fuel? Give four examples.
6 Write down the equations for respiration, and the burning of methane. In what ways are they alike?
7 How would you test a gas, to see if it is oxygen?

12.2 Oxides

On page 166 you saw that oxygen reacts with many substances, to form **oxides**. There are different types of oxides, as you will see below.

Basic oxides

Look at the way these metals react with oxygen:

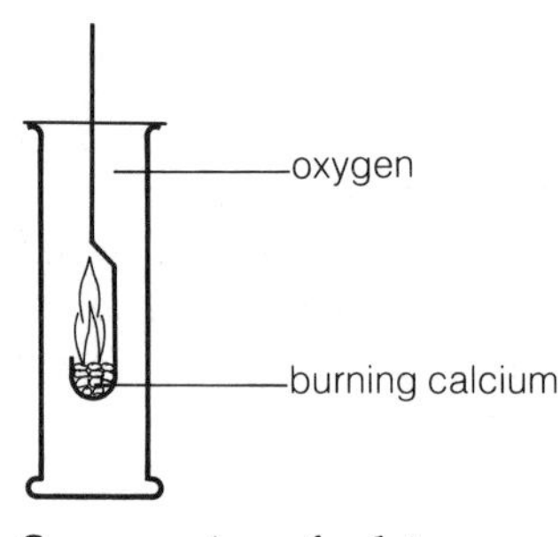

Some grains of calcium are lit over a bunsen flame, then plunged into a jar of oxygen. They burn brightly, leaving a white solid called **calcium oxide**:

$2Ca(s) + O_2(g) \longrightarrow 2CaO(s)$

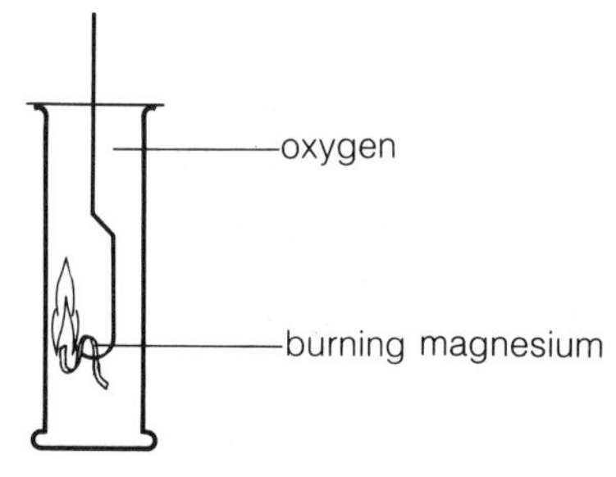

Magnesium ribbon is lit over a bunsen flame, and plunged into a jar of oxygen. It burns with a brilliant white flame, leaving a white ash called **magnesium oxide**:

$2Mg(s) + O_2(g) \longrightarrow 2MgO(s)$

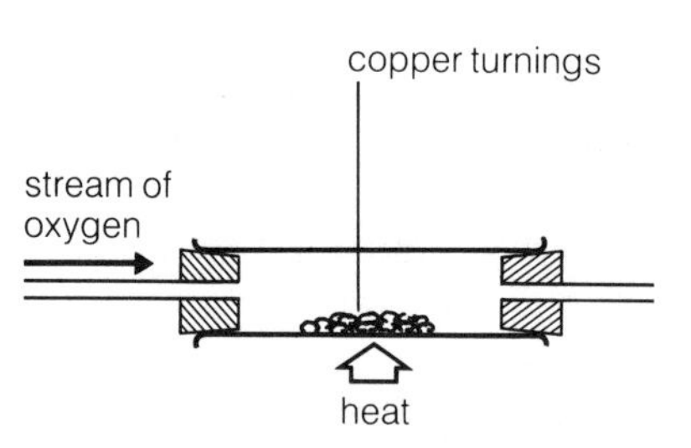

Copper is too unreactive to catch fire in oxygen. But when it is heated in a stream of the gas, its surface turns black. The black substance is **copper(II) oxide**:

$2Cu(s) + O_2(g) \longrightarrow 2CuO(s)$

The way each metal reacts depends on its **reactivity**.
The last reaction above produces copper(II) oxide, which is insoluble in water. But it does dissolve in dilute acid:

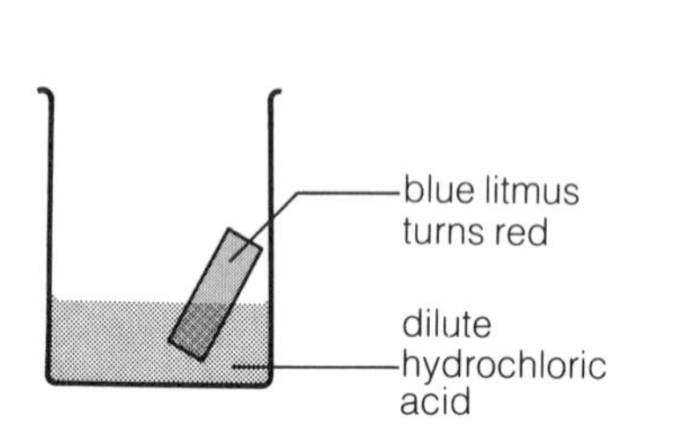

This is dilute hydrochloric acid. It turns blue litmus paper red.

copper(II) oxide

heat

Copper(II) oxide dissolves in it, when it is warmed. But after a time, no more will dissolve.

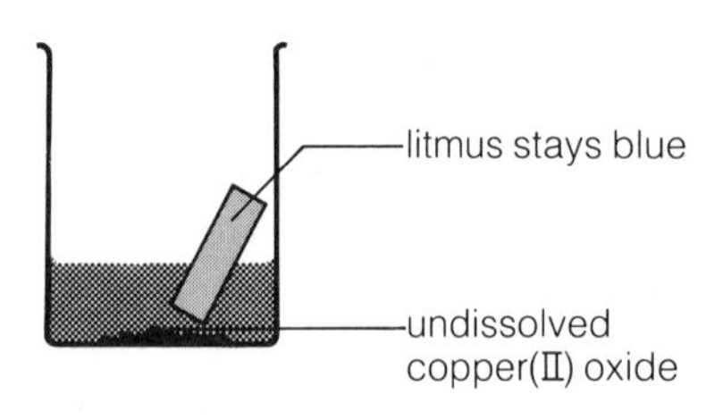

The resulting liquid has no effect on blue litmus. So the oxide has **neutralised** the acid.

Copper(II) oxide is called a **base**, or **basic oxide**, since it can neutralise acid. The products are a salt and water:

base + acid $\longrightarrow$ salt + water
$CuO(s) + 2HCl(aq) \longrightarrow CuCl_2(aq) + H_2O(l)$

Calcium oxide and magnesium oxide behave in the same way—they too can neutralise acid, so they are basic oxides.
In general, metals react with oxygen to form basic oxides.

Acidic oxides

Now look at the way these non-metals react with oxygen:

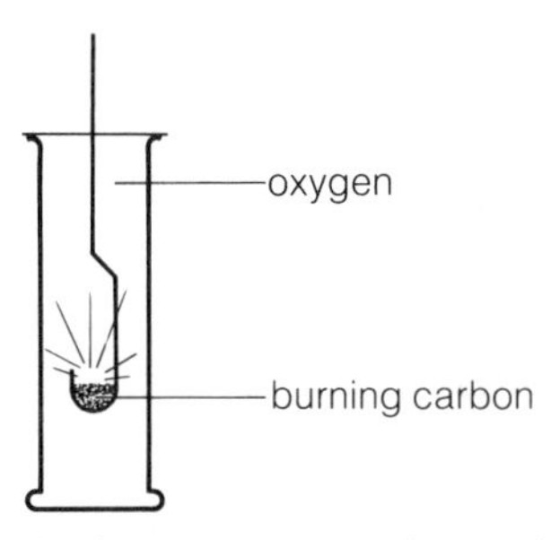

Carbon powder is heated over a bunsen, until it is red-hot. It is then plunged into a jar of oxygen. It glows bright red and the gas **carbon dioxide** is formed:

$C\,(s) + O_2\,(g) \longrightarrow CO_2\,(g)$

oxygen

burning sulphur

Sulphur catches fire over a bunsen, and burns with a blue flame. In pure oxygen it burns even more brightly. The product is **sulphur dioxide**, a gas:

$S\,(s) + O_2(g) \longrightarrow SO_2\,(g)$

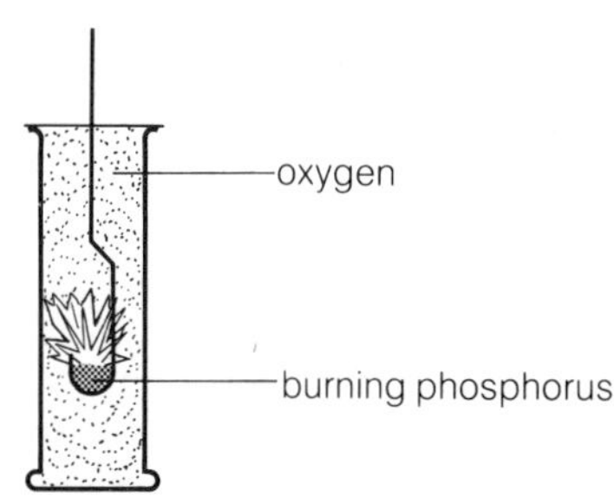

Phosphorus bursts into flame in air or oxygen, without being heated. (That is why it is stored under water.) A white solid called **phosphorus pentoxide** is formed:

$P_4\,(s) + 5O_2\,(g) \longrightarrow P_4O_{10}\,(s)$

The first reaction above produces carbon dioxide, which is slightly soluble in water:

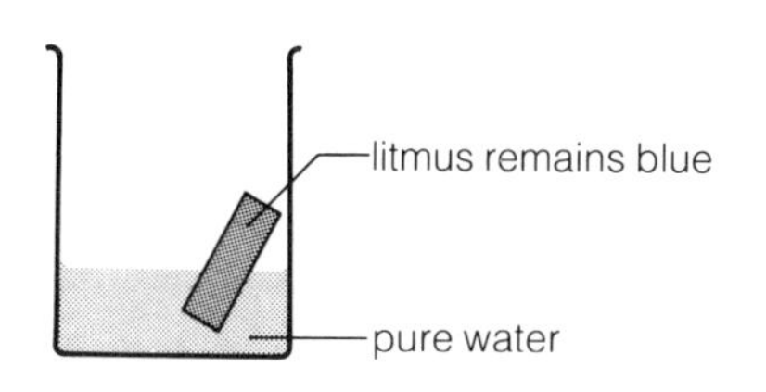

This is pure water. It is neutral. That means it has no effect on red *or* blue litmus paper.

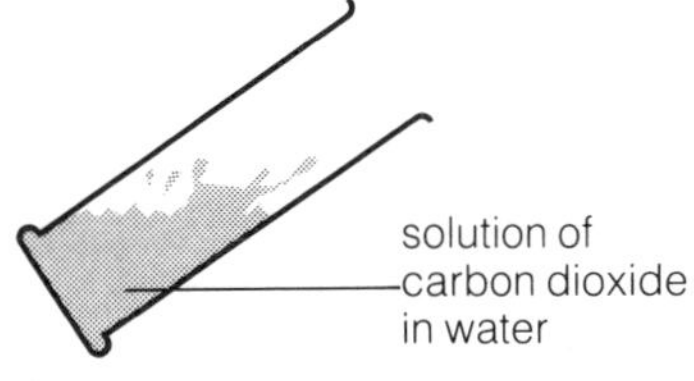

When it is poured into the jar of carbon dioxide, and shaken well, some of the carbon dioxide dissolves in it.

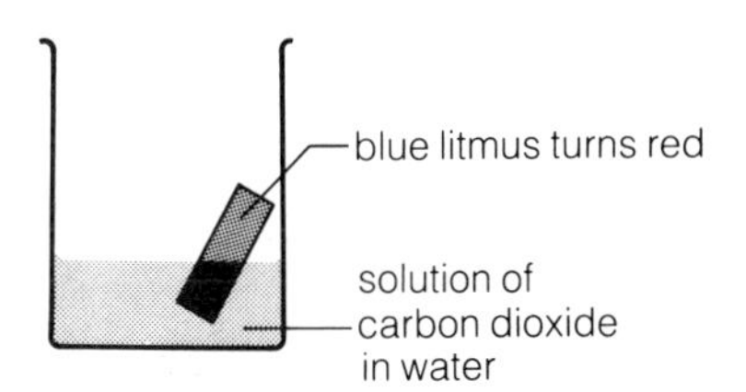

The solution turns blue litmus paper red, so it is **acidic**. It is called **carbonic acid**.

Carbon dioxide is called an **acidic oxide**, because it dissolves to form an acid. Sulphur dioxide and phosphorus pentoxide also dissolve in water to form acids, so they too are acidic oxides.
In general, non-metals react with oxygen to form acidic oxides.

Neutral oxides

Nitrogen forms the acidic oxide **nitrogen dioxide**, NO_2. But it also forms an oxide which is not acidic, called **dinitrogen oxide**, N_2O. Dinitrogen oxide is quite soluble in water, but the solution is neutral—it has no effect on litmus. So dinitrogen oxide is a **neutral oxide**. Another example of a neutral oxide is carbon monoxide, CO.

Questions

1 How could you show that magnesium oxide is a base?
2 Copy and complete: Metals usually form .... oxides while non-metals form .... oxides.
3 Why is phosphorus stored under water?
4 What colour change would you see, on adding litmus solution to a solution of phosphorus pentoxide?
5 Name two non-metal oxides that are *neutral*.

12.3 Sulphur and sulphur dioxide

Sulphur

Sulphur is obtained from two sources:

1 Most comes from large underground sulphur beds found in Poland, Mexico, and the USA.
2 Some is extracted from crude oil and natural gas.

Some crude oil and natural gas contains a high percentage of sulphur. If the oil or gas is burned, sulphur dioxide forms. This causes air pollution and acid rain. So oil companies are forced to remove the sulphur before the oil or gas can be used.

The properties of sulphur

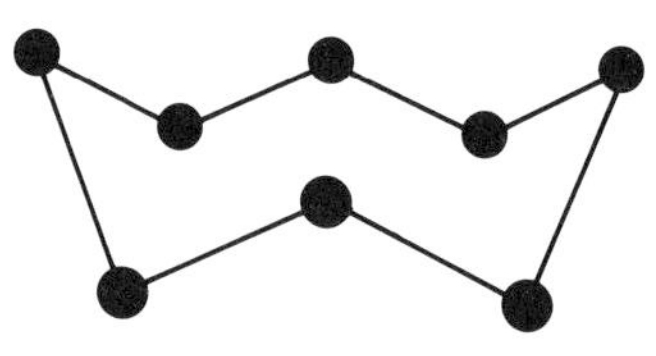

A molecule of sulphur.

1 It is a brittle yellow solid.
2 It is made up of crown-shaped molecules, each with eight atoms.
3 Because it is molecular, it has quite a low melting point (115 °C). It melts easily in a bunsen flame.
4 Like other non-metals, it does not conduct electricity.
5 Like most non-metals, it is insoluble in water.
6 It reacts with metals to form sulphides. With iron it forms iron(II) sulphide:

$$Fe(s) + S(s) \longrightarrow FeS(s)$$

7 It burns in oxygen to form sulphur dioxide:

$$S(s) + O_2(g) \longrightarrow SO_2(g)$$

Uses of sulphur

Most of it is used to make sulphuric acid.
Some is added to rubber, to make it tough and strong.
Some is used to make drugs, pesticides, matches and paper.
Some is used to make a special concrete called **sulphur concrete**. Unlike ordinary concrete, sulphur concrete is not attacked by acid. So it is used for floors and walls in factories where acid may get spilled.

Sulphur has two crystalline forms or allotropes. This is a crystal of **rhombic** sulphur, the form stable at room temperature.

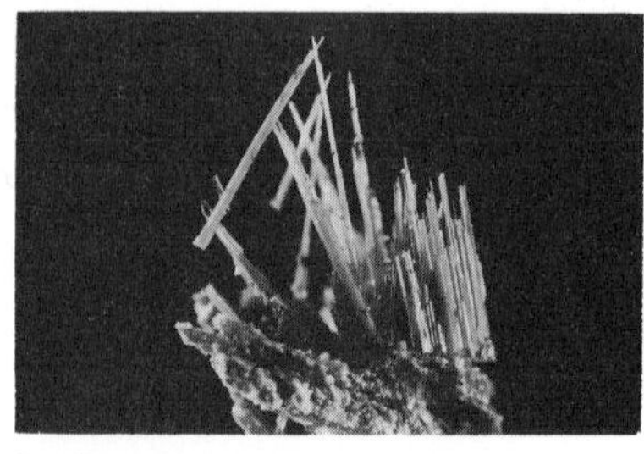

But if you heat rhombic sulphur to above 96 °C, the molecules rearrange themselves to form needle-shaped crystals of **monoclinic** sulphur.

Sulphur concrete is being used here for the floor of a copper electroplating plant. Gallons of sulphuric acid are used for this electroplating process and ordinary concrete would corrode in such an acid environment.

Sulphur dioxide

These trees were killed by sulphur dioxide, which dissolved in rain to give *acid rain*.

Sulphur dioxide is formed when sulphur burns in air. Its formula is SO_2. It has these properties:

1 It is a colourless gas, with a strong choking smell.
2 It is heavier than air.
3 It is soluble in water. The solution is acidic because the gas reacts with water to form **sulphurous acid**, H_2SO_3:

$$H_2O(l) + SO_2(g) \longrightarrow H_2SO_3(aq)$$

Sulphur dioxide is therefore an acidic oxide. The acid easily decomposes again to sulphur dioxide and water.
4 It acts as a bleach, when it is damp or in solution. Some coloured things lose colour when they lose oxygen—that is, when they are **reduced**. Sulphur dioxide bleaches them by reducing them.
5 When it escapes into the air from engine exhausts and factory chimneys, it causes air pollution. It attacks the breathing system in humans and other animals. It dissolves in rain to give acid rain. Acid rain damages buildings, metalwork, and plants.

Uses of sulphur dioxide

Some is used to bleach wool, silk, and wood pulp for making newspaper. Some is used in the preparation of soft drinks, jam, and dried fruit, because it stops the growth of bacteria and moulds. But most is used to make sulphuric acid, as shown on page 172.

Scrap paper being recycled for newsprint. The next stage is to bleach the paper using sulphur dioxide.

This photograph shows two sets of potato strips. The strips on the left have been treated with sulphur dioxide to stop browning.

Questions

1 What are the 2 sources of sulphur?
2 Sulphur reacts with lead to form lead(II) sulphide. Write an equation for the reaction.
3 Write down 3 uses of sulphur.
4 Explain why sulphur dioxide makes rain acidic.
5 Sulphur dioxide is heavier than air. How do you think that might affect pollution?
6 Write down 3 uses of sulphur dioxide.

12.4 Sulphuric acid and sulphates

Part of the ICI sulphuric acid plant at Billingham.

How sulphuric acid is made

More sulphuric acid is produced each year than any other chemical. Most of it is made by the Contact process. The raw materials are sulphur, air and water. This flow chart shows what happens:

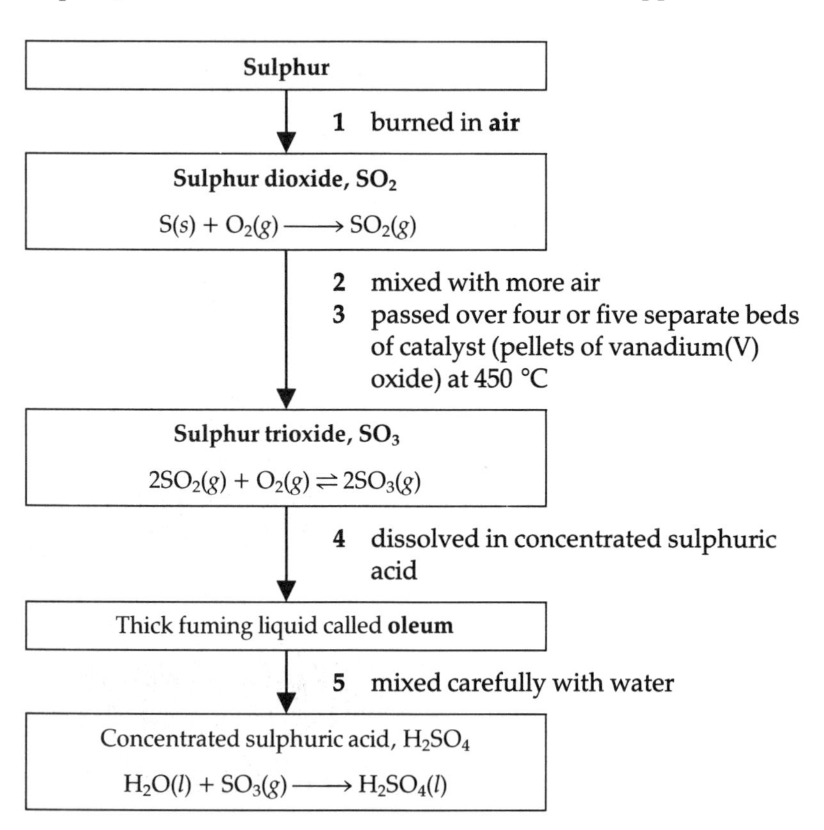

This warning sign is displayed on all containers and tankers containing sulphuric acid. What message does it give?

Things to note about the Contact process

1 The reaction between sulphur dioxide and oxygen is **reversible**. The gas mixture is passed over several beds of catalyst in order to get a high yield of sulphur trioxide.
2 The catalyst will not work below 400 °C. So at lower temperatures the reaction would be too slow. And at higher temperatures the yield of sulphur trioxide drops. So the reaction is carried out at 450°C as a compromise.
3 The burning of sulphur in air is **exothermic**—it gives out heat. So is the reaction between sulphur dioxide and oxygen.
4 The heat given out by these reactions is used to make steam. The steam is then used to make electricity, or sold to nearby factories for heating. This helps to cover the cost of running the plant.
5 The engineers must make sure that no sulphur dioxide or sulphur trioxide or acid escapes from the plant, because that would cause serious pollution.

Sulphuric acid is a hazardous chemical to store and transport. If possible, the acid plant is built close to major users, to keep transport costs down.

Uses of sulphuric acid

Most of it is used to make fertilisers such as ammonium sulphate. Some is used to make paint, soapless detergents, soap, dyes, and plastic. Some is used as battery acid. In fact nearly every industry uses some sulphuric acid. (It is the cheapest acid to buy.)

This is what happened when concentrated sulphuric acid was added to two teaspoons of sugar.

The properties of sulphuric acid

The concentrated acid It has these properties:

1. It is a colourless oily liquid.
2. When it is hot, it acts as an oxidising agent. It will oxidise metals to sulphates. For example it oxidises copper like this:

   $$Cu(s) + 2H_2SO_4(l) \longrightarrow CuSO_4(aq) + 2H_2O(l) + SO_2(g)$$

   When it acts as an oxidising agent, it is itself reduced to sulphur dioxide, which bubbles off as a gas.
3. It acts as a **dehydrating agent**. That means it can remove water. It turns blue copper(II) sulphate crystals into a white powder by removing the water of crystallisation:

   $$CuSO_4.5H_2O(s) - 5H_2O(l) \longrightarrow CuSO_4(s)$$

   It also dehydrates sugar, paper and wood. These substances are all made of carbon, hydrogen and oxygen. The acid removes the hydrogen and oxygen as water, leaving carbon behind.

The dilute acid It has typical acid properties:

1. It turns blue litmus red.
2. It reacts with metals to give hydrogen, and salts called **sulphates**.
3. It reacts with metal oxides and hydroxides to give sulphates and water.
4. It reacts with carbonates to give sulphates, water, and carbon dioxide.

The dilute acid is made by carefully adding concentrated acid to water (never the other way round, because so much heat is produced that the acid could splash out and cause damage).

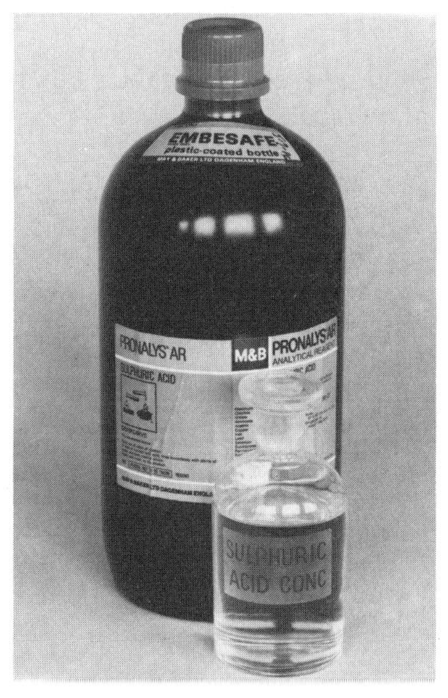

Because it acts as a dehydrating agent, concentrated sulphuric acid can also be used to dry gases, as long as it does not react with them. For example it is used to dry hydrogen chloride (page 176).

Test for sulphates

Most sulphates are soluble in water. But lead, calcium and barium sulphates are insoluble. This provides a way to test for sulphates.

To see if an unknown solution contains a sulphate, this is what to do:

1. First add a few drops of dilute hydrochloric acid, to acidify it.
2. Then add a few drops of barium chloride solution.
3. If a sulphate is present, a white precipitate forms. The white precipitate is barium sulphate, $BaSO_4$.

Questions

1. Name: **a** the process **b** the raw materials **c** the catalyst for making sulphuric acid.
2. What gas is given off when concentrated sulphuric acid acts as an oxidising agent?
3. Explain why concentrated sulphuric acid:
   a turns blue copper(II) sulphate white
   b chars sugar **c** is used to dry gases
4. Describe the test for sulphates.

12.5 Chlorine

How chlorine is made

Chlorine is very reactive, so it is never found as a free element, in nature. It occurs mostly as the compound **sodium chloride**, or rock salt.

In industry In industry, chlorine is made by electrolysing molten sodium chloride (page 140) or brine (page 82). Brine is a concentrated solution of sodium chloride in water.

In the laboratory It is easier to make chlorine from hydrochloric acid, in the laboratory. An **oxidising agent** is used to remove hydrogen from the acid, leaving chlorine:

$$2HCl\,(aq) + \underset{\text{from the oxidising agent}}{[O]} \longrightarrow H_2O\,(l) + Cl_2\,(g)$$

Manganese(IV) oxide is a suitable oxidising agent. The apparatus is shown below. It must be set up in a fume cupboard.

Chlorine is used to sterilise the water in swimming pools.

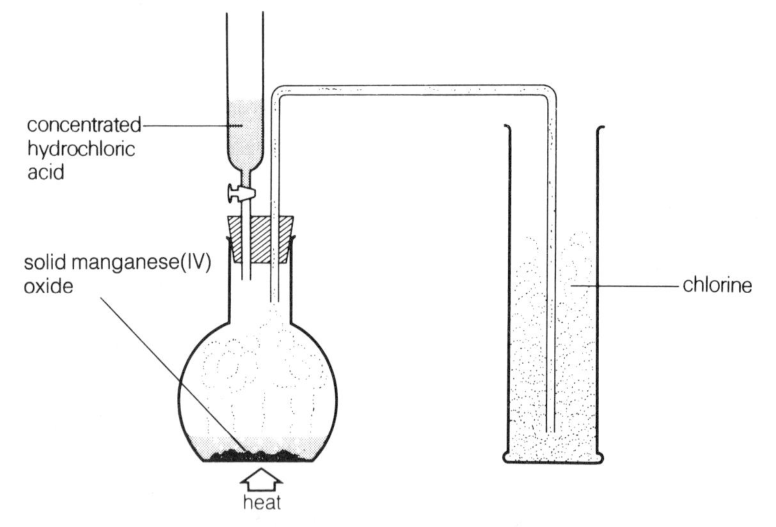

The mixture has to be heated gently as the concentrated acid is dripped onto the manganese(IV) oxide. Look at the way the gas is collected. Is it heavier or lighter than air?

The properties of chlorine

1. It is a greenish-yellow gas with a choking smell.
2. It is heavier than air.
3. It is poisonous to all living things. In fact, it was used as a weapon in World War I.
4. It is soluble in water. The solution is called **chlorine water**. It is acidic because chlorine reacts with water to form *two* acids:

   $$Cl_2\,(g) + H_2O\,(l) \longrightarrow \underset{\text{hydrochloric acid}}{HCl\,(aq)} + \underset{\text{hypochlorous acid}}{HOCl\,(aq)}$$

   Hypochlorous acid slowly decomposes again, giving off oxygen:

   $$2HOCl\,(aq) \longrightarrow 2HCl\,(aq) + O_2\,(g)$$

A soldier in World War I, ready for a chlorine gas attack.

5 Chlorine water acts as a bleach. This is because the hypochlorous acid in it can lose oxygen to other substances—it can **oxidise** them. Some coloured substances turn colourless when they are oxidised.

6 Like other bleaches, chlorine water also acts as a sterilising agent—it kills bacteria and other germs.

7 Chlorine is very reactive. It combines with many non-metals and most metals:

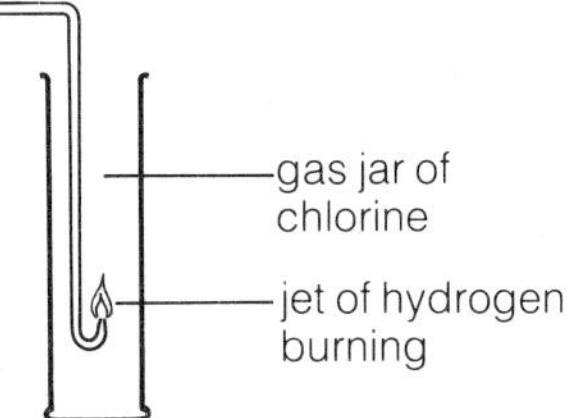

Hydrogen burns in it to form hydrogen chloride:

$H_2(g) + Cl_2(g) \longrightarrow 2HCl(g)$

This can be an explosive reaction.

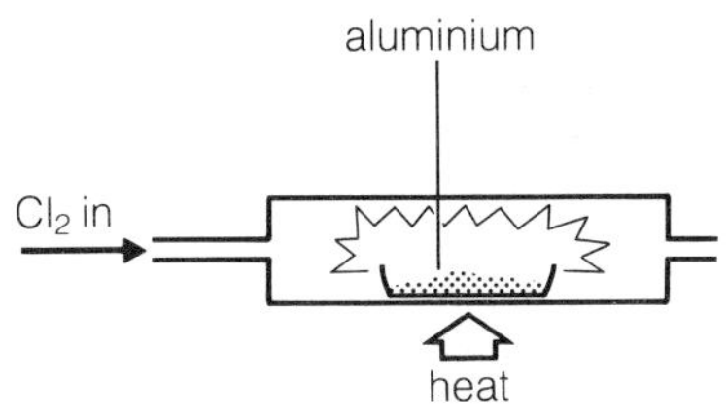

When it is passed over heated aluminium, the metal glows white and turns into aluminium chloride:

$2Al(s) + 3Cl_2(g) \longrightarrow 2AlCl_3(s)$

You can read about its reaction with iron on page 127.

Testing for chlorine in the laboratory. The gas bleaches damp litmus. Can you explain why?

8 If chlorine is bubbled into iron(II) chloride solution, iron(III) chloride is formed:

$2FeCl_2(aq) + Cl_2(g) \longrightarrow 2FeCl_3(aq)$

9 Chlorine displaces other halogens from solutions of their compounds. For example, potassium bromide solution is colourless, but it turns orange when chlorine is bubbled through it, because the bromine is displaced:

$2KBr(aq) + Cl_2(g) \longrightarrow 2KCl(aq) + Br_2(aq)$

colourless ... orange

A solution of potassium iodide also changes colour:

$2KI(aq) + Cl_2(g) \longrightarrow 2KCl(aq) + I_2(aq)$

colourless ... red-brown

These reactions show that chlorine is more **reactive** than bromine or iodine—it can drive them from their compounds.

Uses of chlorine

It is used to sterilise drinking water, and the water in swimming pools. It is used to make polyvinylchloride (PVC). It is used to make bleaches, pesticides, weed-killers, and solvents such as tetra-chloroethane (used in dry-cleaning) and trichloroethane (for typist's correction fluid). But most of it is turned into hydrochloric acid.

All these were made using chlorine. Note that it is dangerous to mix chlorine-based bleaches and germ-killers with other kinds. There could be an explosive reaction.

Questions

1 Describe how you would make chlorine in the laboratory, and draw the apparatus you would use.

2 Explain why a solution of chlorine in water:
a is acidic **b** is able to bleach things
c slowly gives off oxygen

3 How would you test for chlorine in the laboratory?

4 Write equations to show how chlorine reacts with: sodium; iron; sodium iodide solution.
What would you *see*, during the last reaction?

5 List *five* uses of chlorine.

12.6 Hydrogen chloride, hydrochloric acid, and chlorides

Hydrogen chloride

You could make hydrogen chloride in the laboratory by dripping concentrated sulphuric acid onto any chloride. Sodium chloride is the cheapest one to use. The reaction is:

$$NaCl\,(s) + H_2SO_4\,(l) \longrightarrow \underset{\text{sodium hydrogen sulphate}}{NaHSO_4\,(aq)} + HCl\,(g)$$

Concentrated sulphuric acid is also used to dry the gas:

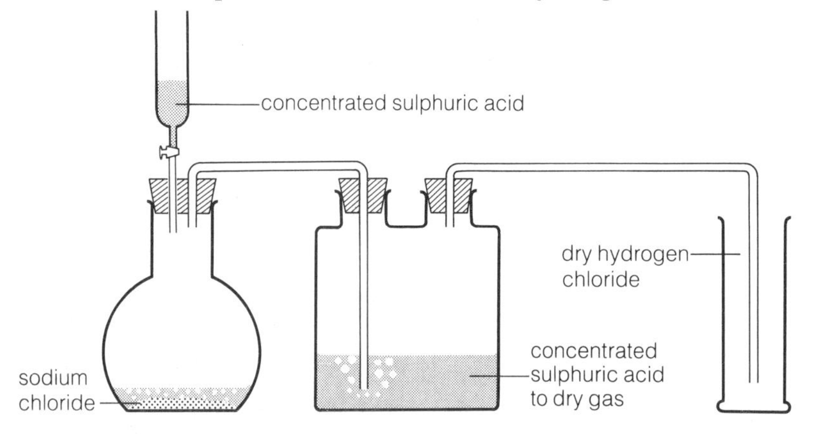

The properties of hydrogen chloride

1 It is heavier than air. (How can you tell, from the diagram above?)
2 It has a choking smell, and it irritates the eyes and lungs.
3 It dissolves very easily in water, to form **hydrochloric acid**. Like all acids, this contains hydrogen ions:

$$HCl(aq) \longrightarrow H^+(aq) + Cl^-(aq)$$

4 Hydrogen chloride reacts with ammonia to form white smoke, made from tiny particles of solid ammonium chloride:

$$NH_3(g) + HCl(g) \longrightarrow NH_4Cl(s)$$

This reaction is used to test for ammonia, or hydrogen chloride.

Hydrogen chloride is so soluble in water that you can do the fountain experiment with it (page 156).

Hydrochloric acid

You saw above that hydrogen chloride dissolves in water to form hydrochloric acid. That is how the acid is made in industry. First hydrogen chloride is made by burning hydrogen in chlorine. Then the gas is passed to absorption towers where it is dissolved in water.

Properties of the acid It shows typical acid reactions:

1 It reacts with metals to give hydrogen and salts called **chlorides**.
2 It reacts with metal oxides and hydroxides to form chlorides and water.
3 It reacts with carbonates to form chlorides, water and carbon dioxide.

Much of Britain's salt comes from underground mines in Cheshire. This shows the salt being mined.

Sodium chloride

Sodium chloride (NaCl) is the most important chloride of all.

1 It occurs naturally as rock salt, and in sea water.
2 It is the starting point for many other chemicals. For example electrolysis of molten sodium chloride gives sodium and chlorine. Electrolysis of a concentrated solution (brine) gives sodium hydroxide, chlorine, and hydrogen. Sodium carbonate and sodium hydrogen carbonate are also made from sodium chloride.
3 It improves the flavour of food. But more important, it provides you with sodium ions, which are essential for body fluids. However, doctors think that too much salt can cause high blood pressure.
4 It is used to melt ice on the roads in winter. But this can have a harmful effect on things like cars, trucks and lamp posts, because sodium chloride makes iron rust faster.

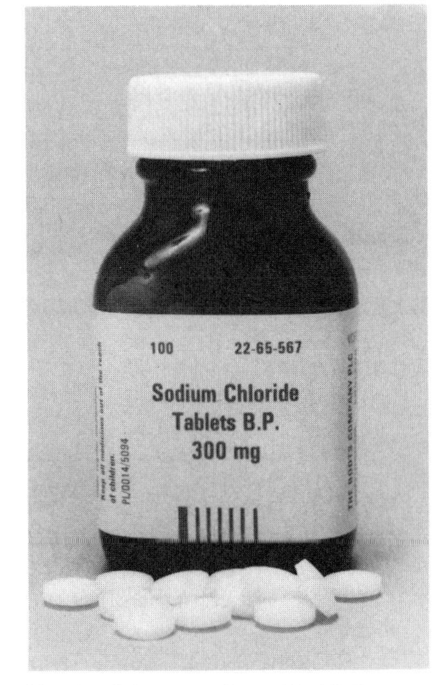

Your body loses sodium chloride in sweat. The result can be exhaustion. So people in hot climates often have to take salt tablets.

Test for chlorides

Only silver and lead chlorides are insoluble in water. All the others are soluble. This is the basis of the test for chlorides.

To see if an unknown solution contains a chloride, this is what to do:
1 First add a few drops of nitric acid, to acidify the solution.
2 Then add a little silver nitrate solution.
3 If a chloride is present, a white precipitate of silver chloride will form.

Questions

1 Give two *physical* properties of hydrogen chloride.
2 Copy and complete: A solution of .... chloride in water is called .... acid. It reacts with metals to give .... These are all soluble except for .... and ....
3 Name three chemicals made from sodium chloride.
4 Explain why sodium chloride is able to melt ice. (Hint: look back at Chapter 1.)
5 Describe how you would test a solution to see if it contained a chloride.

Questions on Chapter 12

1 Look at each description below in turn. Say whether it fits oxygen, or sulphur, or chlorine.
- **a** Quite soluble in water
- **b** Solid at room temperature
- **c** Reacts with metals to form oxides
- **d** Exists in more than one solid form
- **e** When damp, removes the colour from dyes
- **f** Burns in air with a blue flame
- **g** Reacts with hydrogen to form water
- **h** A poisonous gas
- **i** Is added to rubber to make it tough and strong
- **j** Relights a glowing splint
- **k** Is colourless
- **l** Reacts with other elements to form chlorides
- **m** Forms a gaseous oxide which causes acid rain when burnt

2 Oxygen reacts with many different elements.
a Copy and complete the following table to show the results of combustion experiments carried out using jars of oxygen gas.

Element	What you would see in the gas jar
potassium	
calcium	
carbon	
phosphorus	
magnesium	
copper	
sulphur	

b Which element does not need heating?
c Which element does not catch fire?
d In the case of potassium, the product is soluble in water. Which of the other products are soluble in water?
e Which products give aqueous solutions that are acidic?
f Which of the elements form basic oxides?

3 The elements sulphur and oxygen are in the same group of the Periodic Table, so they have similar properties. But there are also some differences between them. Use the information on pages 166–171 to answer these questions.
a Oxygen is a non-metal. Is sulphur a non-metal?
b Do the elements look alike at room temperature? Explain your answer.
c Both the elements are molecular. What type of bonding do their molecules contain?
d In what way are their molecules different?
e Sulphur combines with hydrogen to form the gas hydrogen sulphide, H_2S. Name one way in which this differs from the compound formed between oxygen and hydrogen.

4 The following diagrams represent molecules of different substances that contain sulphur.
a Write the chemical formula for each substance, then name the substance.

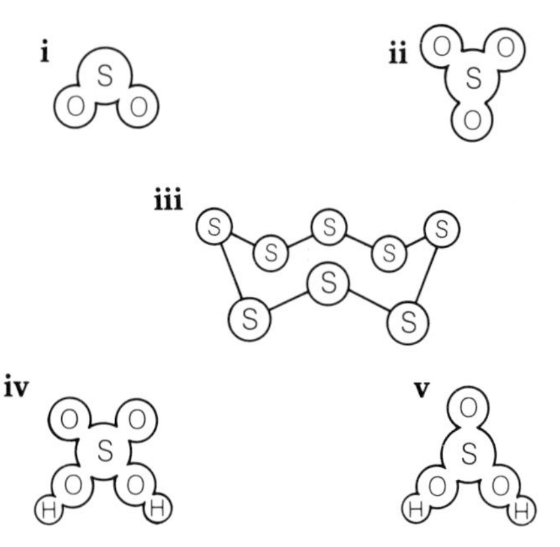

b How would you convert:
substance **iii** into substance **i**?
substance **i** into substance **ii**?
substance **i** into substance **v**?

5 Below is a flow chart for the Contact Process.

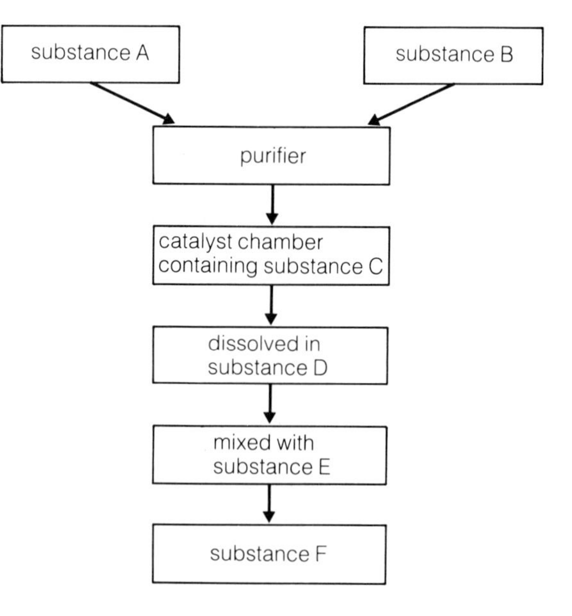

a Name substances A, B, C, D, E and F.
b Why is a catalyst used?
c Write a chemical equation for the reaction that takes place on the catalyst.
d Why is the production of substance F very important? Give three reasons.
e Copy out the flow chart, writing in the full names of the different substances.

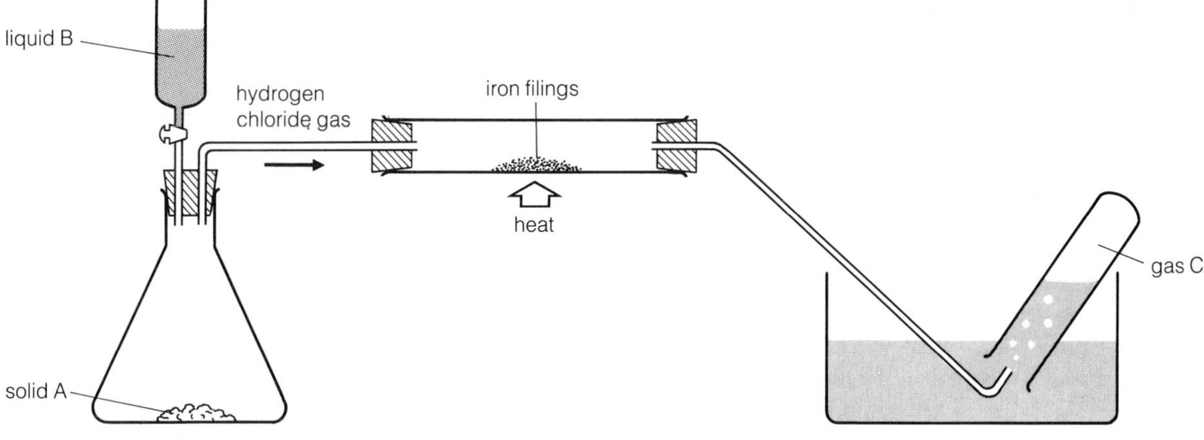

6 The following table shows the properties of certain oxides.

Oxide	State at 20 °C	When added to water: Is energy given out?	pH of solution
Magnesium oxide	solid	no	above 7
Calcium oxide	solid	yes	above 7
Copper(II) oxide	solid	no	—
Sulphur dioxide	gas	yes	below 7
Carbon dioxide	gas	no	below 7

a Why is no pH given for copper(II) oxide?
b Which two compounds react when added to water? Give a reason for your choice.
c Which two compounds contain only non-metals?
d From the information in the table, what conclusions can you draw about:
- **i** the pH of solutions of oxides?
- **ii** the state of oxides at room temperature (20 °C)?

7 Sulphuric acid can act as a dehydrating agent and an oxidising agent, as well as an acid. Decide how it is behaving in each of the following cases:
a It turns blue litmus red.
b It reacts with magnesium to give hydrogen.
c It can be used to dry hydrogen chloride gas.
d It neutralizes sodium hydroxide to form a salt.
e It reacts with copper, releasing sulphur dioxide.
f It turns sugar black.
g It turns blue copper sulphate white.
h It produces carbon dioxide when added to a carbonate.

8 **a** Look again at the reactions in questions 7. For each say whether the sulphuric acid should be concentrated, dilute or either.
b To make dilute sulphuric acid, the concentrated acid must be added carefully to water, and *never* the other way round. Explain why.

9 The apparatus above was used to investigate the reaction between hydrogen chloride and iron filings.
a Name a suitable substance for solid A.
b Name a suitable substance for liquid B.
c The reaction between iron and hydrochloric acid is exothermic. What would you see in the combustion tube?
d Gas C burns with a squeaky pop. What is it?
e Suggest a name for the product left in the combustion tube, after the reaction.
f Write a word equation for the reaction.
g The apparatus could be dangerous. Suggest why.

10 When a piece of damp blue litmus paper is placed in a gas jar of chlorine, the litmus changes colour. First the blue colour turns to red, then the red colour quickly disappears, leaving the paper white.
a Explain why the paper turns red.
b Explain why the red colour then disappears.

11 Chlorine gas is passed through two solutions, as shown below.

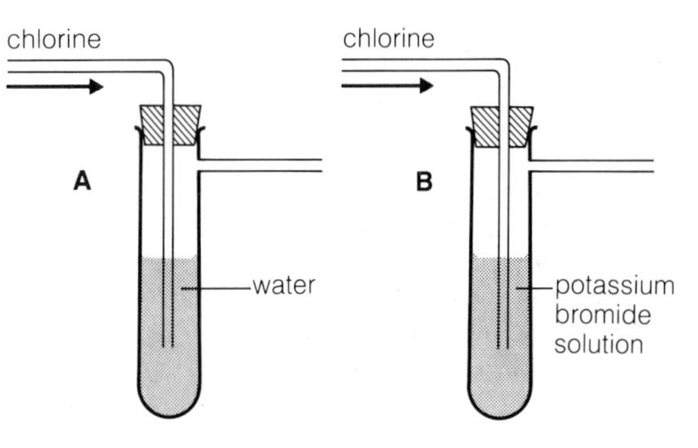

a What would you *see* in each test tube?
b What chemical(s) are formed in each tube?
c Write word and chemical equations for each reaction.
d Which of the reactions is a displacement?
e Write the word equation for a similar displacement reaction, using chlorine.

13.1 Carbon and carbonates

Carbon

The element A small amount of carbon occurs as the free element, in the earth's crust. It has two forms, diamond and graphite. They are the two allotropes of carbon (page 50).
Diamond is a clear, hard substance that sparkles in light. It is rare. It is mostly mined in South Africa. **Graphite** is a dark greasy solid. It is more common than diamond and is mined in Mexico and several other places.
Charcoal and **soot** are also forms of graphite, but they do not occur naturally. They are made by heating coal or wood or animal bones in very little air.

Diamond is so hard that it can be used to cut stone. This cutting wheel is edged with diamond.

Carbon compounds Carbon also occurs naturally in thousands of compounds.

1 It is found as **carbon dioxide** in the air.
2 It is contained in **proteins**, **carbohydrates** and other compounds that make up living things.
3 Coal, wood and natural gas are all mixtures of carbon compounds.
4 It occurs in many rocks, as **carbonates**.

In this chapter we will look at some of these carbon compounds, starting with the carbonates.

Carbonates

Carbonates are compounds that contain the carbonate ion, CO_3^{2-}.
Two examples are sodium carbonate and calcium carbonate.
Sodium carbonate has the formula Na_2CO_3. It is often called **washing soda**, and it is used in making glass, and to soften water (page 110).
Calcium carbonate has the formula $CaCO_3$. It has several forms, including **limestone**, **chalk** and **marble**:

Limestone is hard and strong. St. Paul's Cathedral in London is built from it. It was completed in 1708.

The White Horse at Westbury is scraped out of chalk rocks. Chalk is not so hard as limestone.

Marble is very hard, and can be very smoothly polished. It has often been used for statues.

Limestone is the most common form of calcium carbonate. Besides being used for building, it has several other uses:

1 It is used for extracting **iron**, in the blast furnace (page 146).

2 It is heated with sand and sodium carbonate, to make **glass**.
3 It is heated with clay to make a grey powder called **cement**. When water is added, cement sets to a hard mass.
4 On heating, limestone decomposes to quicklime or **calcium oxide**, a white substance.

$CaCO_3(s) \longrightarrow CaO(s) + CO_2(g)$

This reaction is carried out on a large scale in lime kilns. When water is dripped onto lumps of calcium oxide, the lumps get hot, swell and crumble to a powder. This is slaked lime or **calcium hydroxide**:

$CaO(s) + H_2O(l) \longrightarrow Ca(OH)_2(s)$

Calcium carbonate, calcium oxide, and calcium hydroxide are bases. They are also cheap. So farmers spread them in powder form on soil that is too acidic. Calcium hydroxide is also added to liquid waste from factories, to neutralise any acids in it (page 123).

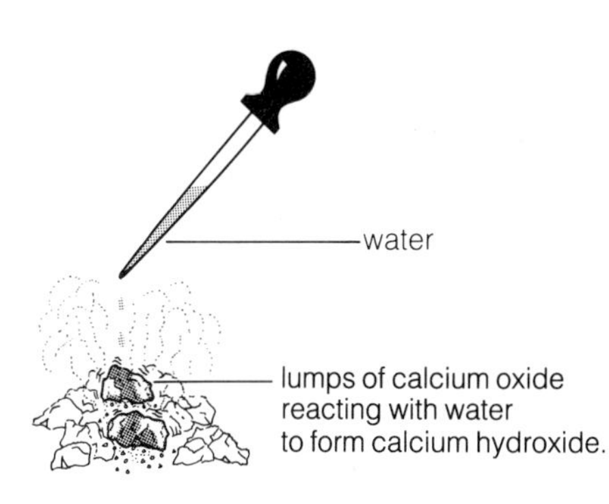

The properties of carbonates

1 They are insoluble in water, except for sodium, potassium and ammonium carbonates.
2 They react with acids to form a salt, water and carbon dioxide.
3 Most of them break down on heating, to carbon dioxide and an oxide. You saw above how calcium carbonate breaks down. Copper(II) carbonate decomposes in the same way:

$CuCO_3(s) \longrightarrow CuO(s) + CO_2(g)$
green → black

But sodium and potassium carbonates are not affected by heat. (The more reactive the metal, the more stable its compounds.)
4 Metal carbonates react with water and carbon dioxide to form **metal hydrogen carbonates**. For example:

sodium carbonate + water + carbon dioxide ⟶ sodium hydrogen carbonate

$Na_2CO_3(s) + H_2O(l) + CO_2(g) \longrightarrow 2NaHCO_3(aq)$

Sodium hydrogen carbonate is often called **bicarbonate of soda**, or **baking soda**. When it is heated it decomposes, giving off carbon dioxide:

$2NaHCO_3(s) \longrightarrow Na_2CO_3(s) + CO_2(g) + H_2O(g)$

Baking soda is used in baking bread. The heat of the oven causes it to decompose, and the carbon dioxide makes the bread 'rise'.

Bicarbonate of soda is sodium hydrogen carbonate. Baking powder is sodium hydrogen carbonate mixed with other things.

This cake was made with baking powder. The tiny holes are caused by carbon dioxide gas.

Questions

1 Name and describe the allotropes of carbon.
2 Name three different forms of calcium carbonate.
3 Write down two names for: **a** CaO **b** $Ca(OH)_2$ How is the second compound made from the first?
4 Why are calcium carbonate, calcium oxide, and calcium hydroxide sometimes spread on fields?
5 Write a balanced equation for the decomposition of magnesium carbonate on heating.
6 Write a word equation to describe the decomposition of baking soda in the oven.
7 What is the formula for potassium hydrogen carbonate?

13.2 The oxides of carbon

Carbon dioxide

Air contains a small amount of carbon dioxide gas. Some of this is produced by **respiration**, which goes on in all living things (page 166). And some is produced by the burning of **fuels**—coal, oil, gas and wood. All these fuels are made of carbon compounds.

How to make carbon dioxide in the laboratory The usual way is to add dilute hydrochloric acid to calcium carbonate:

$$CaCO_3\,(s) + 2\,HCl\,(aq) \longrightarrow CaCl_2\,(aq) + H_2O\,(l) + CO_2\,(g)$$

The apparatus is shown below:

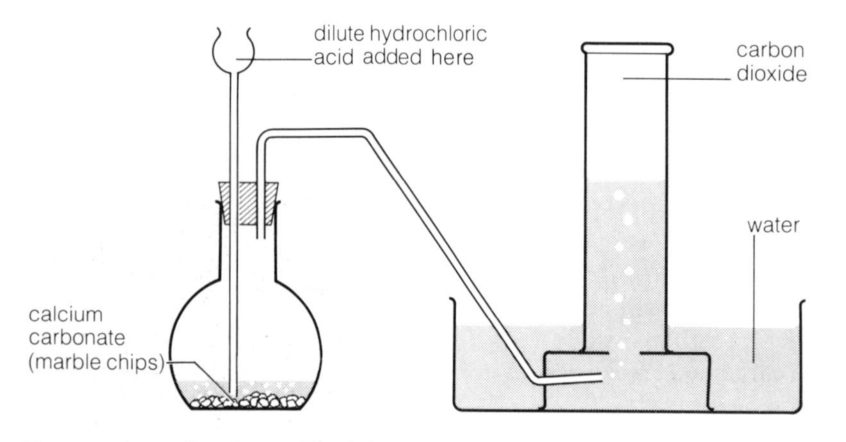

Solid carbon dioxide, or dry ice, is used on aeroplanes to keep food cold.

Properties of carbon dioxide These are the main ones:

1 It is a colourless gas, with no smell.
2 It is much heavier than air.
3 When it is cooled to $-78\,°C$, it turns straight into a solid. Solid carbon dioxide is called **dry ice**. It sublimes when it is heated.
4 Carbon dioxide does not usually support combustion:

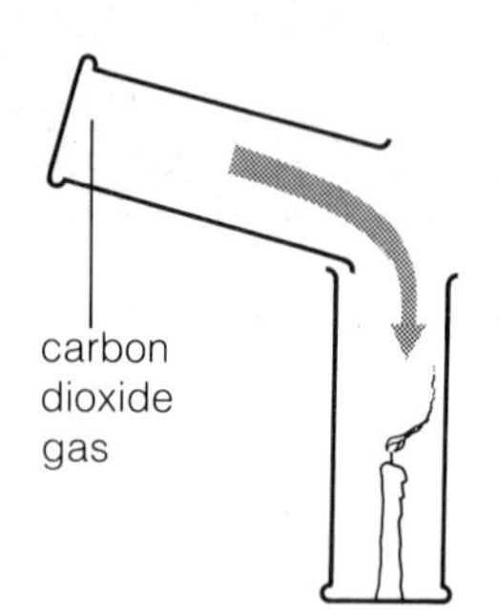

For example, it will put out a candle, and other burning substances. For this reason it is used in fire extinguishers.

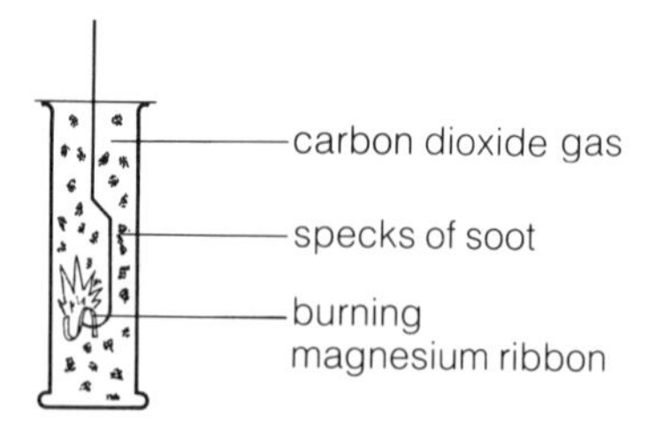

But very reactive metals such as magnesium *will* burn in it:

$$2Mg(s) + CO_2(g) \longrightarrow 2MgO(s) + C(s)$$

Magnesium removes its oxygen, and the carbon is left as soot.

5 It is slightly soluble in water, forming an acidic solution called carbonic acid (page 169). This is a very weak acid.

Carbon dioxide fire extinguishers are used mainly for electrical fires, like the one in this car engine.

6 It reacts with alkalis to form **carbonates**. For example:

$$CO_2(g) + 2NaOH(aq) \longrightarrow Na_2CO_3(aq) + H_2O(l)$$

This reaction can be used to remove carbon dioxide from the air.

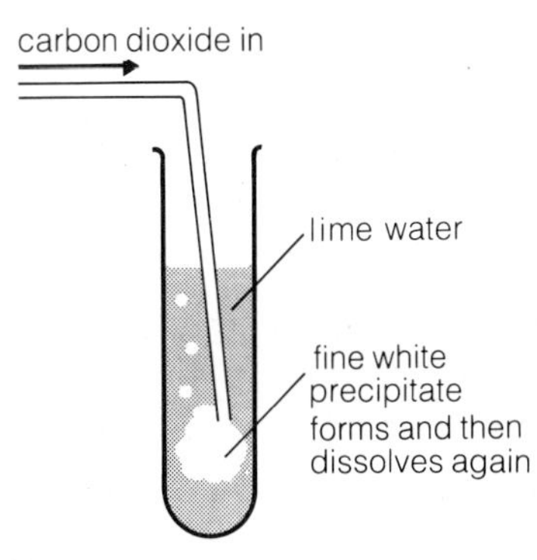

The test for carbon dioxide.

Test for carbon dioxide For this, **lime water** is needed. Lime water is a solution of calcium hydroxide in water. Carbon dioxide makes it go milky, because a fine white precipitate forms. But when more carbon dioxide is bubbled through, the precipitate disappears.
The precipitate is calcium carbonate, from this reaction:

$$Ca(OH)_2\,(aq) + CO_2\,(g) \longrightarrow CaCO_3\,(s) + H_2O\,(l)$$

It disappears again because it reacts with more carbon dioxide to form calcium hydrogen carbonate, which is soluble:

$$CaCO_3\,(s) + CO_2\,(g) + H_2O\,(l) \longrightarrow Ca(HCO_3)_2\,(aq)$$

Uses of carbon dioxide These are the main ones:
1 It is used in fire extinguishers.
2 It is put in drinks like coke and lemonade, to make them fizzy. It is only slightly soluble. But it is bubbled into these drinks under pressure, to make more dissolve. When the bottles are opened it escapes again, and that causes the 'fizz'.
3 Solid carbon dioxide (dry ice) is used to keep food frozen.

Carbon monoxide

When carbon compounds burn in plenty of air they form **carbon dioxide**. But when there is a limited supply of air, **carbon monoxide** (CO) is produced instead. For example it forms when petrol burns in car engines and when cigarettes are smoked.

Properties of carbon monoxide
1 It is a colourless gas with no smell.
2 It is insoluble in water.
3 It is poisonous. If it gets into your blood it reacts with the **haemoglobin** in your red blood cells. Haemoglobin is the substance that carries oxygen round the blood. Carbon monoxide stops it working by forming **carboxyhaemoglobin**. The result could be death by oxygen starvation.
4 It acts as a reducing agent, by removing oxygen. For example it reduces lead(II) oxide to lead:

$$PbO(s) + CO(g) \longrightarrow Pb(s) + CO_2(g)$$

In the blast furnace it reduces iron(III) oxide to iron (page 146).

The air over the British Isles contains about 8½ million tonnes of carbon monoxide, mostly from car engines. Luckily it is not concentrated enough to poison everyone! However it does harm people with heart disease.

Questions

1 Page 182 shows two reactants for making carbon dioxide. Suggest two others that would do instead.
2 Now suggest a different way to make the gas.
3 Carbon dioxide *sublimes*. What does that mean?
4 Describe the test for carbon dioxide.
5 Write down three uses of carbon dioxide, and explain what makes it suitable for each use.
6 Explain why carbon monoxide is poisonous.
7 Carbon monoxide reduces iron oxide (Fe_2O_3) to iron. Write a balanced equation for the reaction.

13.3 Carbon in living things

All living things contain carbon compounds:

Grass is about 4% carbon by weight.

The hard shell of this insect is almost 40% carbon.

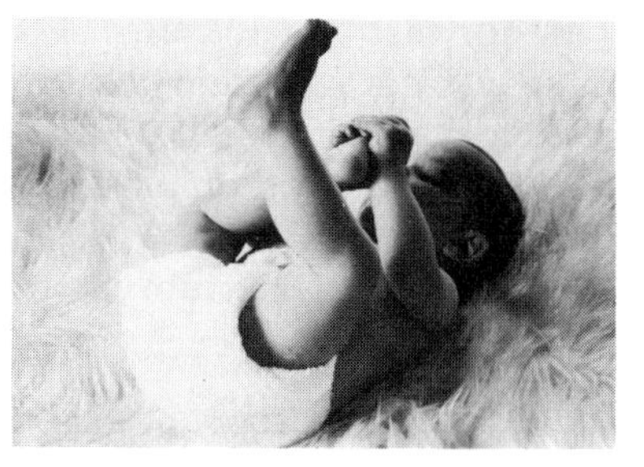

Humans contain hundreds of different carbon compounds. This baby is about 20% carbon.

From air to living things

Carbon dioxide is the source of all the carbon compounds in living things. This is what happens:

1 First, plants take in carbon dioxide from the air, and water from the soil, to make a sugar called **glucose**. This process is called **photosynthesis**. It takes place in the plant's leaves. It needs both heat energy and light energy from sunlight. Chlorophyll, the green substance in leaves, acts as a catalyst for the reaction:

$$\text{carbon dioxide} + \text{water} \xrightarrow[\text{chlorophyll}]{\text{sunlight}} \text{glucose} + \text{oxygen}$$

$$6\,CO_2(g) + 6\,H_2O(l) \longrightarrow C_6H_{12}O_6(s) + 6\,O_2(g)$$

Note that photosynthesis produces **oxygen** as well as glucose. The oxygen goes off into the air. Every year plants put millions of tonnes of oxygen into the air. At the same time we take millions of tonnes of it from the air for respiration (see below) and for burning fuels. Without photosynthesis our supply of oxygen would run out.

Plants take in carbon dioxide and give out oxygen through tiny holes in their leaves. This is a hole in a wheat leaf, enlarged 5000 times.

2 Inside the plant, the glucose is converted to **starch**, **proteins** and other carbon compounds that are needed for roots, stems and leaves. For proteins the plant also needs nitrogen, and small amounts of sulphur, which it obtains from the soil.

3 Animals then obtain carbon compounds by eating plants. And humans eat both animals and plants.

From living things to air

Carbon dioxide gets back to the air, in these ways:

1 Plants and animals give out carbon dioxide during **respiration**. Respiration is the reaction that gives living things the energy they need. It takes place in their cells:

$$\text{glucose} + \text{oxygen} \longrightarrow \text{carbon dioxide} + \text{water} + \text{energy}$$

Note that this is the opposite reaction to photosynthesis.

Tropical forests and jungles play a large part in making sure we do not run out of oxygen.

During photosynthesis heat and light energy are taken in and stored as chemical energy in glucose. Respiration releases the stored energy from the glucose again.

2 Bacteria feed on dead plants and animals, and produce carbon dioxide at the same time.

3 Wood, coal, oil, and natural gas are all made of carbon compounds. (They are all formed from the remains of living things.) They all produce carbon dioxide when they burn in plenty of air.

. . . breathing out carbon dioxide from respiration.

The carbon cycle

You saw above that carbon dioxide is removed from the air by plants, and put back again in other ways. The whole process is called the **carbon cycle**:

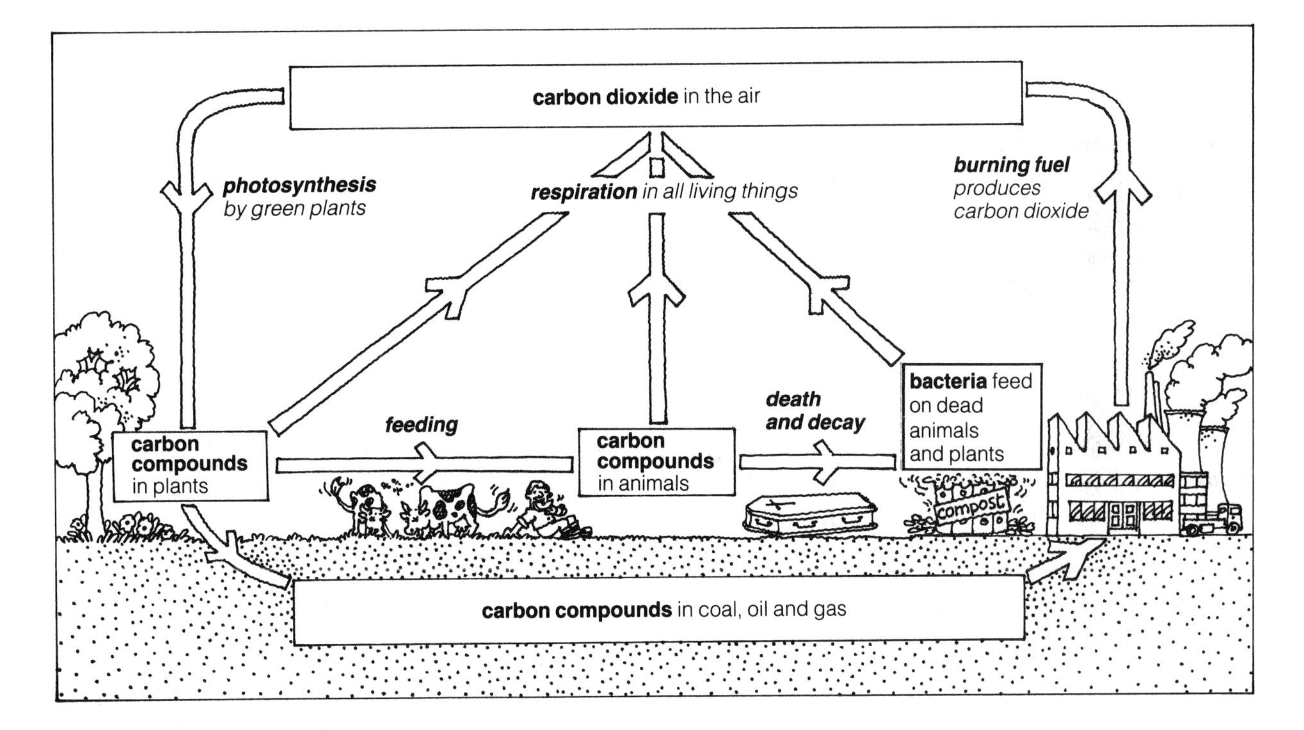

Too much carbon dioxide? Many scientists think that we may upset the carbon cycle, by burning too much fuel. This could put too much carbon dioxide into the air. The gas might then act as a blanket round the earth, keeping in heat. Scientists call this the **greenhouse effect**. It means that the earth could get too warm, so that the weather would change and the ice around the poles would melt, causing huge floods.

Questions

1 What is photosynthesis? In which part of a plant does it take place?

2 Describe how carbon dioxide from the air is converted to a protein in your body.

3 Describe three ways in which carbon dioxide is returned to the air. Give equations if you can.

4 Respiration is the opposite reaction to photosynthesis. Explain why.

13.4 The fossil fuels

Most of the energy used to cook food, drive cars, and keep us warm comes from coal, oil, and gas.

1 Coal, oil, and gas are **fossil fuels**. That means they are the remains of plants and animals that lived millions of years ago.
2 They are all made of carbon compounds.
3 They are used as fuels because they give out plenty of heat energy when they burn.
4 They produce carbon dioxide and water vapour as well as energy when they burn. For example North Sea gas (methane) burns like this:

$$CH_4(g) + 2O_2(g) \longrightarrow CO_2(g) + 2H_2O(g) + \text{energy}$$

Coal

Coal is the remains of trees, ferns, and other plants that lived as much as three million years ago. These were crushed into the earth, perhaps by earthquakes or volcanic eruptions. They were pressed down by layers of earth and rock. They slowly decayed into coal.

Around 105 million tonnes of coal are mined each year in Britain. The table shows that most of it is burned in power stations, to make electricity. The electricity is then used for heating, lighting, and driving machinery.

Coal used for	Amount (million tonnes)
Power stations	80
Coke ovens	10
Industry	9
Home heating	7
Exports	5
Offices, hospitals etc.	4

Oil and gas

Oil and gas are the remains of millions of tiny plants and animals that lived in the sea. When they died, their bodies sank to the sea bed and were covered by silt. Bacteria attacked the dead remains, turning them into oil and gas. Meanwhile the silt was slowly compressed into rock. The oil and gas seeped into the porous parts of the rock, and got trapped like water in a sponge.

Although oil was formed under the sea, many oil wells are on dry land. Millions of years ago, movement of the earth's crust forced some seabeds upwards to form land.

Where does electricity fit in?

Electricity is not itself a fuel. But it is made from fuels:

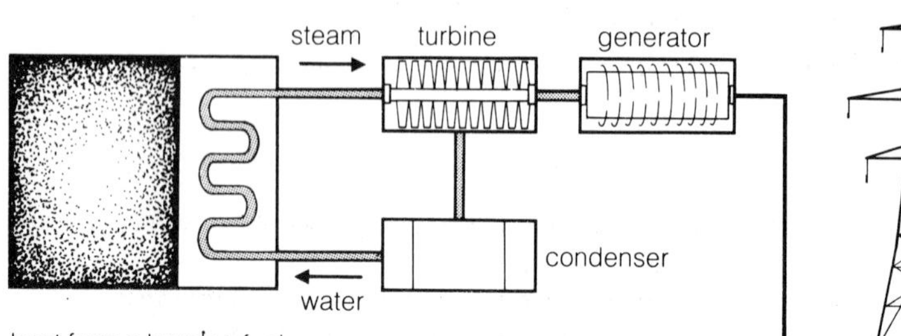

Problems with burning the fossil fuels

North Sea gas burns cleanly, with no smoke or fumes.
But some oil fuels contain sulphur compounds. These form **sulphur dioxide** when they burn, and this pollutes the atmosphere.
Some coal produces a lot of smoke and harmful tarry fumes. It must be made 'smokeless' before it can be used in factories and homes. This is done by heating the coal in the absence of air. **Coal gas** and **coal tar** are driven from the coal, leaving a fuel called **coke** behind.

Smokeless fuels. These burn much more cleanly than ordinary coal. But they still produce sulphur dioxide, because they contain sulphur compounds.

They are not just fuels . . .

Crude oil is a mixture of many carbon compounds. The first step is to separate it into groups of similar compounds (page 197). Some of these are used as fuels. But others are turned into plastics, man-made fibres, drugs, pesticides, insecticides and detergents.
Most coal is used as a fuel. But coke is used in the blast furnace to reduce iron ore to iron. And coal tar is used to make things like inks, detergents, and insecticides. (In fact anything made from oil can also be made from coal. But at the moment it would cost more.)

How long will they last?

This diagram shows how much we depend on coal, oil and gas.

Where Britain got its energy, 1984

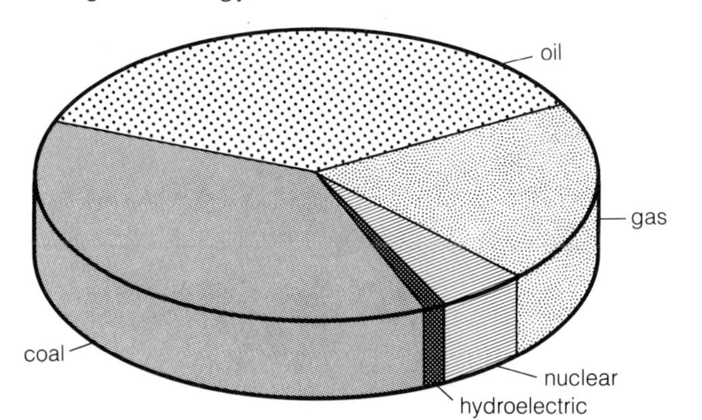

But this state of affairs cannot last. Most of our oil and gas will have run out by the year 2000. And by 2200 most of the coal will have gone. Even the uranium for nuclear power stations is expected to run out in about 30 years. So scientists are searching for alternative sources of energy. You can read about these on the next two pages.

What makes a good fuel? These are the ten key questions . . .

1. Does it light easily?
2. Does it burn steadily?
3. Does it give out much energy?
4. Does it pollute the air?
5. Does it leave much ash behind?
6. Is it plentiful?
7. Does it cost a lot?
8. Is it easy and safe to store?
9. Is it easy and safe to transport?
10. Might it be better used for something else?

Questions

1 What does the term *fossil fuel* mean?
2 Describe how coal was formed.
3 What is the main use of coal at present?
4 Oil and gas are usually found together. Why?
5 How is coke made from coal?
6 Name three products made from coal tar.
7 Although coke is cleaner than coal, it can still cause pollution. Explain why.
8 Why do we need to find other forms of energy?
9 Using the list of ten questions above, decide whether: **a** paper **b** magnesium ribbon would make a good alternative fuel.

13.5 Alternative sources of energy

On the last two pages you read about fossil fuels. These are **finite** or **non-renewable** sources of energy. Sooner or later, they will run out. So scientists are looking for other sources to replace them.

Windmills for generating electricity on Orkney Island. Scientists think that windpower could meet well over half our present electricity needs. But more research is needed, especially for windmills offshore.

Wind power

Windmills were widely used in the past, for grinding corn and pumping water. Now engineers are working on windmills to drive electricity generators. One idea is to build the windmills on man-made islands, several miles out to sea. There they would be out of sight but exposed to plenty of wind.

Solar power

The energy reaching earth from the sun every year is about 15 000 times more than our energy needs. There is a lot of research happening to find ways to use this **solar power**.
One simple way is to fit solar panels onto the roofs of buildings. These panels absorb sunlight and use it to heat water. But that saves energy on just a small scale.
A more ambitious plan is to use giant mirrors to concentrate the sun's rays. The heat from the rays would then be used to produce steam, for driving turbines to make electricity.

Tidal power

There have been many ideas for obtaining energy from waves. One scheme is the **tidal barrage**. The photograph below shows a tidal barrage in France. The tide generates electricity as it flows through turbines set in the concrete wall. There are plans for a similar barrage across the River Severn at Bristol and the Mersey at Liverpool. But they may cost too much to build.

Solar panels in a south-facing roof. They help to cut fuel bills by trapping the sun's warmth to heat water.

The tidal barrage at the mouth of the River Rance in France. Six British rivers have tidal estuaries suitable for barrages.

Hydroelectric power

When rain falls on high ground the water can be trapped in a reservoir and then used to drive turbines in a **hydroelectric station**, as it flows to lower ground. In Britain less than 1% of our energy is supplied by hydroelectric power. It makes economic sense to build more hydroelectric stations in the future.

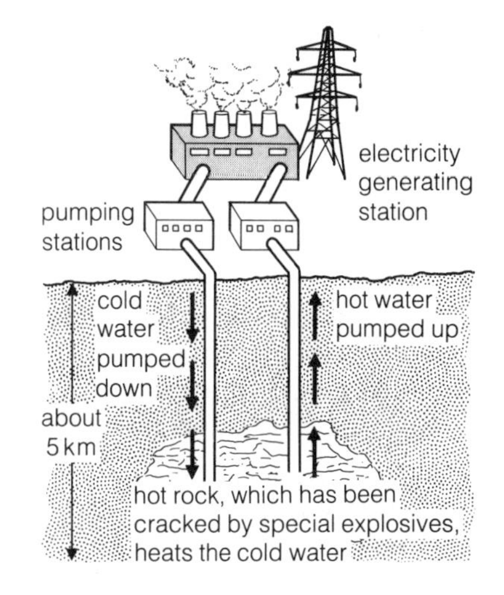

Using hot rocks to heat water. Geothermal energy could meet about 10% of our present electricity needs.

Power from hot rocks

Deep in the earth, the rock is hot. The deeper you go, the hotter it gets. The heat comes from the molten rock that lies below the earth's crust. This **geothermal energy** could be used to heat water for homes and industry, and to make steam for electricity.
Such an experiment is already under way at Rosemanowes in Cornwall. The plan is to pump water down five kilometres onto granite rocks, where the temperature is 200 °C, and then pump the hot water back up again.

Power from the biomass

Biomass means plant material. Plants collect energy from the sun and store it as chemical energy. We release this energy when we burn wood or coal. But that is not the only way.
Plants produce methane when they rot. Animal waste started as plants, and it too rots to give methane. The methane can be collected and used for heating, or to generate electricity.

A sewage digester. Sewage is broken down by bacteria to form methane, which is then used to fuel the sewage works. The same process can now be used to make gas from rotting vegetation.

Another way to use biomass for energy. In Brazil, sugar cane, turned into alcohol, has been used as engine fuel. This helps Brazil to cut down its spending on imported oil.

Questions

1 Coal is a finite energy source. What does that mean?
2 Explain how solar panels could help to save energy in the home.
3 How does a tidal barrage work?
4 **a** What is geothermal energy?
  b Explain how we can make use of it.
5 Explain how animal waste can be used as a source of energy. Write a balanced equation for the reaction that produces the energy.
6 Which of these are renewable sources of energy?
  a the sun **b** rainfall **c** the wind
  d the tide **e** hot rocks **f** biomass

13.6 Alkanes

Organic compounds

On page 184 you saw that living things contain carbon compounds. So do coal, oil and gas, since they were once living things.
These carbon compounds are often called organic compounds. (The word *organic* means *a living thing*.)
There are nearly three million known organic compounds. This huge number is due to the fact that carbon atoms can join together easily, to make chains of different lengths. Like these:

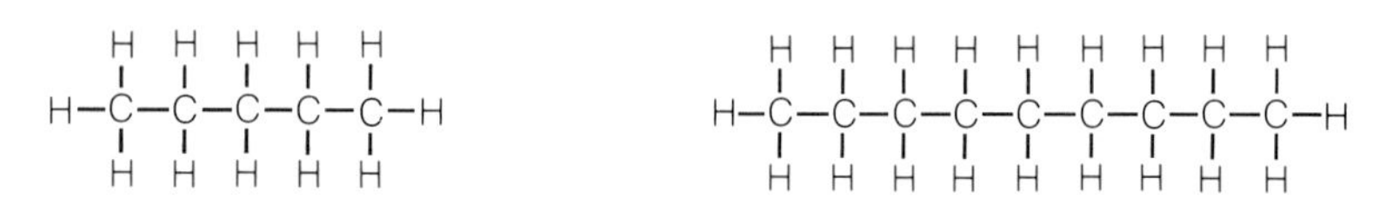

In this molecule, 5 carbon atoms have joined to make a chain.

This has a chain of 9 carbon atoms.

Here 6 carbon atoms form a branched chain.

The carbon atoms join by sharing electrons with each other, to make **covalent bonds** (page 46).
Look again at the molecules above. The first two contain only hydrogen and carbon, so the compounds are called **hydrocarbons**.
Hydrocarbons are organic compounds that contain only hydrogen and carbon.
The last molecule above contains oxygen, so this compound is *not* a hydrocarbon.

Alkanes

Hydrocarbons can be arranged into families of compounds that are alike in some way. The simplest family is called the alkanes. This table shows the first four alkanes:

Name	Methane	Ethane	Propane	Butane
Formula	CH_4	C_2H_6	C_3H_8	C_4H_{10}
Structure of molecule	H H–C–H H	H H H–C–C–H H H	H H H H–C–C–C–H H H H	H H H H H–C–C–C–C–H H H H H
Number of carbon atoms in chain	1	2	3	4
Boiling point	−164 °C	−87 °C	−42 °C	−0.5 °C

boiling point increases with chain length

The number of carbon atoms in the chain increases by 1, each time.
A family of compounds like this is called a **homologous series**.

Things to remember about the alkanes

1. The four alkanes in the table are gases at room temperature. But boiling point increases with chain length, so the next twelve alkanes are liquids and the rest are solids.
2. Alkanes are found in natural gas and crude oil. Natural gas is mostly methane, with small amounts of ethane, propane and butane. Crude oil is a much more complicated mixture of hydrocarbons, and can contain alkanes with up to 100 carbon atoms in their molecules.
3. In all alkane molecules, each carbon atom forms four single bonds. The bonding in ethane is shown on the right.
4. Alkanes are unreactive. For example acids and alkalis have no effect on them. However they do burn well in a good supply of oxygen, forming carbon dioxide and water vapour. The reactions give out plenty of heat, so alkanes are often used as fuels. When propane burns, the reaction is:

   $C_3H_8\,(g) + 5\,O_2\,(g) \longrightarrow 3\,CO_2\,(g) + 4\,H_2O\,(g) + \text{heat}$

   Both propane and butane are used as camping gas, and in gas lighters. Calor gas is mainly butane. And natural gas (North Sea gas) is used for cooking and heating in homes.

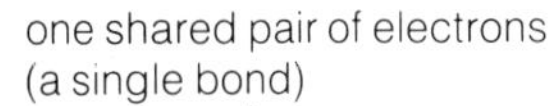

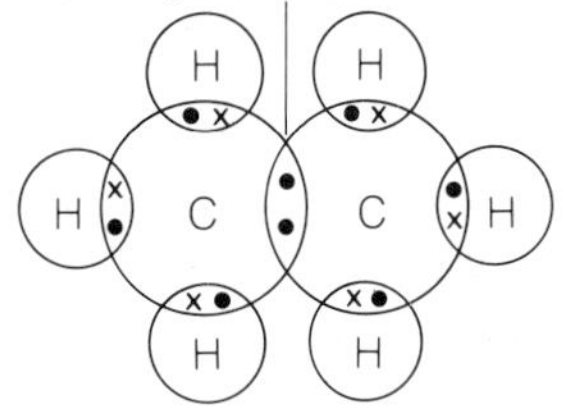

The bonding in ethane.

Isomers

Look at these two molecules:

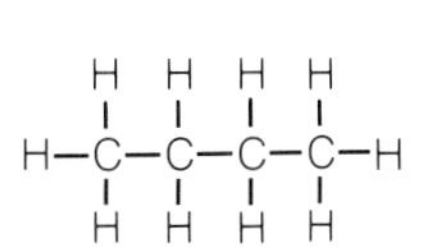

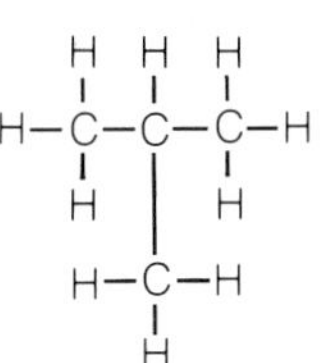

This is a molecule of **butane**, which has the formula C_4H_{10}. It has a straight chain of carbon atoms.

This is a molecule of **methyl propane**, which also has the formula C_4H_{10}. But this time the carbon chain is branched.

So both compounds have the same formula, but their molecules have different structures. They are called **isomers**.
Isomers are compounds that have the same formula, but different molecular structures.
All the alkanes from butane onwards have isomers.

Camping gas being used by a mountaineer.

Questions

1. Write the formulae for the first four alkanes.
2. **a** Suggest formulae for the next two alkanes, and draw the structures of their molecules.
   b Now suggest boiling points for them.
3. Will alkanes conduct electricity? Explain.
4. Draw a diagram to show the *bonding* in:
   a methane **b** ethane
5. Why is propane used as a fuel? Write a balanced equation for its combustion.
6. It is dangerous to burn hydrocarbons in too little air. Can you think why?
7. The next alkane after butane is pentane. Its formula is C_5H_{12}, and it has three isomers. See if you can draw all three.

13.7 Alkenes

On the last two pages you met the **alkanes**. There is another family of hydrocarbons called the **alkenes**. The first two alkenes are **ethene** and **propene**. Their molecular structures are shown below. Compare them with the corresponding alkanes:

ethane, C_2H_6

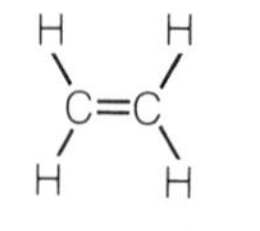

ethene, C_2H_4

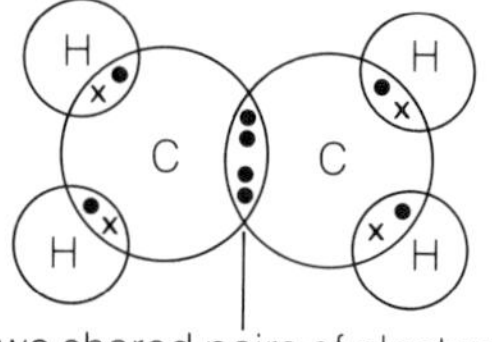

The bonding in ethene.

propane, C_3H_8

propene, C_3H_6

In all the molecules above, each carbon atom forms four bonds. The alkane molecules have only *single bonds* between carbon atoms, but each alkene molecule has one *double bond* between carbon atoms. **Alkene molecules contain one double bond between carbon atoms.**

Alkenes are more reactive than alkanes

Alkenes are made from alkanes (page 197). They are like alkanes in some ways. For example, ethene and propene are both gases, like ethane and propane. They also burn in oxygen, forming carbon dioxide and water vapour, and giving out plenty of heat:

$$C_2H_4(g) + 3O_2(g) \longrightarrow 2CO_2(g) + 2H_2O(g) + \text{heat}$$

However, there is an important difference between the two families of compounds. Alkanes are **unreactive** compounds. But alkenes are **reactive**, because of their double bonds. Ethene reacts with hydrogen like this:

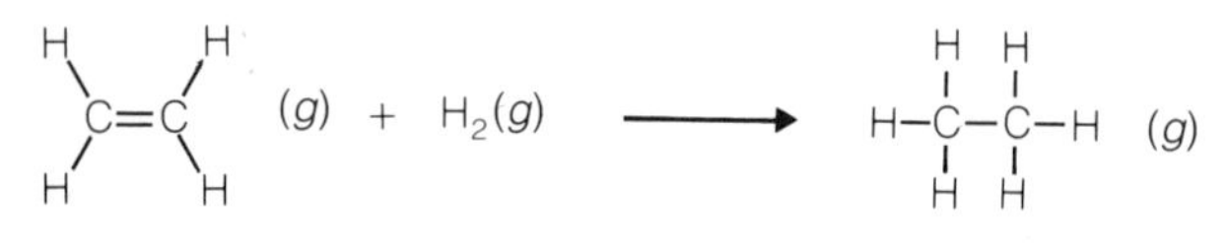

ethene ethane

The ethene molecule is able to 'add on' hydrogen, because the double bond can break to form single bonds. This is called an **addition** reaction. Each carbon atom forms bonds to four other atoms, instead of three.

Ethene is called **unsaturated** (not full) since its molecules can add on more atoms. But ethane is **saturated** (full) since its molecules cannot fit in more atoms—there are no double bonds to break, and each carbon atom already has four single bonds.

Alkanes are saturated. But alkenes are unsaturated, and that makes them more reactive than alkanes.

Alkenes are made from some of the alkanes in oil, by a process called cracking. This is the cracking plant at Esso's Fawley refinery.

Tests for unsaturation There are two ways you could test a hydrocarbon, to see whether it is an alkane or an alkene:

1 You could add some bromine water. Bromine water is an orange solution of bromine in water. It turns colourless in the presence of an alkene, because the bromine adds on to the alkene, to form a colourless compound. For example:

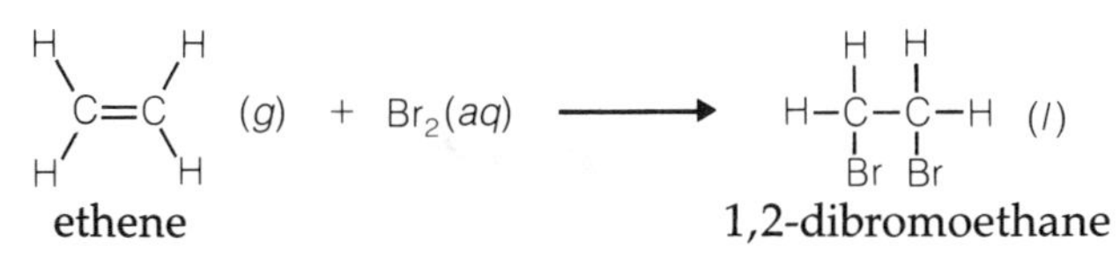

2 You could add acidified potassium manganate(VII) solution. This is purple, but turns colourless if an alkene is present.

Just some of the things that can be made from polythene.

Polymerisation

Because of their double bonds, alkene molecules can also add on to each other, in an important reaction called **polymerisation**. **During polymerisation, many small molecules join together, to make very large molecules.**

The small molecules are called **monomers**, and the very large molecules are called **polymers**. (*Poly-* means *many*.)

When ethene is heated under high pressure, its molecules join together to form **polythene**:

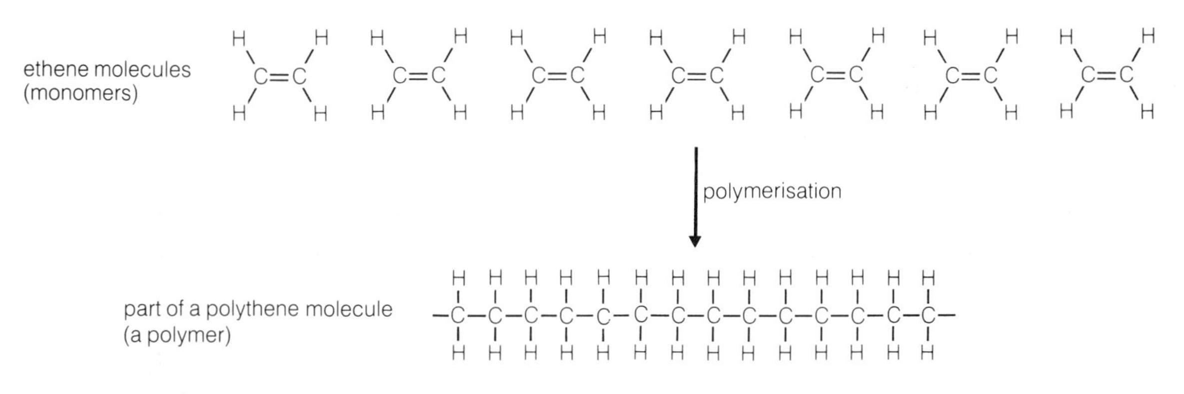

Up to 50 000 ethene molecules can join in this way, to make one molecule of polythene.

Polythene is a solid. It is unreactive, since it contains only single carbon bonds. It can be rolled into thin sheets and moulded into different shapes. Because it is easy to mould, it is called a **plastic**. You can find out more about it, and other plastics, on page 198.

Questions

1 What is the difference between an ethane molecule and an ethene molecule?

2 Alkenes are *unsaturated*. What does that mean? What effect does it have?

3 Write an equation to show how ethene reacts with bromine water. What change would you *see*?

4 Draw a molecule to fit the formula C_3H_6. This substance is called *propene*. What group does it belong to?

5 What is a polymerisation reaction?

6 Ethane will not polymerise. Explain why.

7 Draw a diagram to show how ethene polymerises.

13.8 Alcohols

Alcohols

You already met two groups of organic compounds, the alkanes and alkenes. There is another group called the **alcohols.**
The two simplest alcohols are **methanol** and **ethanol.**

```
   H                 H  H
   |                 |  |
H—C—OH            H—C—C—OH
   |                 |  |
   H                 H  H
```

methanol, CH_3OH ethanol, C_2H_5OH

The two simplest alcohols.

Note that both compounds have an $-OH$ group.
Alcohols are organic compounds that contain an $-OH$ group.

Ethanol

Ethanol is the best-known of all alcohols. In fact it is often just called 'alcohol'.
It is a good solvent—it dissolves many substances that are not soluble in water. It also evaporates quickly. So it is used in glues, paints, varnishes, printing inks, deodorants, colognes and after-shaves. It is also used as a raw material for making other substances, such as synthetic rubbers and flavourings.
Ethanol is well known for another reason too—it makes people drunk. Beer, wine, and all alcoholic drinks contain some ethanol.

Ethanol makes it easy . . . it is used in wine, aftershave and perfume.

Methylated spirits is mainly ethanol, mixed with a little methanol, which is poisonous. Colouring is added, and a horrible taste to stop people drinking it.

The properties of ethanol

1. It is a clear, colourless liquid, that boils at 78°C.
2. The pure liquid is dangerous to drink, and even dilute solutions of it affect the body. At first, a dilute solution makes you feel relaxed, but too much causes 'drunkenness', with headaches, dizziness and vomiting. Over time it can ruin your liver.
3. It burns well in air or oxygen, giving out heat:

   $$C_2H_5OH(l) + 3O_2(g) \longrightarrow 2CO_2(g) + 3H_2O(g)$$

   As methylated spirits, it is often burned as a fuel, in **spirit lamps.** These are used by jewellers, and sometimes by cooks.
4. Ethanol is slowly oxidised to **ethanoic acid,** by bacteria in the air:

   $$C_2H_5OH(l) + O_2(g) \longrightarrow CH_3COOH(aq) + H_2O(l)$$

   This reaction is the reason why wine goes 'sour' when left open in air. But the ethanol can be oxidised much more quickly in the laboratory by heating it with an oxidising agent such as potassium dichromate(VI).

Unfortunately, some people become addicted to ethanol. They turn into alcoholics.

Making ethanol for industry

Ethanol for use as a solvent is made from ethene and water. Ethene is mixed with steam and passed over a catalyst. This addition reaction takes place:

$$\underset{\text{ethene}}{\mathrm{CH_2{=}CH_2}} + \mathrm{H_2O} \xrightarrow{\text{catalyst}} \underset{\text{ethanol}}{\mathrm{H{-}CH_2{-}CH_2{-}OH}}$$

The product is in fact a solution of ethanol in water. Most of the water is then removed by fractional distillation (page 20).

Wine is made from crushed grapes. Unlike beer, no yeast needs to be added — the skins of grapes contain a natural yeast.

Making ethanol by fermentation

The ethanol in alcoholic drinks is made by **fermentation.** During fermentation, glucose from fruit, vegetables or cereals such as barley is turned into ethanol by natural catalysts called **enzymes.** The enzymes are contained in **yeast.**

$$\underset{\text{glucose}}{C_6H_{12}O_6(s)} \xrightarrow[\text{yeast}]{\text{enzymes in}} \underset{\text{ethanol}}{2C_2H_5OH(l)} + \underset{\text{carbon dioxide}}{2CO_2(g)}$$

For example, beer is made from **barley** by fermentation. Barley grains are crushed and soaked in hot water to extract the glucose. Hops are added for flavour. Then the liquid is strained, and yeast added. The yeast converts the glucose to ethanol.
Fermentation is allowed to go on until enough ethanol has formed. Then the beer is carefully heated to kill the yeast.

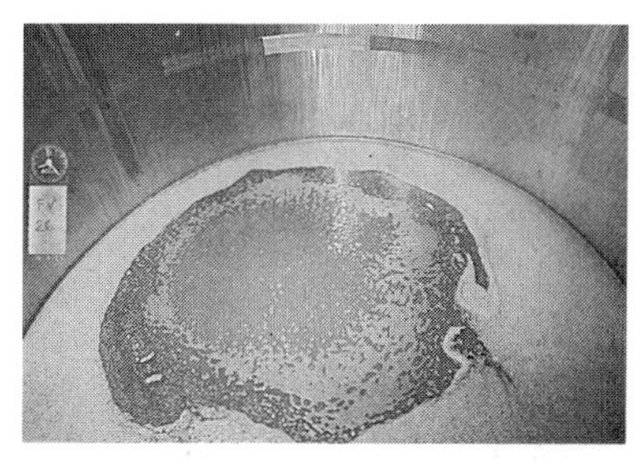

Beer fermenting. The froth is caused by escaping carbon dioxide. Fermentation is carried out at 18–20°C. At lower temperatures the reaction is too slow. At higher temperatures the yeast may get killed.

How alcoholic is it?

The important question about an alcoholic drink is 'How much ethanol does it contain?' A pint of strong beer contains twice as much ethanol as a pint of ordinary beer, for example. Like beer, whiskey is made from barley. But it is more 'alcoholic' because it is distilled after fermentation, which makes the ethanol more concentrated.
The drinks below all contain about the same amount of ethanol:

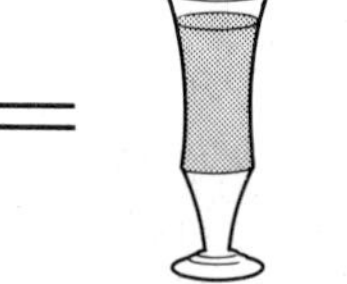

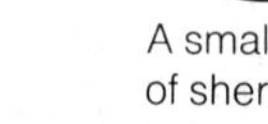

A half pint of ordinary beer = A small glass of sherry = A small whiskey (or gin or vodka) = A glass of wine

Questions

1 List three uses of ethanol.
2 Why does wine go sour when left in air?
3 Describe what happens during fermentation.
4 Why does the temperature need to be carefully controlled during fermentation?
5 Why is whiskey more 'alcoholic' than beer?

13.9 Chemicals from oil

On page 186 you saw that oil is the remains of tiny plants and animals that lived millions of years ago. Oil is in fact a mixture of many compounds. They are mostly hydrocarbons, and they are very useful:

Some of the compounds from oil are used as fuels for central-heating boilers, cars, aeroplanes and ships.

Some are turned into nylon, polythene, PVC and other plastics. These sailors' boats and jackets all started off as oil.

Some are used to make washing-up liquids, washing powders, paints, drugs and cosmetics.

Getting oil from the earth

Oil is usually found about 3 or 4 kilometres below ground. It is trapped in rocks, like water in a sponge.
In Britain, the oil-bearing rocks lie beneath the North Sea.
To get at the oil, a hole is bored in the sea bed, using a giant drill. The drill is hung from a **drilling platform**, like the one on the right.
Next, the hole is lined with steel. Then the oil is pumped up through it. It is carried ashore, to an **oil refinery**, in pipes that run along the sea bed.

An oil platform in the North Sea.

Refining the oil

Some of the compounds in oil have only short carbon chains. But some have long chains, with up to 100 carbon atoms. To make the best use of oil, it is separated into groups of compounds that are close in chain length. This is called **refining** the oil. It is carried out by **fractional distillation**.
The distillation takes place in a tall tower called a **fractionating tower**. There the oil is heated up. The compounds with short chains boil off first, because they have the lowest boiling points. They rise to the top of the tower. The compounds with longer chains have higher boiling points, and are collected lower down. The different groups of compounds are called **fractions**.

Laying down the pipes for carrying oil.

cool (25°C)

fractions out

crude oil in

solid

solid

very hot (over 400°C)

Name of fraction	Length of carbon chain	What the fraction is used for
Gas	C_1 to C_4	Separated into methane, ethane, propane and butane—all fuels. Methane is used to make hydrogen (page 150).
Petrol	C_4 to C_{10}	Fuel for cars.
Kerosene	C_{10} to C_{16}	Jet fuel. Detergents.
Diesel oil	C_{16} to C_{20}	Fuel for central heating. Some cracked (see below).
Lubricating oil	C_{20} to C_{30}	Oil for cars and other machines. Some cracked.
Fuel oil	C_{30} to C_{40}	Fuel for power stations and ships. Some cracked.
Paraffin waxes	C_{40} to C_{50}	Candles, polish, wax-papers, waterproofing, grease.
Bitumen	C_{50} upwards	Pitch for roads and roofs.

Cracking hydrocarbons

Bitumen—one of the fractions from oil.

Look at the equation below. It shows how an alkane molecule can be broken into smaller molecules, using heat and a catalyst:

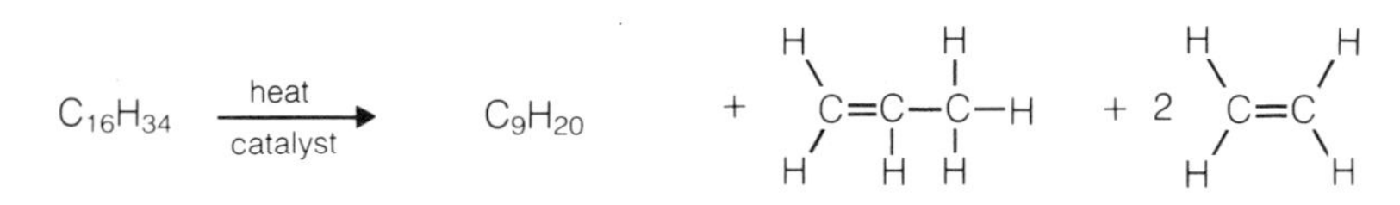

an alkane in diesel oil — an alkane suitable for petrol — an alkene (propene) — an alkene (ethene)

This type of reaction is called **cracking**. It is very important in the oil industry, for two reasons:

1. Some fractions from oil are more useful than others. For example there is a greater demand for petrol than for diesel oil or lubricating oil. So cracking is used to convert part of these fractions to petrol.
2. Cracking produces short-chains **alkenes**, like ethene and propene. These are reactive because of their double bonds, so they can be used to make other substances. For example, ethene is made into polythene and ethanol.

Questions

1. Copy and complete: Oil is the remains of tiny prehistoric .... and .... It is a .... of compounds. These are mostly ...., which means they contain just hydrogen and ....
2. Oil is *refined* before use. What does that mean?
3. Name the oil fraction where the compounds have:
   a the longest chain lengths
   b the lowest boiling points
   c the highest boiling points
4. Which fraction is used for making polish?

13.10 Plastics

What are plastics?

You probably own several plastic things. Records and cassette tapes are made of plastic. So are toothbrushes, combs and nylon socks! There are many different kinds of plastics. Here are some:

Name	Used for
Polythene	Plastic bags, dustbins, plastic basins
Polyvinyl chloride (PVC)	Raincoats, seat covers, records
Polystyrene	Plastic cups, packaging materials
Nylon	Rope, bristle for brushes, tights, clothing
Melamine	'Unbreakable' dishes and mugs, ashtrays
Phenolic	Electric plugs, saucepan handles

All these objects are made of plastic.

They all share these properties:

1. They are all carbon compounds.
2. They can be moulded into different shapes. That is why they are called plastics. (*Plastic* means *easy to mould*.)
3. The starting materials for them are usually obtained from oil. For example polythene and PVC start off as **ethene**. This is made by cracking some of the alkanes in oil (page 197).
4. They are all **polymers**. That means they consist of very long molecules, made by joining many small molecules together. (*Poly* means *many*.) The small molecules are called **monomers**.

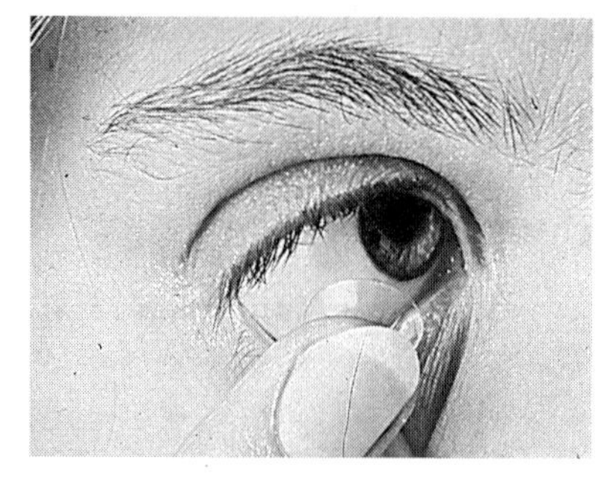

Contact lenses are made of plastic.

On page 193 you saw how ethene molecules join to make polythene. To make polyvinyl chloride, ethene is first reacted with hydrochloric acid, giving the monomer **vinyl chloride**. This is mixed with warm water, under pressure, and the monomers join together or **polymerise**:

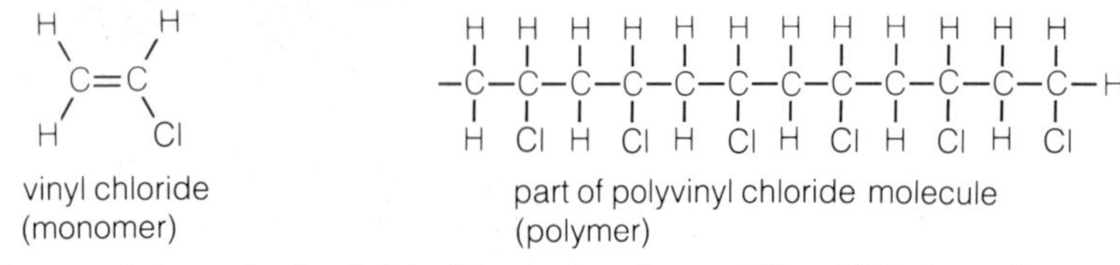

A complete polyvinyl chloride molecule could be 4000 times longer!

Plastics: good or bad?

Good points	Bad points
1 Quite cheap, and easily made	1 Difficult to dispose of. Plastic bags and cartons do not rot when they are thrown away, so they pollute the countryside, but **biodegradable** plastics rot away
2 Lighter than wood, stone or metal	2 Some plastics catch fire very easily
3 Unreactive. They do not corrode in air or water Many are not affected by acids or alkalis	3 When they burn, they often produce harmful gases: For example, PVC gives off fumes of hydrogen chloride when it burns. This would form hydrochloric acid in your eyes and throat
4 Do not conduct heat or electricity, so can be used as insulators	4 They usually do not look as good as wood or stone
5 Can be moulded into any shape	
6 Can be made very strong	
7 Can be coloured, by adding pigments	

Two groups of plastics

All plastics can be sorted into two groups: **thermoplastics**, and **thermosetting plastics**, or **thermosets**. You can tell which group a plastic belongs to, from the way if behaves on heating.

Thermoplastics These plastics get soft and runny when they are heated, and hard again when they are cooled. They can be made soft and hard *over and over again*. This is because of their structure:

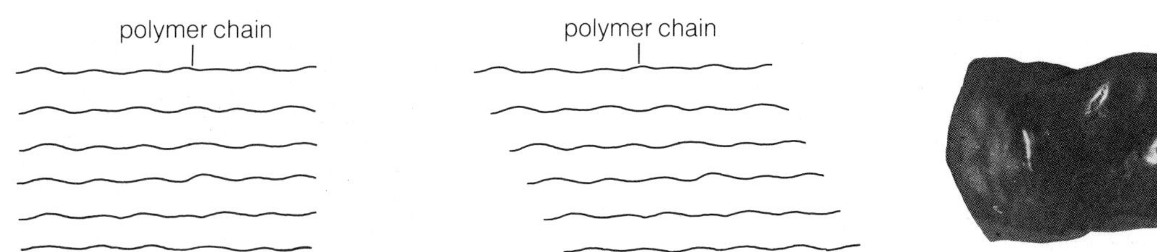

In a thermoplastic, the polymer chains lie next to each other, like this. A thermoplastic gets soft on heating . . .

. . . because the chains can slide past each other. The soft plastic can be moulded into any shape, and the shape can easily be changed again.

Polythene is a thermoplastic. This polythene object was once a bottle! It was put in a hot oven for just a few seconds.

Look again at the list of plastics on the opposite page. The first four are thermoplastics. So they are used only for things that do not get too hot.

Thermosets These get soft only once—the first time they are heated:

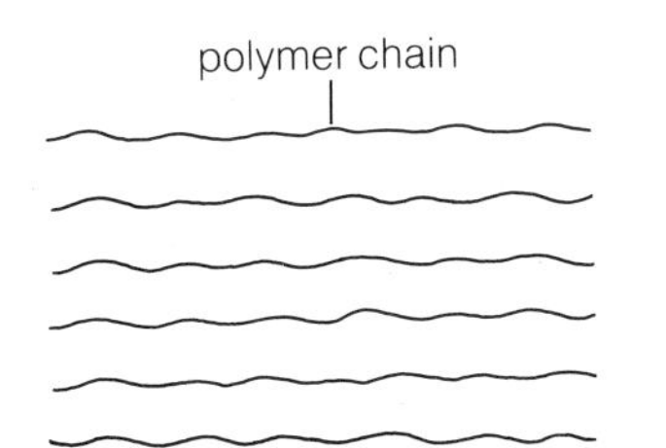

At the start, the polymer chains in a thermoset are like this. The first time a thermoset is heated, if softens. That means it can be moulded into a shape.

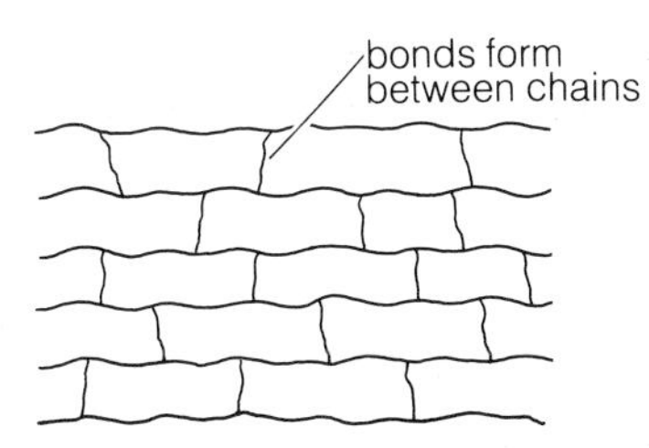

But the heat causes bonds to form between the chains. The plastic sets firmly into its new shape. The bonds keep it hard, even when it is heated again.

Phenolic is a thermoset. That is why it is chosen for saucepan handles. Melamine is a thermoset too.

Questions

1. Name four different plastics.
2. Which monomer is used to make PVC?
3. Draw a row of vinyl chloride molecules. Then show how they join up to make polyvinyl chloride. (The drawing for polythene on page 193 might help you.)
4. Describe three problems about using plastics.
5. What is the difference between a thermoplastic and a thermoset?
6. Why are thermoplastics *not* used for making saucepan handles?
7. Explain why thermosets keep their shape after being heated the first time.

Questions on Chapter 13

1 **a** Copy this diagram and complete it by writing in **i** the common names **ii** the chemical formulae

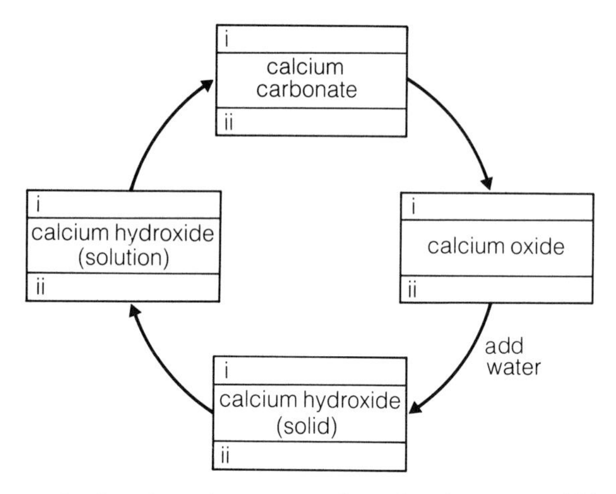

b Beside each arrow say how the change would be carried out. One example is shown.
c Give three reasons why limestone is an important raw material.

2 Carbon dioxide can be converted into carbon monoxide by passing a stream of carbon dioxide over heated carbon.
a Name two chemicals you could react together to produce a steady stream of carbon dioxide.
b Draw a diagram of a suitable apparatus for the reaction, showing:
  i a combustion tube containing carbon.
  ii a flask through which the gases can pass, containing a liquid that will absorb *unreacted* carbon dioxide.
  iii the collection of the carbon monoxide over water.
c Describe what you would see when carbon monoxide is passed over heated copper(II) oxide. Write a balanced equation for the reaction.

3 Sodium hydrogen carbonate is often called sodium bicarbonate or bicarbonate of soda. It is used as baking soda, in baking powder, and in indigestion tablets.
a Write its chemical formula.
b What happens when it is heated?
c Write an equation for this reaction.
d Explain its use in baking.
e Why is sodium *carbonate* no use in baking?
f Hydrogen carbonates react like carbonates, with acids. What products would be formed when sodium hydrogen carbonate reacts with hydrochloric acid?
g Write an equation for this reaction.
h Explain why sodium hydrogen carbonate is used to cure indigestion.

4 Below are ten statements about gases. Which of them describe the gas carbon dioxide?
a Colourless
b Given out during photosynthesis
c Turns lime water milky
d Burns in air
e Insoluble in water
f Heavier than air
g Reacts with bases to form salts
h Used up in burning carbon compounds
i Reduces oxides by removing oxygen
j Does not support combustion

5 Which of the properties in question 4 are properties of carbon monoxide but *not* of carbon dioxide?

6 Put these sentences in the correct order to make a description of the carbon cycle.
a In the reaction, carbon dioxide and water are given out.
b The carbon dioxide and water combine to form sugars.
c The reaction also gives out a lot of energy, which helps to keep the animals warm.
d All green plants take in carbon dioxide and water.
e Animals eat plants for food.
f Sunlight provides the energy for this reaction.
g The sugar from the food reacts with oxygen, in the cells of the animals.

7 Use the information on pages 190 and 191 to answer these questions about the alkanes:
a Which two elements do alkanes contain?
b Which alkane is the main compound in natural gas?
c After butane, the next two alkanes in the series are *pentane* and *hexane*. How many carbon atoms would you expect to find in a molecule of:
  i pentane? **ii** hexane?
d Write down the formulae for pentane and hexane.
e Draw a molecule of each substance.
f Is pentane a solid, liquid or gas at room temperature?
g Suggest a value for the boiling point of pentane, and explain your answer.
h Draw a molecule of a compound which is an isomer of pentane.
i Alkanes burn in a good supply of oxygen. Name the gases formed when they burn.
j Write a balanced equation for the burning of pentane in oxygen.
k Butane, pentane and hexane have different *chain lengths*. Explain what that means.
l What name is given to families of organic compounds such as the alkanes?

8 Ethanol is a member of a family of compounds called alcohols.

a Write the chemical formula of ethanol.

b How does a molecule of ethanol differ from a molecule of ethane?

c Ethanol is one of the products of the fermentation of sugar. The diagram shows apparatus which could be used to study the fermentation.

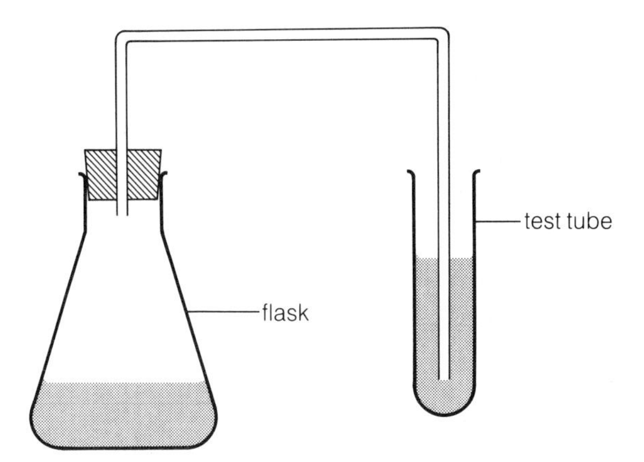

What should go in the flask?

d Which of these temperatures would be best for the reaction?

i 0 °C **ii** 10 °C **iii** 25 °C **iv** 55 °C

Explain your choice.

e The test tube contains water, which prevents air reaching the flask. Explain why this is important.

f Which gas is released during the fermentation?

g Complete the equation for the reaction:

$C_6H_{12}O_6 \longrightarrow$
(sugar)

h How long would you expect the reaction to take?

i 5 minutes **ii** 5 hours **iii** 5 days **iv** 5 months

i What process would you use, to separate a reasonably pure sample of ethanol from the mixture?

9 Propene is one of the many important hydrocarbons made from oil. Like *propane*, it is made up of molecules which contain three carbon atoms. Like *ethene*, it has a double bond.

a Draw a molecule of propene.

b How does it differ from a molecule of propane?

c To which group of hydrocarbons does **i** propane **ii** propene belong?

d Write formulae for propane and propene.

e Which of the two is a *saturated* hydrocarbon?

f i Explain why propene reacts immediately with bromine water, while propane does not.

ii What would you *see* in the reaction?

g Name another reagent that would react immediately with propene but not with propane.

h Propene is obtained by breaking down longer-chain hydrocarbons. What is this process called?

i Propene is the monomer for making an important plastic. Suggest a name for the polymer it forms.

10 Crude oil is a mixture of hydrocarbons, each with a different boiling point.

a It is an important raw material. Why?

b How is the mixture separated?

c A simple separation can be carried out in the laboratory, using this apparatus:

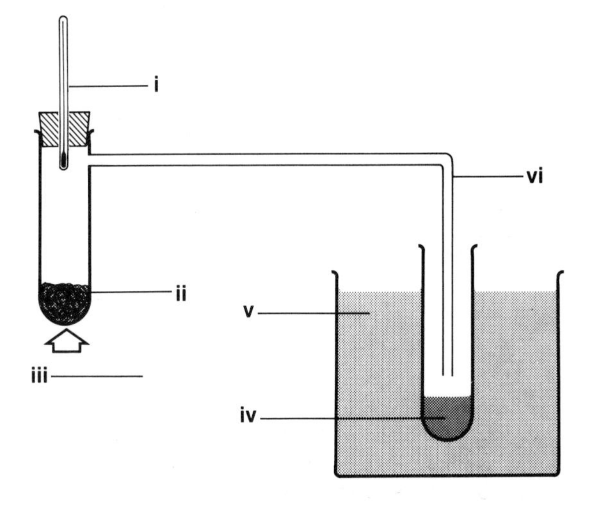

Copy the diagram and label it using these labels:
mineral wool *delivery tube* *thermometer* *heat* *water* *fraction*

d i What is the purpose of the mineral wool?

ii What is the purpose of the water?

iii Why is the thermometer placed where it is?

11 In the experiment in question 10, a crude oil sample was separated into four fractions. These were collected in the temperature ranges shown below:

Fraction	Temperature range/°C
A	25–70
B	70–115
C	115–200
D	200–380

Which fraction:

a has the lowest range of boiling points?

b is the most like petrol?

c has molecules with the longest chains?

12 a Which of these molecules could be used as monomers for making plastics? Explain your choice.

ethene, $CH_2 = CH_2$ ethanol C_2H_5OH,
propane C_3H_8 styrene $C_6H_5CH = CH_2$
chloropropene $CH_3CH = CHCl$

b Suggest a name for each polymer obtained.

c The polymers obtained are all thermoplastics. What special property do thermoplastics have?

d Name one thermosetting plastic. What special property do thermosetting plastics have?

e What problems can the use of plastics cause?

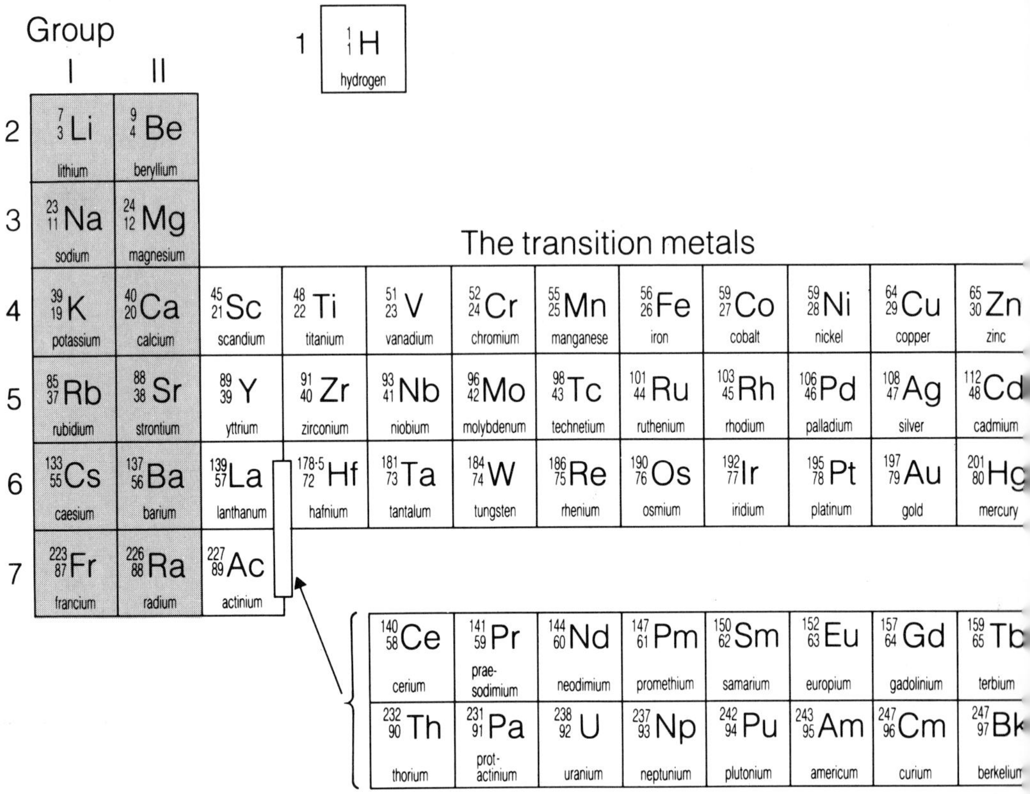

Group

	I	II											
1					$^{1}_{1}$H hydrogen								

The transition metals

	I	II										
2	$^{7}_{3}$Li lithium	$^{9}_{4}$Be beryllium										
3	$^{23}_{11}$Na sodium	$^{24}_{12}$Mg magnesium										
4	$^{39}_{19}$K potassium	$^{40}_{20}$Ca calcium	$^{45}_{21}$Sc scandium	$^{48}_{22}$Ti titanium	$^{51}_{23}$V vanadium	$^{52}_{24}$Cr chromium	$^{55}_{25}$Mn manganese	$^{56}_{26}$Fe iron	$^{59}_{27}$Co cobalt	$^{59}_{28}$Ni nickel	$^{64}_{29}$Cu copper	$^{65}_{30}$Zn zinc
5	$^{85}_{37}$Rb rubidium	$^{88}_{38}$Sr strontium	$^{89}_{39}$Y yttrium	$^{91}_{40}$Zr zirconium	$^{93}_{41}$Nb niobium	$^{96}_{42}$Mo molybdenum	$^{98}_{43}$Tc technetium	$^{101}_{44}$Ru ruthenium	$^{103}_{45}$Rh rhodium	$^{106}_{46}$Pd palladium	$^{108}_{47}$Ag silver	$^{112}_{48}$Cd cadmium
6	$^{133}_{55}$Cs caesium	$^{137}_{56}$Ba barium	$^{139}_{57}$La lanthanum	$^{178.5}_{72}$Hf hafnium	$^{181}_{73}$Ta tantalum	$^{184}_{74}$W tungsten	$^{186}_{75}$Re rhenium	$^{190}_{76}$Os osmium	$^{192}_{77}$Ir iridium	$^{195}_{78}$Pt platinum	$^{197}_{79}$Au gold	$^{201}_{80}$Hg mercury
7	$^{223}_{87}$Fr francium	$^{226}_{88}$Ra radium	$^{227}_{89}$Ac actinium									

$^{140}_{58}$Ce cerium	$^{141}_{59}$Pr prae-sodimium	$^{144}_{60}$Nd neodimium	$^{147}_{61}$Pm promethium	$^{150}_{62}$Sm samarium	$^{152}_{63}$Eu europium	$^{157}_{64}$Gd gadolinium	$^{159}_{65}$Tb terbium
$^{232}_{90}$Th thorium	$^{231}_{91}$Pa prot-actinium	$^{238}_{92}$U uranium	$^{237}_{93}$Np neptunium	$^{242}_{94}$Pu plutonium	$^{243}_{95}$Am americum	$^{247}_{96}$Cm curium	$^{247}_{97}$Bk berkelium

Relative atomic masses based on internationally agreed figures.

Element	Symbol	Atomic number	Relative atomic mass
Actinium	Ac	89	
Aluminium	Al	13	26.9815
Americium	Am	95	
Antimony	Sb	51	121.75
Argon	Ar	18	39.948
Arsenic	As	33	74.9216
Astatine	At	85	
Barium	Ba	56	137.34
Berkelium	Bk	97	
Beryllium	Be	4	9.0122
Bismuth	Bi	83	208.980
Boron	B	5	10.811
Bromine	Br	35	79.909
Cadmium	Cd	48	112.40
Caesium	Cs	55	132.905
Calcium	Ca	20	40.08
Californium	Cf	98	
Carbon	C	6	12.01115
Cerium	Ce	58	140.12
Chlorine	Cl	17	35.453
Chromium	Cr	24	51.996
Cobalt	Co	27	58.9332
Copper	Cu	29	63.54
Curium	Cm	96	
Dysprosium	Dy	66	162.50
Einsteinium	Es	99	
Erbium	Er	68	167.26
Europium	Eu	63	151.96
Fermium	Fm	100	
Fluorine	F	9	18.9984
Francium	Fr	87	
Gadolinium	Gd	64	157.25
Gallium	Ga	31	69.72
Germanium	Ge	32	72.59
Gold	Au	79	196.967
Hafnium	Hf	72	178.49
Helium	He	2	4.0026
Holmium	Ho	67	164.930
Hydrogen	H	1	1.00797
Indium	In	49	114.82
Iodine	I	53	126.9044
Iridium	Ir	77	192.2
Iron	Fe	26	55.847
Krypton	Kr	36	83.80
Lanthanum	La	57	138.91
Lawrencium	Lw	103	
Lead	Pb	82	207.19
Lithium	Li	3	6.939
Lutetium	Lu	71	174.97
Magnesium	Mg	12	24.312
Manganese	Mn	25	54.9380
Mendelevium	Md	101	

Group III	IV	V	VI	VII	O
					$^{4}_{2}$He helium
$^{11}_{5}$B boron	$^{12}_{6}$C carbon	$^{14}_{7}$N nitrogen	$^{16}_{8}$O oxygen	$^{19}_{9}$F fluorine	$^{20}_{10}$Ne neon
$^{27}_{13}$Al aluminium	$^{28}_{14}$Si silicon	$^{31}_{15}$P phosphorus	$^{32}_{16}$S sulphur	$^{35\cdot5}_{17}$Cl chlorine	$^{40}_{18}$Ar argon
$^{70}_{31}$Ga gallium	$^{73}_{32}$Ge germanium	$^{75}_{33}$As arsenic	$^{79}_{34}$Se selenium	$^{80}_{35}$Br bromine	$^{84}_{36}$Kr krypton
$^{115}_{49}$In indium	$^{119}_{50}$Sn tin	$^{122}_{51}$Sb antimony	$^{128}_{52}$Te tellurium	$^{127}_{53}$I iodine	$^{131}_{54}$Xe xenon
$^{204}_{81}$Tl thallium	$^{207}_{82}$Pb lead	$^{209}_{83}$Bi bismuth	$^{210}_{84}$Po polonium	$^{210}_{85}$At astatine	$^{222}_{86}$Rn radon

$^{162}_{66}$Dy dysprosium	$^{165}_{67}$Ho holmium	$^{167}_{68}$Er erbium	$^{169}_{69}$Tm thulium	$^{173}_{70}$Yb ytterbium	$^{175}_{71}$Lu lutecium
$^{251}_{98}$Cf californium	$^{254}_{99}$Es einsteinium	$^{253}_{100}$Fm fermium	$^{256}_{101}$Md mendelevium	$^{254}_{102}$No nobelium	$^{257}_{103}$Lr lawrencium

Approximate atomic masses for calculations.

Element	Symbol	Atomic mass for calculations
Aluminium	Al	27
Bromine	Br	80
Calcium	Ca	40
Carbon	C	12
Chlorine	Cl	35.5
Copper	Cu	64
Helium	He	4
Hydrogen	H	1
Iodine	I	127
Iron	Fe	56
Lead	Pb	207
Lithium	Li	7
Magnesium	Mg	24
Manganese	Mn	55
Neon	Ne	20
Nitrogen	N	14
Oxygen	O	16
Phosphorus	P	31
Potassium	K	39
Silicon	Si	28
Silver	Ag	108
Sodium	Na	23
Sulphur	S	32
Zinc	Zn	65

Element	Symbol	Atomic number	Relative atomic mass
Mercury	Hg	80	200.59
Molybdenum	Mo	42	95.94
Neodymium	Nd	60	144.24
Neon	Ne	10	20.179
Neptunium	Np	93	
Nickel	Ni	28	58.71
Niobium	Nb	41	92.906
Nitrogen	N	7	14.0067
Nobelium	No	102	
Osmium	Os	76	190.2
Oxygen	O	8	15.9994
Palladium	Pd	46	106.4
Phosphorus	P	15	30.9738
Platinum	Pt	78	195.09
Plutonium	Pu	94	
Polonium	Po	84	
Potassium	K	19	39.102
Praseodymium	Pr	59	140.907
Promethium	Pm	61	
Protactinium	Pa	91	
Radium	Ra	88	
Radon	Rn	86	
Rhenium	Re	75	186.2
Rhodium	Rh	45	102.905
Rubidium	Rb	37	85.47
Ruthenium	Ru	44	101.07

Element	Symbol	Atomic number	Relative atomic mass
Samarium	Sm	62	150.35
Scandium	Sc	21	44.956
Selenium	Se	34	78.96
Silicon	Si	14	28.086
Silver	Ag	47	107.868
Sodium	Na	11	22.9898
Strontium	Sr	38	87.62
Sulphur	S	16	32.064
Tantalum	Ta	73	180.948
Technetium	Tc	43	
Tellurium	Te	52	127.60
Terbium	Tb	65	158.924
Thallium	Tl	81	204.37
Thorium	Th	90	232.038
Thulium	Tm	69	168.934
Tin	Sn	50	118.69
Titanium	Ti	22	47.90
Tungsten	W	74	183.85
Uranium	U	92	238.03
Vanadium	V	23	50.942
Xenon	Xe	54	131.30
Ytterbium	Yb	70	173.04
Yttrium	Y	39	88.905
Zinc	Zn	30	65.37
Zirconium	Zr	40	91.22

Answers to numerical questions

page 17 **3** The solubility of X is 16 g at 60°C The solubility of Y is 42 g at 60°C. **4 a** 31.5 g/100 g, 45.5 g/100 g **b** 27 g, 23 g

page 22 **7 a i** copper(II) sulphate 30 g, sodium chloride 36.5 g per 100 g of water **ii** copper(II) sulphate 40 g, sodium chloride 37 g per 100 g of water **b i** B **ii** A **c** 54°C **d** 68°C **e** No

page 23 **8 a** 35 g **b** 2 g **9** 0.09 g per 100 g of water **10 a i** 31 g per 100 g of water **ii** 0.003 g per 100 g of water

page 39 **10 a** The correct values for rubidium are: m.pt. 39°C, b.pt. 688°C; it is very reactive. **11 b i** 1 minute **iii** 50

page 55 **5 a** 32 **b** 254 **c** 16 **d** 46 **e** 132

page 57 **4 a** 1 g **b** 127 g **c** 35.5 g **d** 71 g **5 a** 32 g **b** 64 g **6** 138 g **7 a** 9 moles **b** 3 moles

page 59 **1** sulphur 50%, oxygen 50% **2** hydrogen 5%, oxygen 20% **3** hydrogen 11.1% **4** 0.1 mole per dm^3, or 0.1 M **5 a** 40 g **b** 20 g

page 61 **1 a** 1 **b** 4 g

page 62 **1 a** 64 g **b** 48 g **c** 48 g **d** 60 g **e** 355 g **f** 1.4 g **g** 4 g **h** 0.6 g **i** 24 g **2 a** 2 g **b** 4 g **c** 32 g **d** 35.5 g **e** 248 g **f** 1024 g **3 a** 1 mole **b** 2 moles **c** 1 mole **d** 2 moles **e** 0.2 moles **f** 0.1 moles **g** 0.4 moles **h** 0.2 moles **i** 2 moles **j** 0.05 moles **4 a** 80 g of sulphur **b** 80 g of oxygen **c** 8 moles of chlorine atoms **d** 1 moles of oxygen molecules **e** 4 moles of sulphur atoms **5 a** 1 litre of 2 M sodium chloride **b** 1 litre of 1 M sodium chloride **c** 100 cm^3 of 2 M sodium chloride **d** both the same **e** 40 cm^3 of 1 M sodium chloride **6 a** 18 g **b** 90 g **c** 160 g **d** 250 g **e** 34 g **f** 48 g **g** 30 g **h** 8 g **i** 15.8 g **j** 325 g **8** The missing numbers are: **a** 40, 16, 1, 56 **b** 1.6, 5.6 **c** 0.16 **d** 6, 3 **e** 71.4%, 28.6% **9 b** There is 80% copper in each sample.

page 63 **10 a** 64 g **b** 4 moles **c** 2 moles **d** MnO_2 **11 a** 2.4 g **b** 0.1 mole **c** 1.6 g **d** 0.1 mole **e** MgO **12 a** 106.5 g **b** 3 moles **c** 1 mole **d** $AlCl_3$ **e** 1 M **f** 0.1 M **13 c** NH_3, 17 g; CO_2, 44 g; H_2, 2 g; O_2, 32 g **d** 2 moles **e** 6 moles **f** 22 g **14 c** The missing figures are: group 4, 0.19 g; group 5, 0.20 g **e** 0.16 g **f** 16 g **g** 2 moles **h** 1 mole **i** Cu_2O

page 69 **2 b** 2 moles **c i** 32 g **ii** 8 g **3 b** $CuCO_3$, 124 g; CuO, 80 g; CO_2, 44 g **c i** 11 g **ii** 0.25 **iii** 6 dm^3

page 74 **5 b** NH_4Cl, −10°C; $CaCl_2$, +20°C **c** $CaCl_2$ **d** NH_4Cl **6 a** 217 g **b** 20.1 g of mercury, 1.6 g of oxygen **7 a** 32 g **b** sulphur **c** 11 g of iron(II) sulphide and 6 g of sulphur

page 75 **8 a** 48 dm^3 48000 cm^3 **b** 12 dm^3 12000 cm^3 **c** 0.24 dm^3 240 cm^3 **d** 7.2 dm 7200 cm^3 **9 b i** 2 moles **ii** 168 g **c i** 1 mole **ii** 24 dm^3 **d i** 12 dm^3 **ii** 1.2 dm^3 **10 b** 0.5 moles **c i** 28 g **ii** 22 g **11 d** 24 dm^3 **e i** 3 **ii** 72 dm^3 **f** 48 dm^3 **g** nitrogen 50 cm^3 hydrogen 150 cm^3 **12 c** 160 g **d** 2000 moles **e** 2 moles **f** 4000 moles **g** 224 kg **14 a** 233 g **b** 0.1 mole **c** 0.1 mole of each **d** 1 litre (or 1 dm^3) **14 b** 0.1 mole **c** 0.1 mole **e** $Zn + CuSO_4(aq) \longrightarrow ZnSO_4(aq) + Cu(s)$

page 85 **12 h** 6 g

page 89 **3 a i** 29 cm^3 **ii** 39 cm^3 **b i** 0.6 minutes **ii** 1.5 minutes **c i** 5 cm^3 of hydrogen per minute **ii** 2 cm^3 of hydrogen per minute

page 91 **1 a i** 60 cm^3 **ii** 60 cm^3

page 93 **1 a** experiment 1, 0.55 g; experiment 2, 0.95 g **b** experiment 1, 0.55 g; experiment 2, 0.95 g **c** experiment 1, $\frac{1}{3}$ g per minute; experiment 2, $\frac{1}{2}$ g per minute

page 96 **2 c i** 14 cm^3 **ii** 9 cm^3 **iii** 8 cm^3 **iv** 7 cm^3 **v** 2 cm^3 **d i** 14 cm^3 per minute **ii** 9 cm^3 per minute **iii** 8 cm^3 per minute **iv** 7 cm^3 per minute **v** 2 cm^3 per minute **e** 40 cm^3 **f** 5 minutes **g** 8 cm^3 per minute **h** faster **i** cool the acid

page 97 **8 c i** 25.1 g **ii** 24.7 g **d** 0.4 g **e** 3 minutes **f** 0–$\frac{1}{2}$ min **9 e** 3 moles

page 114 **2 d** 33 cm^3 **e** oxygen 33%, nitrogen 67% **f** oxygen 21%, nitrogen 78% **g** because the gases have different solubilities in water **h** oxygen

Index

If more than one page number is given, you should look up the **bold** one first.